THE ASTROLOGY CODE
Statistical Evidence of Universal Design
2nd Edition

By Michael Bergen

Seven Lights Press

The Astrology Code: Statistical Evidence of Universal Design
© 2014 by Michael Bergen

All rights reserved.
No part of this book may be reproduced or utilized, in any form or by any means, electronic or mechanical, without prior permission in writing from the publisher.

First Edition 2014
Second Edition 2022

Seven Lights Press
Brooklyn, NY 11215

Book and cover design by Michael Bergen
Cover art by Svetlana Shapovalova

The author may be contacted by writing to the publisher,
by e-mail at michael_bergen@msn.com,
or the website www.astrologycode.org.

ISBN 978-0-9899200-6-3

Dedicated to my beloved daughters:

Eva Soleil & Sophie Luna
(my Sun and Moon)

Contents

Prologue	1
Preface to 2nd Edition	3
Introduction	5
Methodology	9
Population Data	12
Summary of Findings	13

Part I: Astrological Techniques

Universal Design	19
Quadrants	27
Elements	31
Modes	35
Sign Pairs	39
Zodiac Signs	41
9th Harmonic	57
Navamshas	61
Libra vs Aquarius	75
Diurnal Cycle	79
4 Axis Points	81
Houses	87
Rising Sign	105
Planetary Aspects	109
Case Studies	125

Part II: The Study Groups

Study Groups	137
Activists	144
Actors	146
Adopted/Abandoned Children	148
AIDS	150
Alcoholics	152
Alzheimer's	154
Artists	156
Astrologers	158
Astronauts	160
Birth Defect	162
Business Owners	164
Cancer	166
Child Abuse	168

Child Performers	170
Child Prodigy	172
Children (4+)	174
Children (none)	176
Comedians	178
Composers	180
Computer Programmers	182
Corporate Tycoons	184
Criminals	186
Culinary	188
Dancers	190
Depression	192
Directors (Film & Stage)	194
Drug Abusers	196
Education (low)	198
Engineers	200
Fashion Designers	202
Fighters	204
Heart Attack	206
Homosexual Men	208
Infant Mortality	210
Inheritance	212
Instrumentalists	214
IQ (high)	216
Journalists	218
Lawyers	220
Lesbians	222
Lifespan: Long (80+ years)	224
Lifespan: Short (29- years)	226
Literature	228
Marriage (15+ years)	230
Marriage (never)	232
Mathematicians and Physicists	234
Medical Professionals	236
Mental Handicap	238
Mental Illness	240
Military Personnel	242
Models	244
Murderers	246
Murder Victims	248
Musicians	250
Mystics	252
Nazis	254

Nobel Prize	256
Novelists	258
Obesity	260
Only Child	262
Out-of-Body (NDE's)	264
Philosophers	266
Poets	268
Police Officers	270
Politicians	272
Presidents	274
Priests	276
Prisoners	278
Professors	280
Psychiatrists	282
Psychics	284
Racers (Cars & Bicycles)	286
Royalty	288
Salespeople	290
Same Job (10+ years)	292
Scientists	294
Sex Abuse Victims	296
Sex Offenders	298
Sex Workers	300
Singers	302
Songwriters	304
Spiritual Teachers	306
Sports	308
Sports Coaches	310
Stroke	312
Suicide	314
Teachers (K-12)	316
Technical Professions	318
Transvestites	320
Twins	322
Wealthy	324
Widowers	326
Writers	328

Personality Trait Groups

Active	332
Aggressive-Brash	334
Brilliant Mind	336
Eccentric	338
Emotional	340
Extrovert	342
Gracious-Sociable	344
Hard Worker	346
Humorous-Witty	348
Introvert	350
Pioneer	352
Principled	354
Sex Drive (high)	356
Sex Symbols	358

Gauquelin Study Groups

Actors	362
Alcoholics	364
Business Executives	366
Journalists	368
Medical	370
Mental Illness	372
Military	374
Murderers	376
Musicians	378
Painters	380
Politicians	382
Priests	384
Scientists	386
Sports	388
Writers	390

Appendices

Appendix 1 – Lunar Phases	395
Appendix 2 – Moon's Nodes	396
Appendix 3 – Parallels of Declination	399
Appendix 4 – Retrograde Planets	400
Appendix 5 – Midpoints	402
Appendix 6 – Planetary Rulership of Signs	403
Appendix 7 – Southern Hemisphere	404
Appendix 8 – Sidereal Zodiac	406
Addendum – Metaphysics	409
About the Author	457

Prologue

How is it possible that the entire Universe can be so organized as to perfectly represent the consciousness of every single organism at every moment of its life?

Astrology asserts that the Universe revolves around each one of us.

However, Copernicus discovered, many centuries ago, that we are not even remotely the center of the Universe, making astrology seem an antiquated part of an outmoded idea.

But, what if the Universe itself is centered around a higher dimension, a Mental Plane which organizes and contains the Universe?

Evidence of astrology is evidence of the Mental Plane.

If so, the Universe itself must be one single organism, which is the body of one single self-aware consciousness who is creating a multitude of simultaneous, interacting experiences that exist within it as a projection of the Mind.

Preface to 2ⁿᵈ Edition

In the eight years since The Astrology Code was first published, I have continued to learn and grow in many different ways. In addition to doing personal analysis, writing, and visiting shamans in South America, I attended the Gestalt and C.G. Jung psychoanalytic institutes in NYC for 3 years. During this phase, I became exposed to a whole new vantage point for understanding the human psyche and soul.

Much of this has been shared through my growing Facebook audience. I have written about metaphysical topics as well as the many issues that astrologers struggle with: the Houses, the Southern Hemisphere, the Great Ages, and the Sidereal Zodiac to name a few.

I have also learned how to improve my statistical research, and study the macrocosm under a better microscope. In the original book, I mainly used raw data to evaluate the study groups. In this edition, every piece of data is compared with a well designed control group. In addition, all Southern hemisphere charts have been removed from the study groups. Since the zodiac is formed by the seasons, it became apparent to me that the signs are most likely reversed "down under", since their seasons are reversed.

I also added many new study groups, such as Astronauts, Film Directors, Computer Programmers, High IQ individuals, and Nazis. I also decided to include 14 of Lois Rodden's personality trait study groups. I did not include them in the original book because of their subjective nature, but decided that it would be useful for readers to see them.

Due to the improved statistical methodology, I was able to take a much more involved look at the Gauquelin data as well. I obtained 15 of his main study groups, comprising over 23k charts, and applied the same methodology he used to analyze their house placements. I also analyzed them in the same manner I used for all of the Rodden groups. The results were profound!

Overall, the book has undergone a major renovation and, in my opinion, has improved in accuracy dramatically. Most of my original findings have stood up, and some new techniques have been verified as well, the most notable being the houses. Originally, I was suspicious of the houses due to the mass of confusion surrounding them. But, using Gauquelin's Diurnal Arc technique, in the traditional zodiac sign order, I have been able to reveal findings that greatly support their usage.

Astrologers often say that there are many different systems of astrology that all work equally well. However, through statistics we can easily observe the effectiveness of any astrological technique. It is time to hold astrology to the same standards as every other scientific field in order to help bring forth a more clear understanding of its truth.

The Astrology Code seeks to begin this process through the study of almost 37,000 people in 122 study groups. Many astrological techniques were validated, but others were found to be questionable. In order to raise astrology to the level of accepted science, astrologers everywhere must be willing to adjust their methods in response to the best research available. This improves the both the credibility of the field and the quality of our work.

Traditional rule-based astrology systems such as Hellenic and Vedic over-simplify astrological symbolism and propose to know a future which does not even exist.

Sun sign astrology is the ultimate over-simplification, proposing that no other planets or cycles have any importance. It grew an audience but lost astrology's soul.

The majority of people are wise to these things and scorn astrology, as a result. In order to make astrology an accepted field of study, it must make sense and be verifiable. It also must serve people's spiritual needs and not imprison their minds in fatalistic thinking.

Many astrologers feel uncomfortable taking instruction from a person outside of the profession. It is important to know that my goal has always been to be an astrologer, but I needed to verify the techniques for myself before using them with clients. In my research, I never had any intention to prove any particular techniques right or wrong. In fact, I was rooting for each one as I studied them. My only intention was to derive accurate findings and present my best assessment of the truth.

Will any of my findings be proven wrong? Quite possibly. But a great deal will stand up under scrutiny and, as any mathematician knows, we only need one proof of a thing to prove it to be true. This was the principle that motivated Michel Gauquelin to focus on the Diurnal Arc method of daily planetary motion. But I am going further by trying to test every single astrological technique in order to develop a comprehensive system of chart interpretation. I believe that I have accomplished this goal.

In addition, I have tied many other metaphysical systems into astrology such as the Jungian Cognitive Functions, MBTI types, the Enneagram, Numerology, and the Chakras.

In our modern era, people are increasingly seeking metaphysical systems of thought to improve self-awareness, connect to our life's purpose, and serve spiritual needs.

The system shown in The Astrology Code understands that astrology is an evolving field of knowledge and that astrologers need not have all the answers. Our exact individual expression can not be known by our birth chart alone. We determine our choices and self-expression in each moment of the present. The art of being an astrologer is to help people find their truth and eternal connection with the divine.

INTRODUCTION

This book aims to present the results of a statistical research project in terms that any student of astrology can understand and apply. It reveals my statistical findings as well as a model of the nature of the planets and the structure of the zodiac. To my knowledge, this is the first project of its kind. Its contents are original and can enhance the way people view astrology. It can also improve the knowledge and accuracy of astrologers, giving them a solid basis for their interpretations. My hope is that this book will stir enthusiasm in many others to do similar research into astrology's effects on human and planetary existence and the many ways that it can be used to better our lives.

BACKGROUND

As a child, I was raised as a half-Jewish atheist in a Catholic neighborhood in 1970's and 80's Brooklyn. I played soccer and baseball on the street with other kids living in the area. We would listen to music, practice dancing, hang-out and flirt. Life was good.

In my teen-age years, most of my friends turned to drugs, sex, and crime as outlets and dropped out of school. I did not go down this road due to the support of my parents, the excellent specialized public high school I attended, and an internal voice filling me with self-belief.

I did not have any particularly spiritual experiences until I was 17, when I was shocked by the sudden realization that everyone and everything revolves around each of us, simultaneously. This caused a flood of thoughts about the nature of life, death, and God. I realized that all of us are parts of God and that it could never be otherwise. At 18, I was seized by a vision of the world entering a new era and that I would participate in it through sharing knowledge gained in my adulthood.

Four years later, I was introduced to astrology by a friend and knew immediately that my life would revolve around this great field of study. I taught myself how to construct charts by hand, since there were no computer programs at the time, and excitedly attempted to analyze the charts of friends and family members. But I started to notice that the standard techniques in books did not always work so well. So, I slowed down on my astrology studies and started studying spirituality.

I devoured every book that seemed to offer any insight at all into the meaning of existence. I learned how to meditate, developed a yoga practice, and used psychoactive substances in attempts to gain communion with God. And, on a couple of occasions, I did get glimpses of the loving presence, which encouraged me greatly to develop myself into a good and helpful person who could make a lasting contribution to life.

Many years later, while working as a high school teacher, I started tinkering around with some new ideas I was having about astrology. I noticed that at the time that I moved in with my girlfriend, got engaged, then married, I had absolutely no astrological influences in my chart. I thought that there must be something I wasn't seeing. I had learned about how astrologers in India use the 9th harmonic chart. They call it the Navamsha chart and give it great importance for seeing the inner nature of a person. So the thought of looking at transits to my 9th harmonic chart positions occurred to me. Sure enough, Pluto was conjunct with my 9th harmonic Moon position exactly at these times. Of course, the Moon represents the home and women. I thought: "This is big!" That was in the year 2005.

I spent a few years working with my thoughts, studying all the charts I could find and eventually gained full confidence in the use of the 9th harmonic chart. In 2008, I bought the birth data of 15,000 people from Lois Rodden, which contains full biographies, and organized as many of them as I could into categories. Then I purchased software that could compile the astrological data of these people (JigSaw) and I started interpreting the tables of information to see if there were any real patterns or if it was just a mirage.

After months of devoting most every day to this statistical project, while waiting for my first daughter to be born, I discovered that the patterns were real. The statistical findings were very strong! But how could I get this information to people? Who could understand? Who would believe me or care?

So I posted my findings to the Internet and naively expected the whole world to applaud. Of course, nothing of the sort happened. So, I patiently continued with my experiments and found that there were many other patterns that I could statistically verify.

This book is a compilation of those patterns that I found to be effective when interpreting an astrology chart. All together, they form an entire system, which leads to an accurate interpretation of a person or event. The patterns in the 1st harmonic chart show the surface personality of a person, while the patterns in the 9th harmonic chart reveal the person's inner nature. Combined, the detailed dynamics of the person's consciousness are revealed.

Philosophy

So what is the real use of astrology after all? Is it to find out if someone is compatible with you or if you will have a good career or many children?

These uses are relevant but inconsequential compared to its real treasure -- the revelation that the universe is alive, conscious, and self-willed! The physical universe is the expression or projection of a mind, which precedes it and is, therefore, contained within this mind. We are a part of this mind and exist whether in physical bodies or not.

So then, if the universe is all the consciousness of one great mind, why is life so chaotic and why do we seem so incomplete and feel unfulfilled? Consider, that before the universe existed there was nothing … except self-awareness. And perhaps before that some state of semi-awareness. The point is that all this is new … a constantly evolving creation of self-expression, of which we are the focal points. We have not done this before and are, therefore, making it up as we go along. We make the best choices we can as we gain knowledge and wisdom about living as physical individuals. There is no final goal because we are in a constant state of becoming something new and having entirely new goals, as a result. Of course, this seems to be happening very slowly for most of us. But it is happening nonetheless. It is what we call evolution and as we evolve we see increasing order in our lives and the universe.

But how can astrology help us out specifically in our goal of expressing ourselves more intelligently and lovingly? The birth chart of an individual is a symbolic map of its consciousness. If understood well enough, the chart shows the tendencies that manifest in every aspect of our lives. The planets of the solar system represent the phases of our consciousness and how they interact, the zodiac signs show how our consciousness expresses itself, and the houses represent the outer world conditions in which our consciousness finds expression.

An astrologer can help interpret a birth chart, but only you will ever be able to reveal everything there is to be learned. True knowledge can only be gained through the individual self's awareness, through contemplation and insight. The birth chart is your special mandala, which will provide surprising revelations to you as life experiences show you how the symbols of your inner world manifest in your life. Knowing the symbolism of your birth chart will give you an objective basis for observing yourself and making the necessary corrections towards becoming who you want to be. The more effectively you do this, the more peaceful and fulfilling your life will become.

Astrology, the New Science

Unbeknown to most people, astrologers have been attempting to discover new techniques for the last few hundred years, and have not stagnated with the advent of science. Astrology has grown from a mere predictive art into a form of psychological analysis with widespread applications. But, up until now, there has been very little evidence, in the form of statistics, to verify that the universe expresses itself through astrological symbolism. As a result, faith and personal experience have been the bedrock of astrologers, which has led to astrology's disrepute amongst scientifically minded people.

Astrology is a science which attains personal and subjective information from the conditions of objective reality. In academic circles, this has been thought to be impossible. However, we are discovering, through quantum physics, that our consciousness plays a vital role in the objective conditions of life. In other words, how we see life influences what we see in life. This can be understood on a personal, collective, or universal level. Further, astrology proves the quantum theory that the phenomena of the universe form a cohesive whole, which is organized in an intelligent, meaningful manner. How else could planetary phenomena be used to gain personal information if all these phenomena were separate, random and not part of a synchronous cohesive system?

In order to establish astrology as a science, we need a set of techniques that can be consistently applied and statistically verified. I believe that this project accomplishes this task and should set the foundation for more research into the many ways that the planetary bodies of our solar system relate to life on Earth. Such knowledge could revolutionize the way we think, behave, and interact.

Summary

In the first chapters of the book, I attempted to show how the universe is organized and how the planets and zodiac reflect this pattern in our solar system. Every feature of this pattern is analyzed statistically. The 12 sign zodiac came to exist by projecting consciousness, represented by the planets, into time and space. I studied the hemispheres, which represent duality, the quadrants, which represent the four planes of being, the four elements, which represent space, and the three modes, which represent time. All calculations were done in the Tropical zodiac, which uses the Sun's position at the Spring equinox as its starting point for Aries.

In the next chapters, the full statistical results and descriptions are shown for each sign. Students can use the tables to analyze directly how planets express through the 1st and 9th harmonics as well as each $1/9^{th}$ portion of every sign. Most results are consistent with traditional meanings for the signs. The next two chapters show the results of the studies on planetary aspects and planets in the houses and on the 4 Axis points (Ascendant, Descendant, Midheaven, and Imum Coeli). Following this are a few examples of how to apply this study to chart reading. There are the charts of some famous sports figures, a comparison of Bill Gates and Steve Jobs, and studies of O.J. Simpson and Adolf Hitler.

In the final section, each group's findings are shown in detail and include a summary of the major findings. The interpretations are meant only to be suggestions for analysis since I do not presume to be an expert about each group of people. Every group's section has a summary table listing all relevant findings. Most groups have a very noticeable theme, which strongly validates the entire study. For example, Actors are strong in Aquarius in both the 1st and 9th harmonics, have frequent Venus aspects, have Venus frequent on the 4 Axis points, and are strong in the Fixed mode and Fixed sector (Fixed represents love and stability).

METHODOLOGY

I used 13,225 birth charts, which were collected by Lois Rodden. She took meticulous notes on each person and listed their careers, hobbies, and life events as well as the reliability of the source of the birth data. I organized the birth charts into as many groups as possible with at least 80 people in them.

I ended up with 93 different groups, some with over 1,200 people. Many individuals are in more than one study group (less than two on average). For instance, an actor who had a heart attack and 4 children would be in all three study groups.

These groups are based on objective facts about a person's life, which helped to eliminate subjective decision-making on my part. I also added 14 study groups that are based on subjective personality traits, such as extroversion and introversion. However, they are not included in the first section of the book which is used to assess the validity of the many astrology techniques.

I also included Michel Gauquelin's 15 study groups to use as a point of reference, since all of these groups have the same categories as the Rodden study groups. The 23,743 people in the Gauquelin groups are primarily French from the 19th century, while the Rodden groups are primarily Americans from the 20th century. Most of the people in the Gauquelin groups have birth times rounded to the nearest hour, while most of the birth times in the Rodden groups are exact to the minute.

The statistical analysis for this project was done by creating a control group, used for comparison with every single point in the study. It was created by calculating over 200,000 evenly spread times from the years 1850 to 2000, when the majority of the population in this study was born, then replicating the birth time-of-day of the entire population.

This technique revealed the exact distribution of planets by sign and aspect, while also reflecting the human birth trend of being born predominantly in the morning hours, thus allowing for studying the houses and 4 Axis points.

Each planet has its own unique distribution through the signs, most displaying sine wave patterns (except Venus), with the Moon being the most evenly distributed. The Sun, Mercury, and Venus are heavily skewed towards the houses on the Eastern horizon, while the other planets are more evenly distributed.

The following tables show the distribution of the Sun, Moon, and Mercury through the 12 signs and houses for the control group:

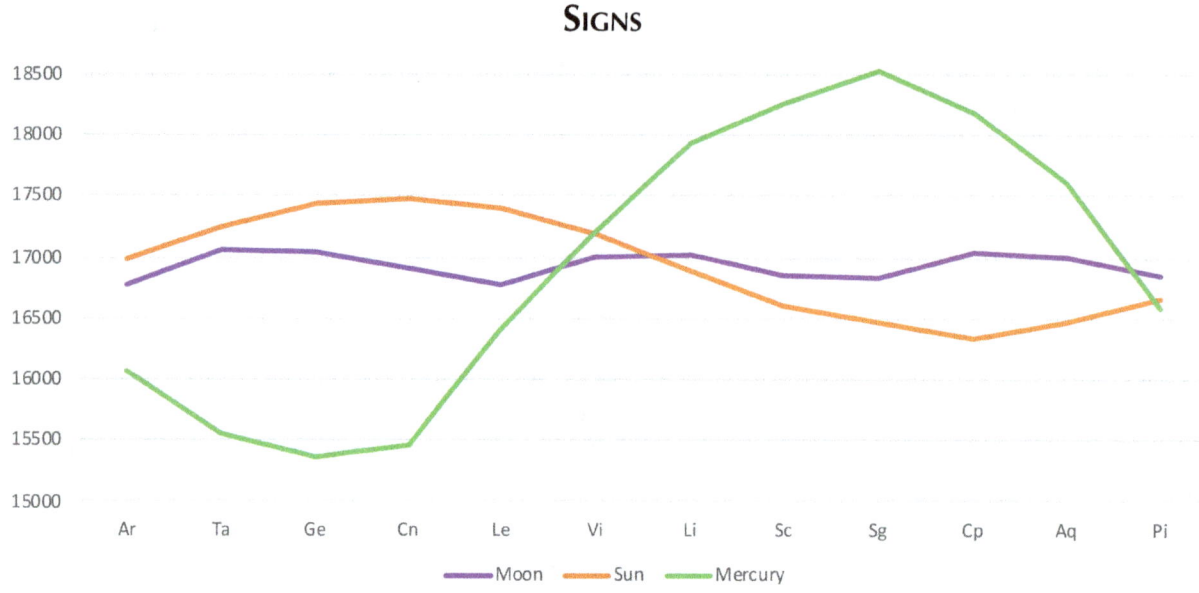

The method used to calculate the probability of any astrological event is to compare the actual results to the expected results. The actual results are calculated by summing the amount of times a planet is in a given position for an entire study group. The expected results are calculated with the control group.

The importance of each result is decided by finding its position on a normal curve, using the expected result as the middle point of the curve. If its position is in the upper 5%, then the probability is at least 1 in 20 and is considered statistically significant. If the result is in the upper 1%, the probability is at least 1 in 100. In this study, there are over 170 findings with odds above 1 in 1,000, which is equivalent to tossing a coin on its head 10 times in a row.

This method is used throughout the book to show the signs and planets in which each study group is dominant. For instance, there are 214 people in the Military study group. The expected number of people with the Moon in each zodiac sign is 18. In this case, 25 people with the Moon in any sign would be in the upper 5% of the Normal Curve. In the Military study group, 29 people have their Moon in the sign of Capricorn and it is, therefore, considered a significant finding with a 1 in 311 chance of being a random accident.

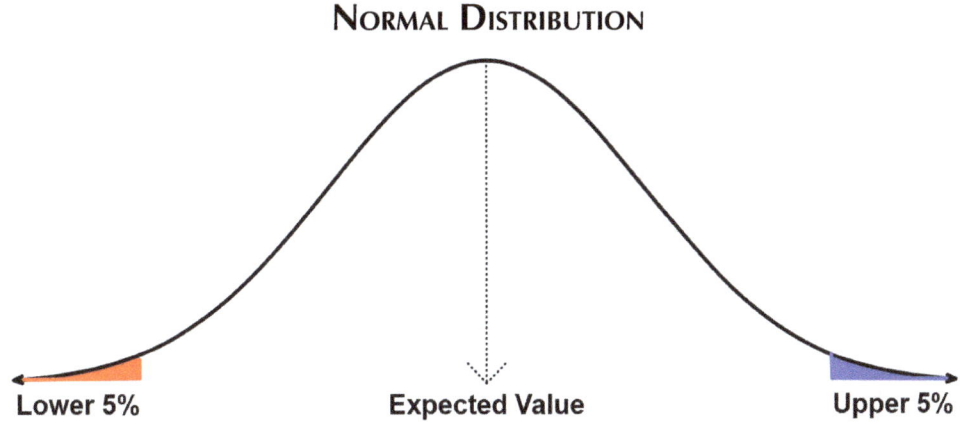

In the study on the zodiac signs, I analyzed the 1st harmonic, the 9th harmonic, and the 108 1/9th portions of the signs. Outer planets (Uranus, Neptune, Pluto, and Chiron) were not studied by sign since they move slowly causing all the study groups to be strong in just a few signs.

For aspects, I used the 36th harmonic, which is the first division of the zodiac containing all of the major traditional aspects as well as the nonile (40 degrees). This study shows that two planets have a meaningful relationship when in any of the 36 points spaced every 10 degrees from each other around the zodiac.

The influence of planets in the diurnal cycle was also studied. The most significant findings were found when planets are near the 4 Axis points: the Ascendant, Midheaven, Descendant, and Imum Coeli, as well as in the 12 Houses. The effect of the Rising sign was less significant. The Lunar Phases, Sun/Moon midpoints, Moon's Nodes, and Parallels of Declination were also studied with clear, meaningful results.

Population Data

The following data is gathered from the entire population of the Rodden study groups.

Rodden Source Ratings

AA –	Birth Record	7,074 – 53.5%
A –	Memory	4,129 – 31.2%
B –	Biography	603 – 4.6%
C –	Other	1,419 – 10.7%

Century of Birth

<1,400	25
1,400's	55
1,500's	62
1,600's	58
1,700's	122
1,800's	1,977
1,900's	10,897
2,000's	29

Decade of Birth

1,900's	642
1,910's	795
1,920's	1,459
1,930's	1,772
1,940's	2,465
1,950's	1,720
1,960's	1,133
1,970's	572
1,980's	229
1,990's	110

Gender

Male	8,495 – 64.2%
Female	3,547 – 26.8%
Undetermined	1,183 – 8.9%

Hemisphere

Northern	12,910 – 97.6%
Southern	315 – 2.4%

Country

United States	6,974 – 52.7%	Canada	218 – 1.6%
France	1,614 – 12.2%	Netherlands	190 – 1.4%
Italy	907 – 6.9%	Brazil	184 – 1.4%
England	770 – 5.8%	India	95 – 0.7%
Germany	564 – 4.3%	Australia	90 – 0.7%
Scotland	364 – 2.8%	Spain	83 – 0.6%
Belgium	227 – 1.7%	Austria	77 – 0.6%

Summary of Findings

1) The solar, lunar, daily, and inter-planetary cycles all have a four quadrant structure.

2) The Tropical zodiac signs likely reverse in the Southern hemisphere, since the seasons are reversed.

3) Traditional planetary rulerships of signs have good correlation to the signs where planets are actually strongest and weakest.

4) Every planet has its own diurnal arc (daily path) and its own unique set of 12 houses. The 9th house is the most beneficent, while the 4th house is the most problematic.

5) The Rising sign is a weak technique for assessing the persona. Its distribution is highly skewed in locations North and South of the Equator.

6) Libra and Aquarius should have their meanings changed. Libra is common with active, assertive thinkers and weak with many of the people-oriented and charismatic types that are strongly influenced by Aquarius and Venus.

7) Virgo is the most sexually active sign, yet shows the most health difficulties.

8) Scorpio is the most sexually magnetic sign, yet is the most deep thinking.

9) The 9th harmonic chart in the Tropical zodiac is highly influential. It represents the inner self, while the 1st harmonic chart represents the outer self.

10) The dominant inter-planetary aspect pattern is the 36th harmonic (every 10 degrees).

11) The Sun is combusted by other planets. The Sun functions more powerfully when it is alone in the sky and not in any aspect relationship with other planetary bodies.

12) Venus expresses its positive attributes best when retrograde and its negative attributes most when direct.

13) Mars is influential for people whose life span was under 29 years and is weak for people who have lived over 80 years.

14) Neptune is more often seen with materialism and deception than spirituality.

15) The Lunation cycle is most descriptive when divided into 12 sections symbolized by the zodiac signs.

Summary of Group Findings by Sign/House

Aries (1st)	Taurus (2nd)	Gemini (3rd)	Cancer (4th)	Leo (5th)	Virgo (6th)
Children (4+)	Alcoholics	Astrologers	Alzheimer's	Homosexuals	AIDS
Dancers	Children (4+)	Business	Cancer	Lesbians	Birth Defect
Marriage (15+)	Comedians	Heart Attack	Child Abuse	Police	Child Abuse
Math-Physics	Directors	Scientists	Child Prodigy	Same Job (10+)	Children (none)
Nazis	Life (80+ yrs)	Singers	Corporate	Songwriters	Engineers
Politicians	Murderers	Sports	Culinary		Fighters
Racers	Nobel Prize		Depression		Infant Mortality
Salespeople	Psychiatrists		Fashion Designer		Journalists
Sports	Same Job (10+)		Infant Mortality		Marriage (none)
Sports Coaches	Sex Abuse		Murderers		Medical
	Singers		Sex Offenders		Military
			Teachers (K-12)		Out-of-Body
					Sex Offenders
					Sex Workers

Libra (7th)	Scorpio (8th)	Sagittarius (9th)	Capricorn (10th)	Aquarius (11th)	Pisces (12th)
Activists	Child Abuse	Instrumentalists	Adopted	Actors	Artists
Astronauts	Child Performers	Marriage (15+)	Composers	Astronauts	Astrologers
Child Performers	Comedians	Scientists	Computers	Comedians	Cancer
IQ (high)	Computers	Twins	Homosexuals	Lawyers	Dancers
Lawyers	Criminals		Math-Physics	Literature	Education (low)
Lesbians	Drug Abuse		Military	Marriage (15+)	Fashion
Priests	Fighters		Murder Victims	Mental Illness	Life (80+ yrs)
Sex Workers	Homosexuals		Out-of-Body	Novelists	Mental Handicap
Stroke	Life (<29 yrs)		Technical	Only Child	Poets
Teachers (K-12)	Mental Handicap		Wealthy	Presidents	Politicians
	Mental Illness			Prisoners	Priests
	Models			Suicide	Psychics
	Murderers			Twins	Royalty
	Mystics			Writers	Spiritual Teacher
	Philosophers				
	Police				
	Priests				
	Prisoners				
	Professors				
	Psychiatrists				
	Spiritual Teacher				
	Transvestites				

Summary of Group Findings by Planet

Moon	Sun	Mercury	Venus	Mars	Jupiter
Alzheimer's	Child Performers	Child Prodigy	Actors	Criminals	Engineers
Child Abuse	Drug Abuse	Children (4+)	AIDS	Drug Abuse	Homosexuals
Comedians	Songwriters	Computers	Child Abuse	Math-Physics	Marriage (15+)
Culinary		Education (low)	Child Performers	Nazis	Nobel Prize
Life (80+ yrs)		Engineers	Child Prodigy	Politicians	Obesity
Marriage (none)		Literature	Children (4+)	Prisoners	Presidents
Mental Illness		Mental Handicap	Culinary	Suicide	Royalty
Obesity		Mental Illness	Inheritance	Transvestites	Sex Workers
Scientists		Mystics	Only Child	Wealthy	Spiritual Teacher
Transvestites		Psychics	Out-of-Body		Wealthy
Widowed		Technical	Salespeople		

Saturn	Uranus	Neptune	Pluto	Chiron
Adopted	Activists	Corporate Tycoons	Activists	Lesbians
Birth Defect	Composers	Instrumentalists	Adopted	Military
Children (none)	Computers	Lawyers	Business Owners	Nazis
Heart Attack	Engineers	Mental Illness	Lesbians	Psychiatrists
Math-Physics	Suicide	Military	Medical	Sex Abuse
Novelists		Police		Stroke
Professors				Widowed
Sex Abuse				

These tables are a summary of the findings by sign and planet. For every group, the overall statistical strength of each sign was assessed by observing the results of the 1st and 9th harmonic signs, houses, Rising sign, Sun/Moon midpoint, Moon's nodes, and Lunar Phase studies. The strength of each planet was assessed by observing the results of the studies on planetary aspects, planets on the 4 Axis points, parallels of declination, Sun/Moon midpoint, Moon's nodes, and retrograde motion. Full details are shown in the study group section of the book.

Part I

Astrological Techniques

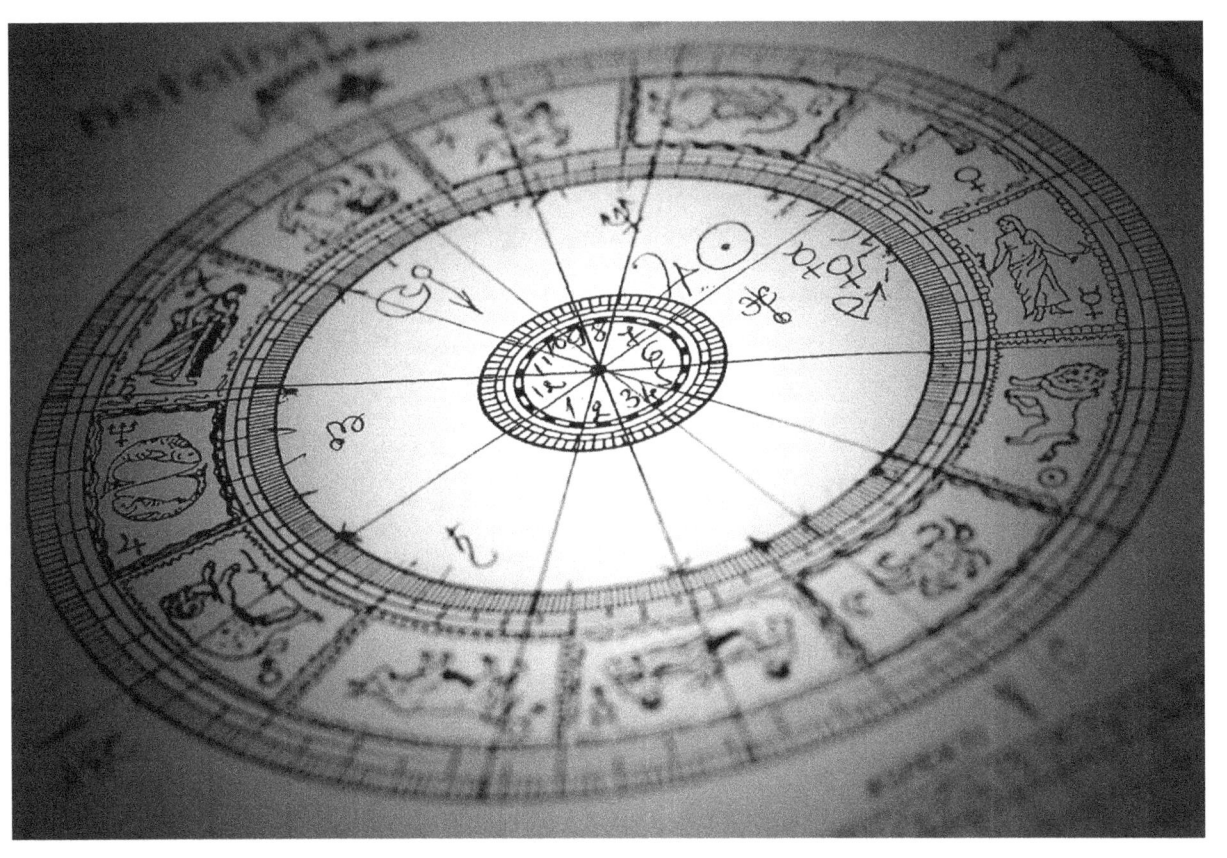

The Universal Design

Our inner self, in its solitude, has no expression and, therefore, cannot be described with symbols. However, our inner self has a body which is composed of consciousness. We call it the soul. This represents the primordial duality, the inner and outer aspects of mind - awareness and consciousness, and is the basic principle or building block with which we create and experience our conscious reality.

The Yin-Yang symbol represents this most basic duality, from which all creation develops. The Moon's cycle is symbolic of this original duality as well. The New Moon represents Yin, subjective awareness, and the Full Moon represents Yang, objective consciousness.

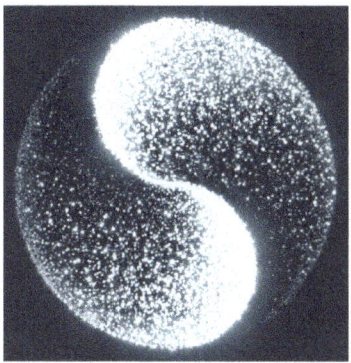

Duality and the process of creation can also be symbolized by a wave. The upward arc represents an outward movement and the downward arc represents an inward movement.

<u>Duality</u>

Outward	Inward
Yang	Yin
Positive	Negative
Active	Passive
Electric	Magnetic

Duality takes place in four major areas of life: the individual, relations, groups, and the universal. The individual area of life represents a single unit and involves the development of a being's abilities. The relations area of life represents the interaction of two beings and involves learning how we contrast with others, how we fit in and how we stand out. The groups area of life represents the functioning of an entity composed of many beings and involves taking on responsibilities by learning how we can contribute to a group such as our family and community. The universal area of life represents the individual being's connection to the whole and involves seeking knowledge about the nature of life and finding one's higher purpose.

The Planets

The planetary bodies of our solar system represent our consciousness as they express in the four major areas of life. They do this by forming duality pairs within each area. Each planetary pair expresses the polarity between outward expansion (yang) and inward unification (yin) within their field of influence. The equality and balance of each pair explains why both planets have approximately the same size (Sun and Moon have the same apparent size). The inner planets represent the personal aspects of our consciousness. The outer, gaseous planets represent the impersonal, collective aspects of our consciousness.

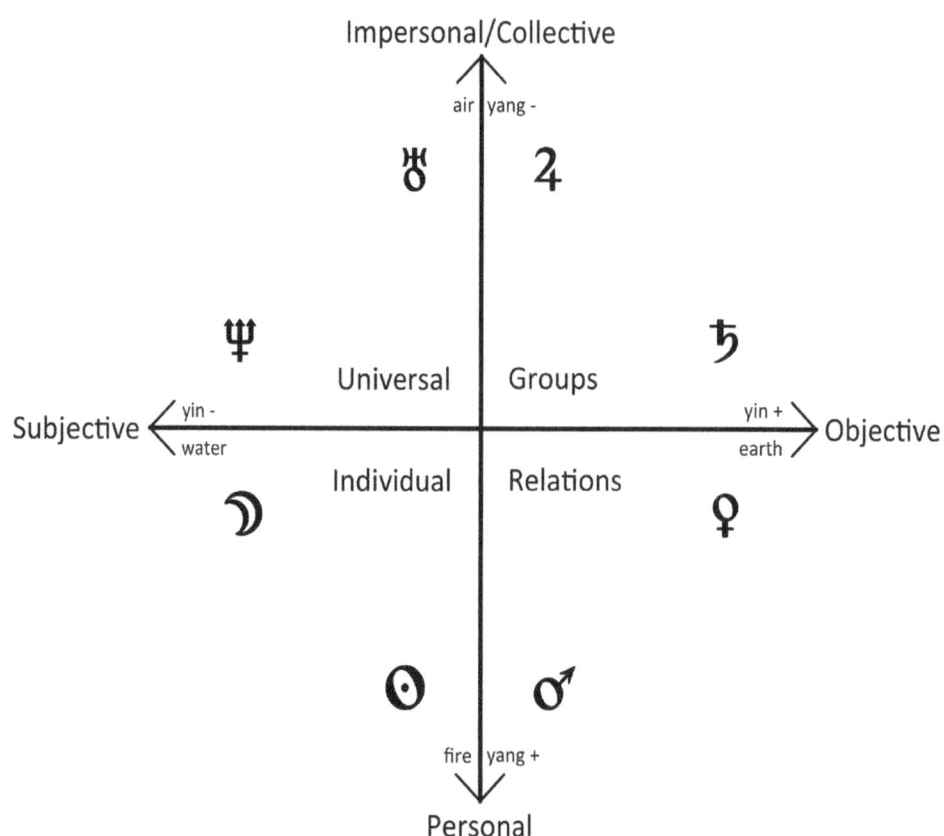

Individual
The Sun (yang) and Moon (yin) represent the outward and inward movements of the individual. They represent the polarity between seeking growth, through new experiences, and security, through family and home. They show how we develop our personalities and form emotional stability in childhood, and together, depict our subjective experience of being. The Sun lights our path, showing our identity and sense of purpose. It represents the father figure who inspires our life goals. The Moon warms our soul, showing our emotional needs and attachments. It represents the mother figure who secures us and nurtures our development.

Relations
Mars (yang) and Venus (yin) represent masculinity and femininity. They demonstrate the youthful dilemma of asserting our individuality vs. getting along with others and fitting in. Mars shows our physical aptitude and how we seek to separate ourselves from others in order to have freedom and differentiate ourselves. It is energetic and productive. Venus shows how we enjoy our senses and attract others toward us in order to have happy and productive relationships. It is sensual and artistic.

Groups
Jupiter (yang) and Saturn (yin) represent the expansive and contracting forces of the beliefs and values shared by groups of people. They represent our dilemma of seeking the freedom to find our own truths vs. accepting the conventional wisdom from our surroundings. They show how we take part in shaping our cultural environment and depict our experience of adulthood. Jupiter represents a person's need to grow and find ways to explore one's potential and impact social movements. Saturn represents the need to conform to our cultural expectations and assume responsibilities, becoming a productive member of society.

Universal
Uranus (yang) and Neptune (yin) represent the universal process of dreaming into and awakening from collective illusions of reality. They show we how we relate to our Universal Self which, like these two planets, is not visible to the naked eye. Together, they depict our experience of spiritual awakening, which peaks in old age. Neptune's dreaming is the Universal Self's way of imagining itself into being as a multitude of expressions in an objective universe. It represents the mystic dreamer. Uranus' awakening is the process of individuation into a unique expression of our one true self. It represents the illuminated mind with intuitive awareness.

THE ZODIAC'S QUADRANTS

The zodiac, through which planets travel and find expression, is also a representation of the four major areas of life.

The first quadrant, which is composed of Aries, Taurus, and Gemini, represents the individual area of life. Its season is the Spring, which is a seeking and expansive time. In the life cycle, it begins at birth and represents childhood.

The second quadrant, which is composed of Cancer, Leo, and Virgo, represents the personal relations area of life. Its season is the Summer, which is a creative and joyful time of ripening. In the life cycle, it begins at puberty, and represents youth. Together, the first two quadrants represent the process of developing personal qualities and talents.

The third quadrant, which is composed of Libra, Scorpio, and Sagittarius, represents the group activities area of life. Its season is the Fall, which is a time of maturing, gaining influence and insight, and reaping the fruits of one's efforts. In the life cycle, it begins when many of us get married, start a family, and pursue a career and represents adulthood.

The fourth quadrant, which is composed of Capricorn, Aquarius, and Pisces, represents the areas of life of universal concern. Its season is the Winter, which is a contemplative and contractive time. In the life cycle, it begins at retirement, represents old age and ends at death. Together, the last two quadrants represent the process of finding one's place within the collective and gaining knowledge of life.

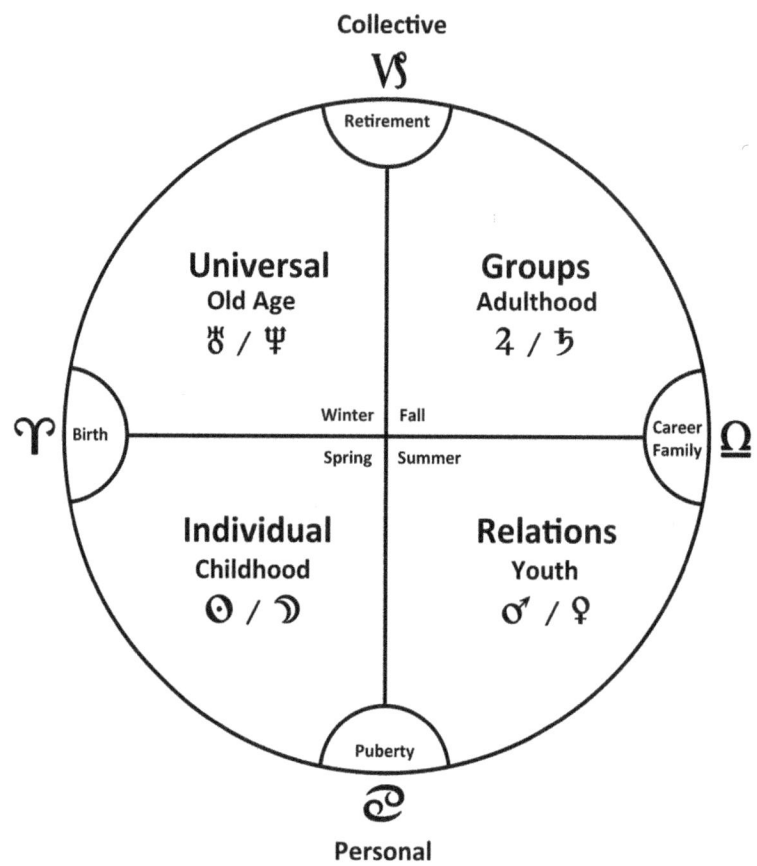

(Childhood lasts 12 years, ending at the first Jupiter return. Youth lasts 18 years, ending at the first Saturn return. Adulthood lasts 30 years, ending at the second Saturn return. Life's end occurs near the Uranus return and Neptune opposition at age 84.)

THE ELEMENTS

The four elements, Fire, Earth, Air, and Water, represent the prime substances of life. In the zodiac, they are represented by the signs. The Fire signs, Aries, Leo, and Sagittarius, are personal and inspired. The Earth signs, Taurus, Virgo, and Capricorn are physical and productive. The Air signs, Gemini, Libra, and Aquarius, are intellectual and sociable. The Water signs, Cancer, Scorpio, and Pisces, are healing and spiritual.

Fire, the spark of life, represents light and warmth. It represents the warm seasons, Spring and Summer. Its planets are the Sun and Mars. Earth, physical matter, is the manifester of life. It represents the dry seasons, Summer and Fall. Its planets are Venus and Saturn. Fire and Earth are masculine elements that provide us with the vast majority of our energy.

Air, representing sound and thought, is used to communicate information. It represents the cool seasons, Fall and Winter. Its planets are Mercury, Jupiter and Uranus. Water, the bringer of life, forms the rhythms of nature. It represents the wet seasons, Winter and Spring. Its planets are the Moon and Neptune. Air and Water are feminine elements which provide the spaces in which beings live and breath.

DUALITY IN THE ZODIAC

There are four types of duality in the zodiac which represent the four major areas of life:

1) The 12 signs alternate every sign from extroverted to introverted, which corresponds with the Sun/Moon duality in individuals. The Fire and Air signs are extroverted, and the Earth and Water signs are introverted.

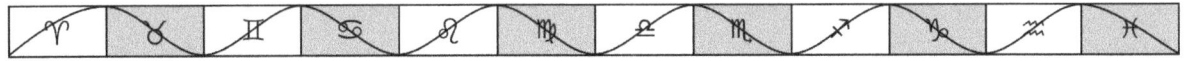

People whose birth charts have a powerful Sun or are dominant in Fire and Air signs are outgoing and seek to express their talents in public. People with a powerful Moon or are dominant in Earth and Water signs are withdrawn and pursue their interests in private.

2) The signs alternate every two signs from masculine to feminine which corresponds with the Mars/Venus duality in relations. The Fire and Earth signs are masculine and Air and Water signs are feminine.

People whose birth charts have a powerful Mars or are dominant in Fire and Earth signs are assertive in relationships. People with a powerful Venus or are dominant in Air and Water signs are submissive and supportive in relationships.

3) The signs alternate every three signs from expansive to contractive which corresponds with the Jupiter/Saturn duality in groups. In this way, the 1st and 3rd quadrants are expansive and the 2nd and 4th quadrants are consolidating and contractive.

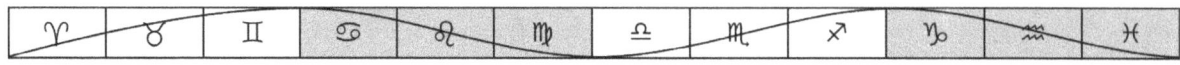

People whose birth charts have a powerful Jupiter or are dominant in the 1st and 3rd quadrants follow their cultural beliefs and values and may impose them on others. People with a powerful Saturn or are dominant in the 2nd and 4th quadrants are abiding of their culture's laws and values and are responsive to the opinions of others.

4) The 1st hemisphere of the zodiac, Aries through Virgo, represents personal goals and outward development. The 2nd hemisphere, Libra through Pisces, represents collective integration and the development of maturity and wisdom. This corresponds with the Uranus/Neptune duality in the universal process.

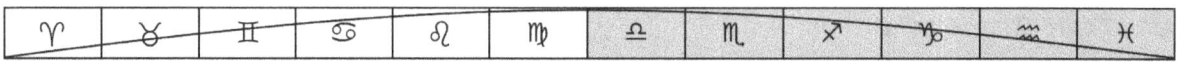

People with a powerful Uranus or are dominant in the 1st hemisphere are focused on individual development. People whose birth charts have a powerful Neptune or are dominant in the 2nd hemisphere are focused on collective integration.

THE MODES

The three modes, Cardinal, Fixed, and Mutable, represent the process of life. The Cardinal mode represents the power of creation and represents forward movement. The Fixed mode is the desire to experience love and fulfillment and represents the preservation of the present. The Mutable mode is the contemplating mind which learns from the past, decides what knowledge needs to be retained and plans for the future.

The modes cycle four times through the zodiac in the order Cardinal, Fixed, Mutable. The four quadrants of the zodiac are each composed of one mode cycle. In the first quadrant, Aries is Cardinal, Taurus is Fixed, and Gemini is Mutable. Every stage of life, represented by the quadrants, goes through the same process of the three modes.

In addition, the entire zodiac can be divided into modes. The first third of the zodiac represents the Cardinal mode, the second third represents the Fixed mode, and the final third represents the Mutable mode.

The Universal Design

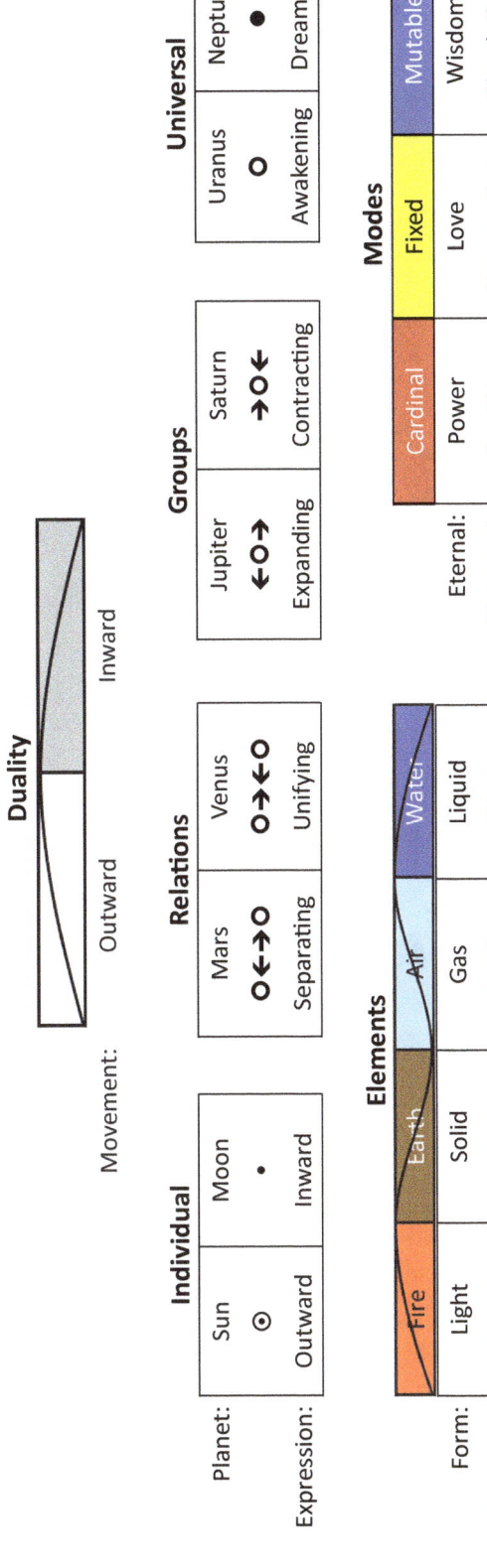

Notes:
a) The planetary pairs represent our consciousness which finds expression in time and space. Mercury is excluded since it represents awareness itself.
b) Awakening (Uranus) and Dreaming (Neptune) refer to the process of arising into the many unique beings of the universe, then dissolving back into the one dreaming mind.
c) The mode's colors are the three primary colors. Each zodiac sign's color is derived from blending its mode color with its element color.

The Quadrants

The solar, lunar, daily, and inter-planetary cycles all share the same basic structure. They are each divided by four axis or cardinal points into four quadrants which correspond with the four major areas of life expression: Individual, Relations, Groups, and Universal.

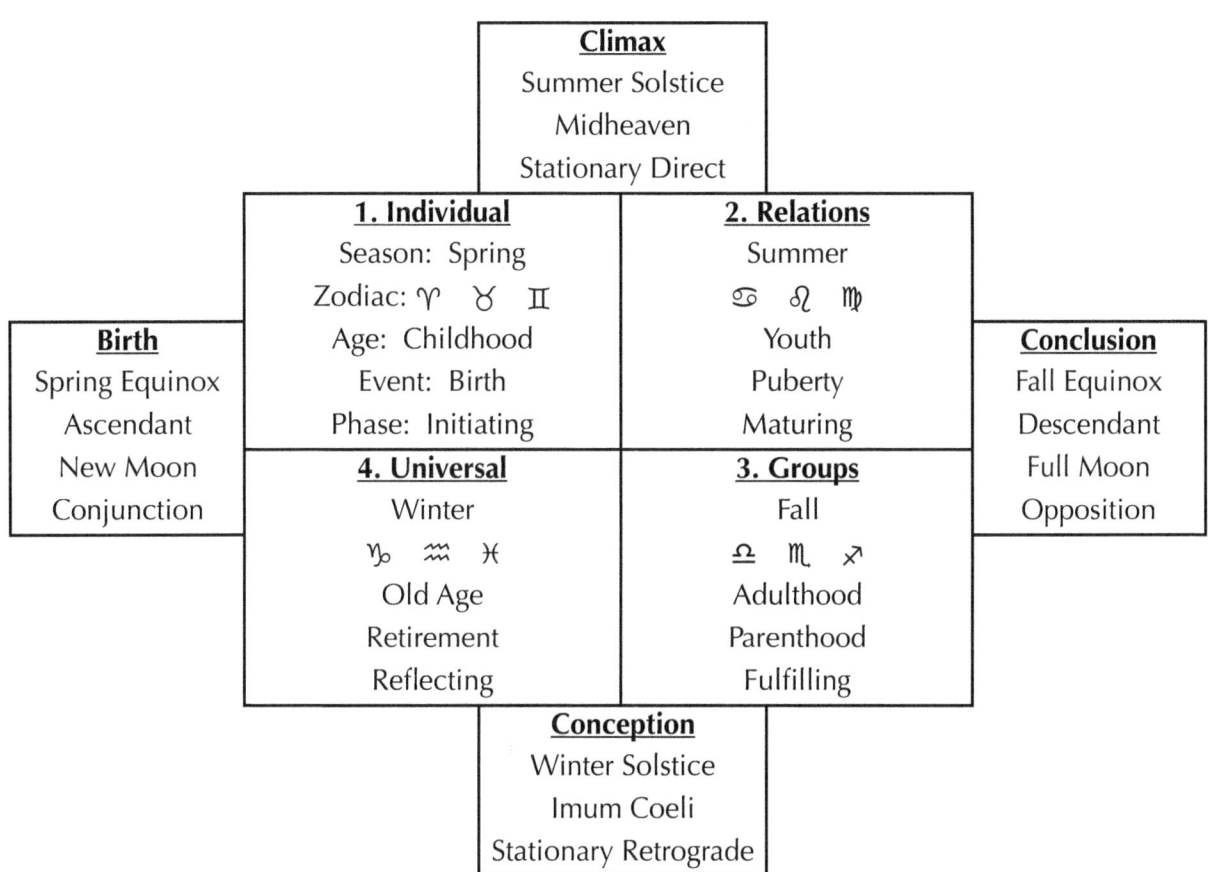

Birth represents the first noticeable beginning of a process, as opposed to the Trough of the cycle, which represents the conception of an idea. The Climax represents the outermost expression of the idea and the Conclusion represents the results of the process.

These four moments represent the critical points of any cycle and each initiates their own phase or stage of existence. Birth initiates the Individual stage of existence, where we take action on our motivations. The Climax initiates the Relations stage of existence, where we fulfill our goals and interact with others. The Conclusion represents the end of personal focus and initiates the Group stage of existence, where we make use of our efforts for the collective well-being of our family and community. The Trough of the cycle initiates the Universal stage of existence, where we resolve the process and look forward to the future.

The Sun is at its trough at the Winter Solstice and Imum Coeli, where the Sun is least powerful. It is symbolically born when its light exceeds darkness at the Spring Solstice, and appears at the Ascendant. The Sun is at its peak at the Summer Solstice and Midheaven, where it shines brightest and symbolically dies when darkness exceeds light at the Fall Equinox and disappears at the Descendant.

The zodiac divides into four quadrants: 1) Aries, Taurus, and Gemini; 2) Cancer, Leo, and Virgo; 3) Libra, Scorpio, and Sagittarius; 4) Capricorn, Aquarius, and Pisces. Each quadrant shows a clear pattern, which holds to expectations. The first quadrant pertains to the individual self and personal will. The second quadrant is where the self interacts with others to achieve its will. The third quadrant is where we seek to understand others and increase our responsibilities through group interaction. The fourth quadrant is where we seek to understand life as a whole and be of service to the world.

Quadrant 1 - Individual
The main focus of these people is on developing personal qualities in order to express themselves effectively. They place great faith in their native abilities and are less inclined to trust others with tasks that they feel they can do better on their own. They tend to be self-centered and highly subjective in their judgments. They place a great value on maintaining their privacy. They see themselves as independent individuals, capable of organizing their own lives.

Quadrant 2 - Relations
These people are focused on their personal relationships. They are strongly influenced by others, for better or worse, and by acts that they do not initiate. They seek fulfillment through the development and expansion of their talents and abilities in an attempt to experience positive, fulfilling interactions with others.

Quadrant 3 - Groups
These people are highly perceptive and seek to understand others in order to have fulfilling interactions with them. They seek to achieve self-realization through giving of themselves. However, others may prove to be a drain on their energies and resources. They look for relationships with an objective, rather than personal outlook. They also want to be public and to play their part in society.

Quadrant 4 - Universal
These people seek to have a strong influence on the world at large. They are highly inspired individuals who are very much in control of their destiny. They have a tendency to seek self-fulfillment through sharing with others. Being focused on the goal of public service, they measure success by their contribution to the world. They are totally willing to consider others in life situations due to a sense of membership with humanity.

In order to see if there is any statistical evidence of the quadrants, I analyzed them using the study groups. The Sun, Moon, Mercury, Venus, Mars, Jupiter, and Saturn were used. I left out the outer planets (Chiron, Uranus, Neptune, and Pluto) because they are much slower moving and tend to show more of a generational (rather than personal) effect in the zodiac. I also did not use the rising sign because its distribution is heavily skewed and, therefore, has questionable value.

Through observing this study of the quadrants, we can see that people with largely personal goals are strong in quadrants 1 and 2. Sports, Fighters, and Racers showed up very strongly in Quadrant 1, verifying the idea that this sector is largely about personal ambition and asserting one's will. Quadrant 2 was strongest for sex-related groups, showing its correspondence with personal relationships. Quadrant 3 is strong with Activists and Spiritual Teachers, showing its correspondence with groups activity. Most of the groups that are strong in Quadrant 4 demonstrate creativity and mental aptitude and seek to affect the way we understand the world, thus having a broad and lasting influence on society.

Personal		Collective	
Individual	**Relations**	**Groups**	**Universal**
Artists	AIDS	Activists	Artists
Business Owners	Astrologers	Child Performers	Comedians
Children (4+)	Fashion Designers	Computer Programmers	Composers
Engineers	Lawyers	Criminals	Inheritance
Fighters	Lesbians	IQ (high)	Life (80+ yrs)
Infant Mortality	Military	Mental Illness	Literature
Marriage (15+ yrs)	Models	Mystics	Math Physics
Mental Handicap	Out-of-Body	Police	Mental Handicap
Nobel Prize	Sex Abuse	Sex Abuse	Musicians
Racers	Sex Offenders	Spiritual Teachers	Only Child
Salespeople	Sex Workers	Transvestites	Racers
Same Job (10+ yrs)		Twins	Scientists
Singers			Songwriters
Sports			Technical
Stroke			Writers

(Highlights show the groups that fit with the character traits of each quadrant.)

Elements

The four elements are the building blocks of manifestation and embody the basic characteristics of our physical experience. Fire represents light, warmth, sight, and the directions down and South. The Fire signs, Aries, Leo, and Sagittarius, are extroverted, masculine, individualistic, sensuous, emotional, and subjective. Earth represents matter, dryness, touch, and the directions right and East. The Earth signs, Taurus, Virgo, and Capricorn, are introverted, masculine, individualistic, sensuous, rational, and objective.

Air represents the atmosphere, coolness, sound, hearing, and the directions up and North. The Air signs, Gemini, Libra, and Aquarius, are extroverted, feminine, communal, intuitive, rational, and objective. Water represents liquids, wetness, smell, taste and the directions left and West. The Water signs, Cancer, Scorpio, and Pisces, are introverted, feminine, communal, intuitive, emotional, and subjective.

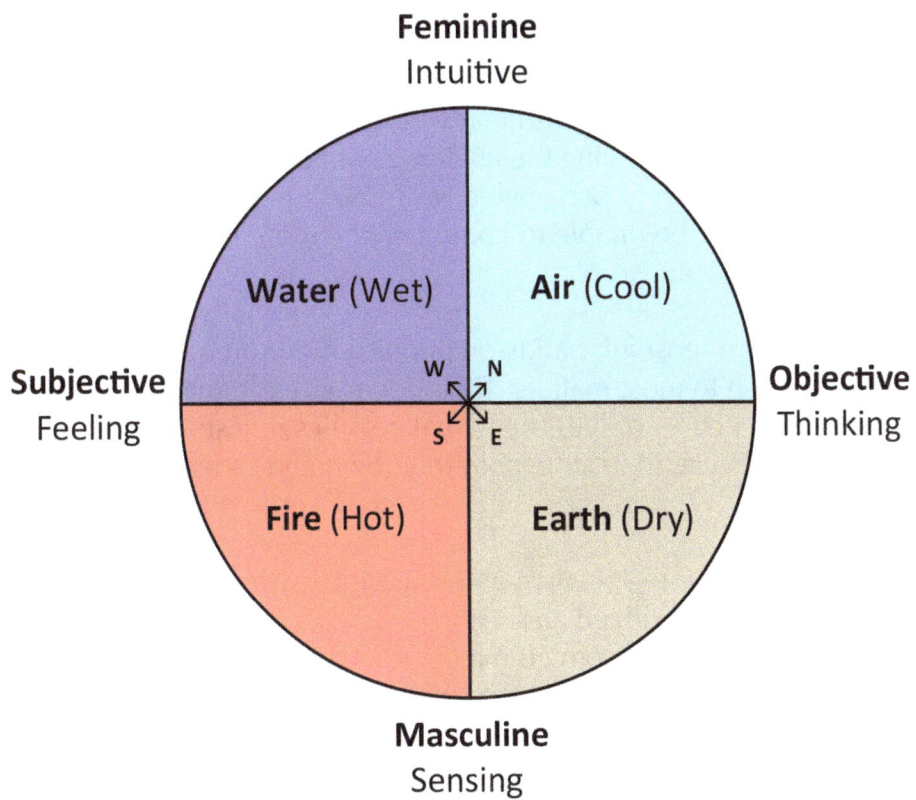

The study on the four elements did not reveal too much about the nature of Fire and Earth. However, the groups that were strong in Air and Water are very representative of these elements. The groups strong in Air are highly communicative people. The groups strong in Water were largely contemplative types, which is consistent with this element's private, sensitive nature.

Fire	Earth	Air	Water
Lesbians	Actors	Activists	Astrologers
	Alzheimer's	Actors	Astronauts
	Musicians	Birth Defect	Child Prodigy
	Out-of-Body	IQ (high)	Comedians
	Priests	Salespeople	Corporate Tycoons
	Technical	Teachers (K-12)	Criminals
		Twins	Infant Mortality
			Mystics
			Psychiatrists
			Psychics
			Royalty
			Spiritual Teachers

Fire (Sensing-Feeling)
These are warm enthusiastic people with many personal goals and aspirations. They are vital and spontaneous, often enjoying the challenge of traveling down new and adventurous roads in life. They are very personable, with much feeling. Their weakness lies in a tendency to exaggerate and be unable to cope with the more mundane activities in life.

Earth (Sensing-Thinking)
These are practical and purposeful, hands-on people. They can be relied on to take a common-sense approach to most matters. Their down-to-earth manner stands them in good stead for the daily chores of life. Their weakness lies in a stubborn refusal to make changes in their lives. They tend to stifle both their own and other people's enthusiasm.

Air (Intuitive-Thinking)
People dominated by Air like to socialize and share ideas with many people. They have a high sense of fairness, are civilized and well-mannered. They are objective and philosophical, preferring to view life from an intellectual perspective. Their rational mind outweighs feelings in decision making, since they consider the world of emotions to be irrational. As a result, they can be out of touch with their feelings.

Water (Intuitive-Feeling)
These people speak from the heart and are compassionate and caring people with a deeply intuitive, spiritual nature. They value personal relationships in which they often take on a caretaker role. They are such sympathetic and understanding people that they are often caught in the role of mother to their loved ones. Their weakness lies in the fact that they fail to take an objective perspective when necessary. They often react emotionally without thinking things through.

ELEMENT PAIRS

Combining the elements into pairs forms six dual-element temperaments. The following table shows the groups that are dominant in each particular element pair.

Fire-Earth	Fire-Air	Fire-Water	Earth-Air	Earth-Water	Air-Water
Cancer	Salespeople	Astrologers	Actors	Composers	Activists
Lawyers		Fashion	Birth Defect	Life (80+ yrs)	Child Performer
Math-Physics		Infant Mortality	Culinary	Military	Child Prodigy
Out-of-Body		Politicians	Murder Victims	Musicians	Comedians
Technical		Psychiatrists		Out-of-Body	Corporate
				Philosophers	IQ (high)
				Psychiatrists	Psychics
				Royalty	Salespeople
				Spiritual Teacher	Spiritual Teacher
					Wealthy

Fire-Earth (Masculine-Sensing)
These are masculine, self-centered, aggressive people who like to be the captains of their own ships. They are concrete thinking, practical enthusiasts with tremendous self-confidence, but can be dictatorial and insensitive to others. They have a pronounced sensuality and enjoy the good life. They have strong constitutions and stay warm easily.

Air-Water (Feminine-Intuitive)
These are the most feminine, helpful people who like to make their lives about others. They have a good sense of humor and creative imaginations. They are highly intuitive, sensitive people who often find the world and its responsibilities difficult to handle. They have sensitive constitutions and get cold easily.

Fire-Water (Subjective-Feeling)
These are the most personable people since they have such great emotional enthusiasm. They enjoy passionate, emotional involvement with others but can be volatile and moody. They are highly intuitive, inner directed and creative, but find it difficult to be detached and impartial.

Earth-Air (Objective-Thinking)
These are the most impersonal, intellectual types who think out everything in advance and take life seriously. They are practical idealists who bring together abstract thought with real world know-how. They have difficulty understanding their emotions and expressing feelings.

Fire-Air (Extroverted)
These people are the most active, adventurous types who are inspired by great ideas. They are visionary, future-oriented people who lack the practicality and work ethic to complete projects. They are highly entertaining and outgoing, but get bored easily.

Earth-Water (Introverted)
These are the most self-contained, reliable types who like to accomplish long-term goals, such as family and careers. They are past-orientation people who need both emotional and material security and view change as threatening. They make trustworthy friends, but have difficulty showing enthusiasm.

DIRECTIONS

The directions of the elements often tell us what the culture is like in a given area. In the Northern Hemisphere, the Air element represents the North, the Fire element represents the South, the Earth element represents the East, and the Water element represents the West.

In Europe, people living in Northern areas, such as the Britain and Scandinavia, tend to be cool, reserved and intellectual (Air). People living in Southern areas, such as Italy and Spain, tend to be much more passionate and in the moment (Fire). People living in Eastern Europe tend to be very practical and conservative (Earth). People living in Western Europe such as the Netherlands and France tend to be very liberal and creative (Water).

In the Unites States, the Northeast is where the majority of our elite universities exist. This corresponds with the objective-thinking type (Air-Earth). The Southwest, especially the Los Angeles area, is where entertainment industry is strongest and the people are the most charismatic, subjective and creative. This corresponds with the subjective-feeling type (Fire-Water).

The Northwest is where the most progressive, spiritual people live, which corresponds with the feminine-intuitive type (Air-Water). The Southeast is where the central government and military complexes are located and its people are known to be conservative and practical thinking, which corresponds with the masculine-sensing type (Fire-Earth).

(In the Southern Hemisphere, the directions are reversed. Fire represents the North, Air represents the South, Earth represents the West, and Water represents the East.)

MODES

While the elements represent the substances of creation, the three modes represent the process of creation. Each experience brought into manifestation (Cardinal) is then experienced (Fixed) and eventually disappears (Mutable).

Prior to creation, there is an idea which finds lodgment in the mind, which we think about. This represents the Mutable mode. Then, we measure in our hearts how the idea feels to us. If it feels good, we build up desire for its fulfillment. This represents the Fixed mode. When desire builds to an intense level, we act upon our desire for the completion of the idea and bring it into manifestation. This represents the Cardinal mode.

The Cardinal mode represents the power of creation and the individual's ability to act upon its will through the physical body. The Fixed mode represents love and a person's heartfelt desires. The Mutable mode represents wisdom and the mind's ability to reflect.

Cardinal	Fixed	Mutable
Power	Love	Wisdom
Creation	Preservation	Dissolution
Action (Body)	Desire (Heart)	Thinking (Mind)

Aries, Cancer, Libra, and Capricorn represent the Cardinal mode. Taurus, Leo, Scorpio, and Aquarius represent the Fixed mode. Gemini, Virgo, Sagittarius, and Pisces represent the Mutable mode.

Cardinal (Power)
These people are strong and forceful in their dealings. They are domineering, inherently courageous and believe in fair play. When in a good mood they exude exuberance, but when angry can be cruel and hurtful. They make friends easily, especially if they perceive that such friends will be useful to them. They have practical intelligence and tend to be impatient with anyone whose intelligence is not equally acute.

They are dedicated to their own self-development, which sometimes becomes a sort of ego-expansion. Their opinions are strongly held, and they can fall into fanaticism. They are visually oriented, and visualize almost everything they think about. They plan methodically and efficiently, and enjoy implementing new ideas that are dreamed up by more theoretical types. They love to engineer ideas into practical uses, but the extroverted, Cardinal types (Aries and Libra) have little direct interest in the day-to-day details of running a project or business.

Fixed (Love)
These people are predominantly magnetic, attractive people who most enjoy the plea-

sures of the senses and relationships. Patience, kindness and humility are common virtues. However, they often struggle with vanity, attachment, possessiveness and greed. They usually have very stable personalities. They study each project carefully before committing themselves. Once committed to a course of action, though, they see it through.

Due to a strong likability, they often attract friendships easily, but prefer to enter them slowly, ensuring that their relationships last. Innate self-satisfaction makes them less motivated for self-development than others may be and, therefore, can lack mental acuity or agility (Taurus and Leo). Their faith is steady and unshakable in whatever they believe, though it is often motivated by a desire to maintain the status quo. They do tend to be innately more compassionate than others. They often think with their emotions, which influence them more than their ideas or beliefs.

Mutable (Wisdom)
These people are highly intelligent, curious, and mentally agile. They think predominantly in words and have an acute sense of hearing. They are life-long learners who enjoy sharing knowledge. They are good original theorists, but tend to jump from idea to idea. This makes it difficult for them to convert their theories into realities, and often fail to complete the projects they start. They recognize the need for self-development but are rarely consistent with any one program.

They are sensitive and perceptive people who react quickly to changes in their environment. They are exceptionally changeable, and resist regularity in their lives because their active minds demand continual stimulation. When their energy is high, they can be the life of the party, but may burn out quickly. Sometimes they crave companionship and other times demand solitude. They usually make friends easily, but their friendships are often short-lived. Their hyper-adaptability gives them flexibility and a potential for detachment, but also tends to make them chaotic and spacey.

Cardinal	Fixed	Mutable
Alzheimer's	Actors	Cancer
Homosexuals	Children (4+)	Fashion Designers
Salespeople	Comedians	Marriage (none)
	Education Low	Only Child
	Life (<29 yrs)	Out-of-Body
	Marriage (15+ yrs)	Poets
	Mental Handicap	Scientists
	Models	Sex Offenders
	Murderers	Singers
	Philosophers	
	Same Job (10+ yrs)	
	Suicide	

There are two ways of dividing the zodiac into thirds or modes. The traditional way is to take every third sign as being in the same mode. The second way of dividing the zodiac into modes is to take the first third (sector) of the zodiac as being Cardinal: Aries to Cancer; the second third as being Fixed: Leo to Scorpio; and the last third as being Mutable: Sagittarius to Pisces.

Mode Sectors

Cardinal (♈-♋)	Fixed (♌-♏)	Mutable (♐-♓)
Business Owners	Activists	Artists
Children (4+)	Astrologers	Astronauts
Children None	AIDS	Comedians
Corporate Tycoons	Child Performer	Composers
Fighters	Drug Abuse	Computers
Infant Mortality	Lesbians	Fighters
Marriage (15+ yrs)	Medical	Inheritance
Racers	Models	Life (80+ yrs)
Salespeople	Mystics	Literature
Same Job (10+ yrs)	Out-of-Body	Math-Physics
Sports	Police	Mental Handicap
Stroke	Prisoners	Musicians
	Sex Abuse	Only Child
	Sex Symbols	Scientists
	Sex Workers	Suicide
		Technical
		Writers
		Twins

The results in both tables look quite similar in spite of being entirely different methods. The Cardinal groups are full of initiative, leadership skill, and power. The Fixed signs represent qualities such as steadfastness and stubbornness (which are true of Taurus and Scorpio, but not as much of Aquarius and Leo), but also represents love in its many positive and negative expressions. The Mutable mode is shown here to represent intelligence and mental activity.

Interestingly, they are very similar to the three doshas of Ayurveda. Therefore, assessing the balance of modes could quite possibly be used to assess a person's body type and dietary needs. (The Cardinal mode is associated with Pitta, the Fixed mode with Kapha, and the Mutable mode with Vata.)

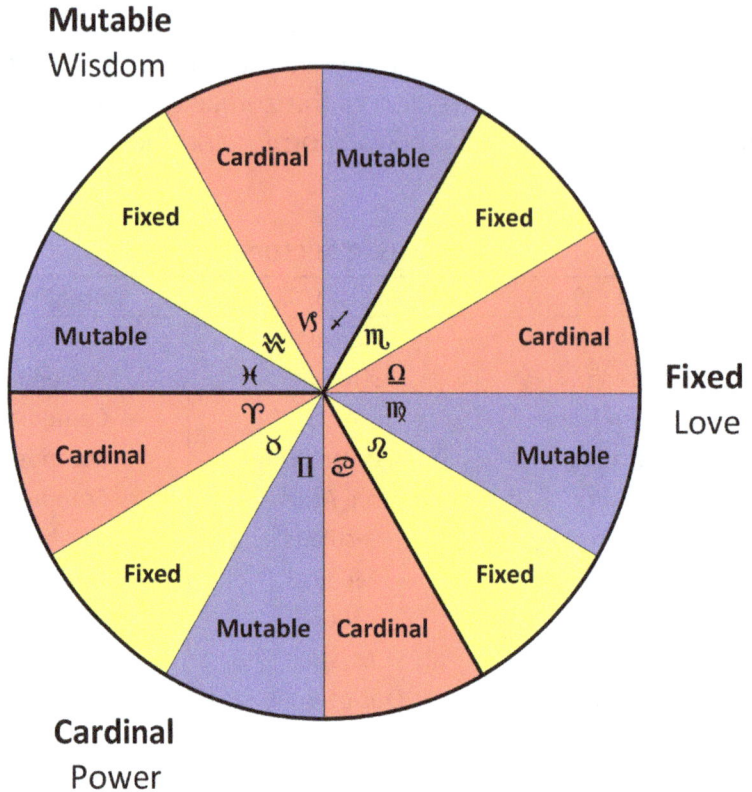

THE LIFE CYCLE

The Cardinal phase, which starts in Aries, represents childhood and early youth, when our lives are full of change and growth. At this time, we are almost totally self-centered, full of adventure, and seek constant discovery.

The Fixed phase starts in Leo at about 18 years of age, when many people leave home to either attend college or start a vocation. In this time period, we are concerned with establishing our place in society, building a career, falling in love, and having children.

The Mutable phase starts in Sagittarius at about 50 years of age, when we are finished having children and free to pursue our higher purpose. In this last stage of life, we utilize what we have developed and learned in order to benefit our family and community. It is also a time of reflection and letting go.

(These time periods were found by trisecting the time-span of each quadrant to find the length of each sign. The 1st quadrant lasts 12 years, 4 years/sign. The 2nd quadrant lasts 18 years, 6 years/sign. The 3rd quadrant lasts 30 years, 10 years/sign. The Fixed phase begins at the 1st return of the Moon's Node, which represents life purpose and karmic relationships. The Mutable phase begins at the 1st Chiron return, at the end of the midlife crisis.)

Sign Pairs of the Zodiac

There are six pairs of opposing signs, which share strong connections with each other. Each pair is called an axis. The Aries-Libra axis represents freedom of action and thought. The Taurus-Scorpio axis represents depth of commitment and perception. The Gemini-Sagittarius axis represents the gaining of information. The Leo-Aquarius axis represents charisma. The Virgo-Pisces axis represents health and healing.

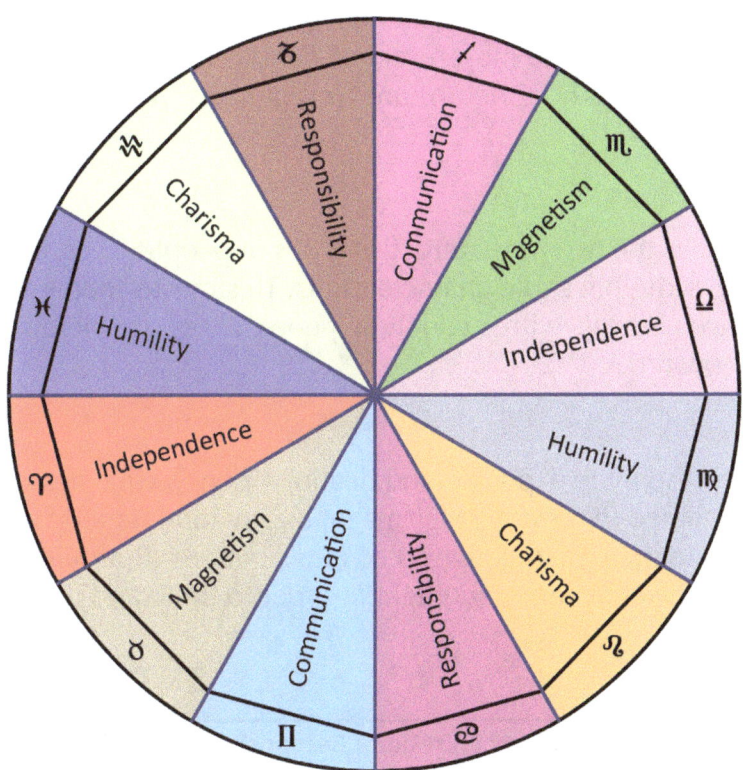

Axis of Independence
The Aries-Libra Axis marks the beginning of Spring and Fall and represents the moment when the Sun's daylight begins to gain or to lose dominance over darkness, the moment of gaining true freedom or the loss of it. Both signs are highly independent, seeking to do things their own way. Aries more care-free than Libra, since it is self focused while Libra seeks freedom for all.

Axis of Magnetism
Taurus and Scorpio represent the positive and negative magnetic poles of the zodiac. They draw people toward them with magnetism and sensuality. Like the trunk and roots of a tree, they are strong and stable and support the efforts of more adventurous types. People with this influence are deep thinkers who are extremely difficult to influence and will change only when they see fit. They get attached to their goals, which are pursued with determination and patience.

Axis of Communication
Geminis and Sagittarians are highly curious people who love to learn and share information. They are highly intelligent and articulate and quite comfortable in front of an audience. They are versatile, humorous people who can be the life of a party. Geminis are interested in concrete information, while Sagittarians are more concerned with beliefs.

Axis of Responsibility
The Cancer-Capricorn Axis represents the beginning of Summer and Winter and represents the moment of the Sun's greatest strength and weakness. Cancer is the most warm, loving sign while Capricorn is the most reserved, serious of signs. As a result, Cancers are highly effective caregivers and Capricorns are well-built for taking on hard work and the burdens of others.

Axis of Charisma
Leos and Aquarians are the two most attractive types of people. Even if not beautiful, they are very distinctive, with noticeable character traits. This makes them well-suited for public positions and involvements with a variety of people. They are likable and know how to put their best foot forward.

Axis of Humility
Virgo and Pisces represent the exposure and healing of one's weaknesses. The humility and understanding gained in these signs is meant as a natural remedy for the pride and vanity often accrued in the previous signs, Leo and Aquarius. Both Virgos and Pisceans seek to remove themselves from the spotlight in order to think and reflect upon issues that concern them.

Independence (♈-♎)	Magnetism (♉-♏)	Communication (♊-♐)	Responsibility (♋-♑)	Charisma (♌-♒)	Humility (♍-♓)
Obesity	Actors	Singers	Adopted	Comedians	Alcoholics
Salespeople	Business	Stroke	Alzheimer's	Education (low)	Cancer
	Child Abuse	Twins	Homosexuals	Lawyers	Fashion
	Children (4+)		Military	Literature	Lawyers
	Computers			Models	Life (80+ yrs)
	Criminals			Only Child	Marriage none
	Mental Illness			Politicians	Medical
	Murderers			Suicide	Only Child
	Philosophers				Out-of-Body
	Royalty				Poets
	Same Job (10+)				Sex Offenders
	Spiritual Teacher				Sex Workers
					Writers

THE ZODIAC SIGNS

The seasonal shifts of light and heat on Earth are measured by the ecliptic, which modulates like a sine wave. The zodiac is the division of the ecliptic into 12 equal sections which represent the phases of every light wave. And, since all things are made of light, the zodiac shows the process of all things. (Divine ideas cause the repetition of light waves to appear as stable objects in our minds.)

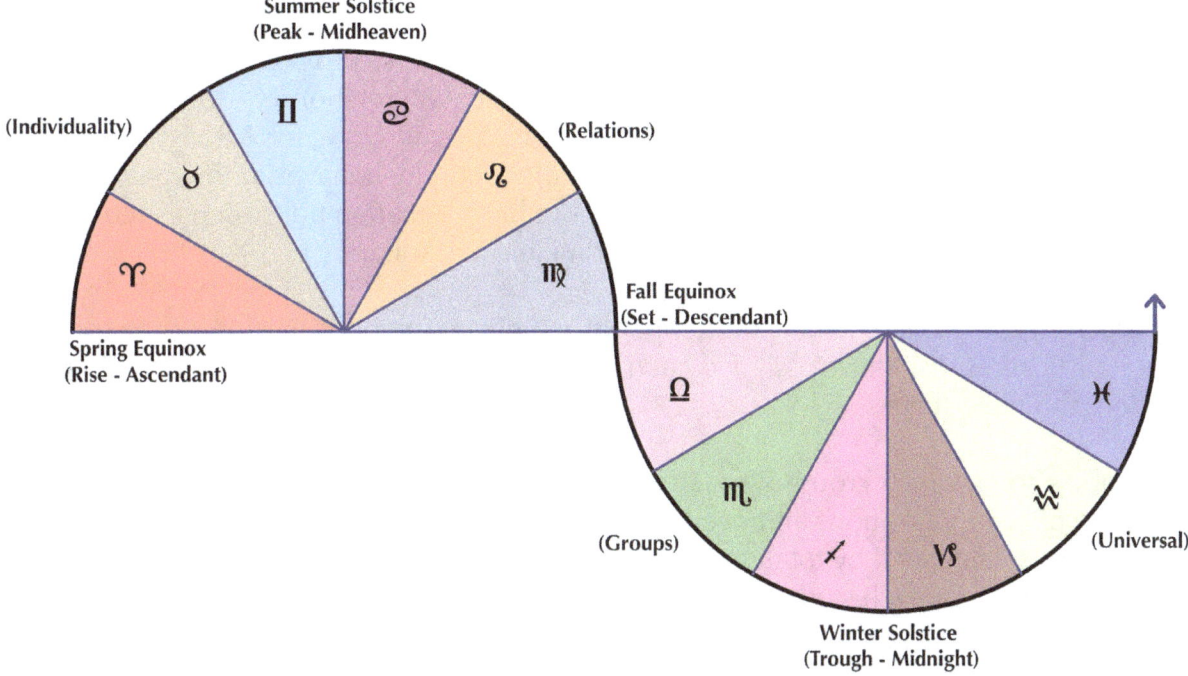

Aries through Virgo represent the upper half of the wave. Cancer is at its peak and, for this reason, our Presidents and Corporate leaders are often born with Cancer dominant in their birth charts. Libra through Pisces represent the lower half of the wave. Capricorn is at its trough and, for this reason, our Military, who sublimate themselves for the needs of the group, are born with Capricorn most dominant in their birth charts.

Since the Earth is divided into two hemispheres, their inhabitants experience the seasons in opposite order (just as opposite sides of Earth reverse day and night). Therefore, the signs of the ecliptic are opposite in opposing hemispheres and two people born at the same time in opposing hemispheres will have opposite influences from the planets. If one person has their Sun in Aries, the other will have their Sun in Libra.

For this reason, I excluded Southern Hemisphere birth charts from the study groups, except for the Obesity group, which is entirely Southern Hemisphere charts. In Appendix 6, groups of Actors and Politicians from both hemispheres are compared.

In this study, most of the signs revealed themselves to have strength in the type of groups that astrologers would have predicted. For instance, Aries is strong for Sports figures. Taurus is strong for the groups Same Job 10+ years, Life Long 80+ years, and Marriage 15+ years. Gemini is strong for Singers. Cancer is strong for the Teachers K-12. Leo is strong for Models. Virgo is strong for Medical practitioners. Libra is strong for Activists. Scorpio is strong for Mystics and Psychiatrists. Sagittarius is strong for Scientists. Capricorn is strong for Military and Technical professions. Pisces is strong for Artists.

However, some of the signs showed surprising results. For instance, Virgo is strongest for sex related groups such as Sex Abuse victims, Sex Offenders, and Sex Workers. In addition, Aquarius is strong for groups of attractive people with relationship success such as Actors, Comedians, and Marriage 15+ years, while Libra is strong for Activists and High IQ people. This led to the conclusion that Aquarius actually behaves in the manner traditionally used for Libra and that planets in Libra have Aquarian traits such as rebelliousness. (A full discussion is shown in the section Libra vs. Aquarius.)

For each sign, there are compact descriptions which are intended as aids for interpreting the statistical results. I have assigned each sign with a new name, giving a more modern understanding to them.

The tables show which groups were strongest in each of the 12 signs. The strongest individual planetary positions and overall sign strengths are both shown. The overall sign strength was calculated by totaling the sum of findings for all seven planetary bodies (Sun through Saturn) for each study group. I did not apply any sort of weighting to the Sun or Moon as is typical among astrologers, since I did not want to influence the study with preconceptions. Study group findings that show a positive or neutral common theme are highlighted in blue or yellow, while results with a negative theme are highlighted in red.

Top Findings (by probability of being an accident)

1.	Same Job (10+ yrs)	Jupiter	Cancer	136,092
2.	Artists	Mars	Taurus	13,710
3.	Only Child	Jupiter	Virgo	11,514
4.	Fighters	Mercury	Capricorn	10,135
5.	Artists	Sun	Pisces	9,894
6.	Sex Workers	Venus	Virgo	9,401
7.	Scientists	Sun	Capricorn	6,165
8.	Racers	Venus	Taurus	4,280
9.	Prisoners	Mars	Virgo	3,787
10.	Child Abuse	Venus	Virgo	3,701
11.	Only Child	Mars	Aquarius	3,639
12.	Philosophers	Mars	Scorpio	3,567
13.	Astronauts	Sun	Pisces	3,407
14.	Artists	Venus	Pisces	3,070
15.	IQ (high)	Moon	Sagittarius	2,738
16.	Prisoners	Mars	Aquarius	2,393
17.	Racers	Venus	Aquarius	2,240
18.	Lesbians	Jupiter	Sagittarius	2,214
19.	Comedians	Sun	Aquarius	2,191
20.	Marriage (15+ yrs)	Mars	Gemini	2,054
21.	Transvestites	Venus	Gemini	1,928
22.	Mental Handicap	Mercury	Aries	1,914
23.	Nazis	Moon	Libra	1,862
24.	Child Abuse	Venus	Scorpio	1,737
25.	Nazis	Jupiter	Aries	1,702
26.	Fighters	Sun	Capricorn	1,267
27.	Inheritance	Mercury	Capricorn	1,254
28.	Scientists	Moon	Sagittarius	1,247
29.	Philosophers	Jupiter	Taurus	1,225
30.	Wealthy	Moon	Capricorn	1,209
31.	Instrumentalists	Jupiter	Cancer	1,190
32.	Sports	Moon	Gemini	1,125
33.	Only Child	Mercury	Pisces	1,096
34.	Life (80+ yrs)	Mercury	Pisces	1,090
35.	Inheritance	Venus	Capricorn	1,072
36.	Spiritual Teachers	Mars	Capricorn	1,052
37.	Child Abuse	Moon	Pisces	1,026
38.	AIDS	Jupiter	Virgo	990
39.	Comedians	Venus	Aquarius	987
40.	Same Job (10+ yrs)	Moon	Taurus	947

ARIES ♈ "THE ACHIEVER"
AXIS OF INDEPENDENCE

QUADRANT: INDIVIDUAL   **ELEMENT:** FIRE   **MODE:** CARDINAL

TEMPERAMENT: EXTROVERT   **GENDER:** MASCULINE   **PERSPECTIVE:** SUBJECTIVE

Aries individuals are self-reliant and ambitious in nature. They are born leaders who love freedom. They hate to be subordinate to people and fare well only when they are their own bosses. Impatient and restless, Arians are quick-tempered and can get offended easily. However, their temper is short-lived and they will cool down just as quickly as they get angry. Their courageous and hard-working nature is the key to their success.

They are regarded as champions, who can turn any situation into a victory. They are brave and always ready to fight for a good cause. Arians have both moral and physical courage to fight against unfair and unjust situations. They have a candid nature and are quite affectionate. Arians are very expressive, outspoken, alert, analytical, quick to make decisions, generous in nature, and are always glowing with energy.

The Arian enthusiasm and aggression are seen with Nazis, Racers, Salespeople, Sports. The analytical side of Aries, which is associated with Mars, is most apparent with the groups Math-Physics, Professors, and Technical. The majority of groups with the Sun strong in Aries are very deep thinking people. The Sun is exalted in Aries and is a position of great success. The Moon's most difficult position is in Aries.

Sun		Moon	Mercury	Venus	
Children (4+)	Priests	AIDS	Children (4+)	Artists	Lesbians
Life (80+ yrs)	Professors	Dancers	Education (low)	Cancer	Marriage (15+)
Math-Physics	Racers	Drug Abuse	Math-Physics	Child Prodigy	Math-Physics
Mystics	Salespeople	Nazis	Mental Handicap	Children (4+)	Mental Illness
Nobel Prize	Same Job (10+)	Philosophers	Racers	Depression	Sports Coaches
Police	Technical	Suicide		Education (low)	Widowed
				Engineers	

Mars		Jupiter	Saturn	TOTAL	
Children (none)	Journalists	Infant Mortality	Sports	Artists	Salespeople
Computers	Marriage (none)	Life (80+ yrs)		Children (4+)	Sports
Fashion	Sports Coaches	Nazis		Math-Physics	Technical
Homosexuals		Salespeople		Nazis	
Inheritance				Professors	
Instrumentalists				Racers	

TAURUS  "THE NATURAL"
AXIS OF MAGNETISM

QUADRANT: INDIVIDUAL	**ELEMENT:** EARTH	**MODE:** FIXED
TEMPERAMENT: INTROVERT	**GENDER:** MASCULINE	**PERSPECTIVE:** OBJECTIVE

Taureans are slow and methodical individuals with a relaxed and laid-back temperament. They are warm, sensual, loving by nature and can easily display care and affection for others. They love peace and hate quarreling. They are generally noted to be quite good looking and have attractive personalities. They usually have an artistic bent of mind and are great lovers of music, art, and beauty.

Taureans are practical and materialistic and known for their down-to-earth approach. They have tremendous patience, immense will power and self-discipline. Once they make up their minds, there is nothing in the world that can deter them from their decisions. Taureans are not risk-takers and always weigh their decisions carefully.

The Taurean stability is evidenced through the groups Children 4+, Life Long 80+ years, and Same Job 10+ years. Taureans are also known for their deep, rich voices, which is seen with Singers being prominent in this sign. Taurus is especially frequent with musicians.

Taurus is ruled by Venus and is its best position. Mars is in detriment in Taurus and is its most difficult position (along with Virgo).

.

Sun	Moon	Mercury	Venus	
Directors	Comedians	Children (4+)	Business Owners	Nobel Prize
Education (low)	Journalists	Drug Abuse	Children (4+)	Novelists
Fashion Designers	Same Job (10+ yrs)	Fashion Designers	Children (none)	Priests
Life (80+ yrs)		Murderers	Composers	Racers
Murderers		Musicians	Instrumentalists	Royalty
Scientists		Singers	Life (80+ yrs)	Same Job (10+ yrs)
Singers		Stroke	Math-Physics	Technical
Widowed			Musicians	
Mars	**Jupiter**	**Saturn**	**TOTAL**	
Artists	Actors	Birth Defect	Alzheimer's	Murderers
Child Abuse	Cancer	Computers	Artists	Musicians
Directors	Criminals	Life (<29 yrs)	Business Owners	Nobel Prize
Homosexuals	Dancers	Marriage (15+ yrs)	Children (4+)	Priests
Life (80+ yrs)	Infant Mortality	Mental Handicap	Children (none)	Same Job (10+ yrs)
Military	Journalists	Models	Fashion Designers	Scientists
Murderers	Murder Victims	Songwriters	Life (80+ yrs)	Singers
Murder Victims	Philosophers	Sports	Mental Handicap	Sports
Priests				
Sex Abuse				

Gemini ♊ "The Inquirer"
Axis of Communication

Quadrant: Individual   **Element:** Air   **Mode:** Mutable

Temperament: Extrovert   **Gender:** Feminine   **Perspective:** Objective

Geminis are bright, witty, entertaining individuals who are adaptable in all situations. They have an optimistic outlook on life and are always ready to take up new activities. Curiosity is the driving force for Geminis, who want to satisfy their quest for knowledge. However, they rarely get deeply involved in one particular task or interest since their attention span is very limited and they get diverted easily.

Geminis are egocentric and restless individuals who are often unstable and unpredictable, especially in relationships. A popular belief amongst Geminis is that life must be lived to the fullest, which makes them enjoyable companions.

Gemini's talent for thinking is mainly seen with the Moon, which is strong with Journalists, Poets, and Writers. Business Owners are strong with many different planets in Gemini, showing their cleverness and skill working with people.

Mars is the strongest planet in Gemini. Mercury, thought to rule Gemini, is strong for Astrologers and Mystics.

Sun	Moon		Mercury	Venus
Business Owners	Activists	Mental Handicap	Astrologers	Adopted
Corporate Tycoons	Directors	Out-of-Body	Business Owners	Birth Defect
Medical	Fighters	Poets	Mystics	Sports
Salespeople	Journalists	Priests	Prisoners	Transvestites
Sex Abuse	Life (<29 yrs)	Sports	Sports	
	Marriage (none)	Writers		

Mars		Jupiter	Saturn	TOTAL
Business Owners	Politicians	Alcoholics	Infant Mortality	Business Owners
Child Performers	Same Job (10+ yrs)	Astronauts	Inheritance	Corporate Tycoons
Corporate Tycoons	Singers	Police	Instrumentalists	Salespeople
Engineers	Stroke	Widowed	Musicians	Singers
Marriage (15+ yrs)	Wealthy		Presidents	Sports
IQ (high)			Same Job (10+ yrs)	Stroke
			Singers	
			Sports	

Cancer ♋ "The Caretaker"
Axis of Responsibility

Quadrant: Relations	**Element:** Water	**Mode:** Cardinal
Temperament: Introvert	**Gender:** Feminine	**Perspective:** Subjective

Cancers are good listeners and are caring by nature, with a streak of possessiveness. They are kind and sympathetic to the problems of others. Cancers are reliable, sincere, dedicated, responsible and determined. They are, by and large, good looking, with a very pleasing personality. They love the comfort and security of their homes and will usually be the most pampered ones in the family.

Childhood memories play an influential role in the minds of Cancer individuals, who can have an affinity for living in the past. They are introspective but also imaginative and can become victims of fantasy. They are very moody and can get aggressive at times. While Cancers are usually emotional, romantic and sentimental, when feeling defensive they can be thick-skinned, unemotional, uncompromising, and stubborn.

The nurturing nature of Cancer is strongly shown by the strength of the Culinary and Teachers K-12 groups. Cancer's association with childhood is seen with Child Prodigy and Infant Mortality.

The Sun is the most beneficent planet in Cancer, while the Moon is surprisingly uncomfortable in its ruled sign. Saturn is weak in its sign of detriment.

Sun	Moon	Mercury	Venus
Birth Defect	Alcoholics	Dancers	AIDS
Child Prodigy	Alzheimer's	Infant Mortality	Astrologers
Culinary	Corporate Tycoons	Salespeople	Fashion Designers
Depression	Culinary	Teachers (K-12)	Lesbians
Fashion Designers	Heart Attack		Teachers (K-12)
Journalists	Police		
Philosophers	Stroke		
Royalty			

Mars	Jupiter	Saturn	TOTAL
Child Prodigy	Alcoholics	Astrologers	Birth Defect
Depression	Infant Mortality	Heart Attack	Child Prodigy
	Instrumentalists	Infant Mortality	Corporate Tycoons
	Mental Illness	Lesbians	Depression
	Salespeople	Out-of-Body	Fashion Designers
	Same Job (10+ yrs)	Sex Abuse	Infant Mortality
		Widowed	Military
			Salespeople
			Teachers (K-12)

Leo ♌ "The Leader"
Axis of Charisma

Quadrant: Relations   **Element:** Fire   **Mode:** Fixed

Temperament: Extrovert   **Gender:** Masculine   **Perspective:** Subjective

Leos are enthusiastic individuals, who are quite affable and pleasing in nature. They have a creative bent of mind, are self-expressive, intelligent, and broad-minded. Leos love to be the center of attention and crave flattery. They have a flair for luxury, love to be pampered, and are generous in nature. But, they are also lazy, arrogant, and very possessive of loved ones.

Born leaders, Leos are known to be ambitious, courageous, strong-willed, positive, independent and self-confident. When placed in commanding situations, Leos tend to fare extremely well. However, they can become short-tempered and blunt, if aggravated. They are prone to short periods of depression, but also know how to bounce back to become their cheery and jovial selves.

The attention loving aspect of Leo is seen through the strength of the groups Models and Politicians. Leo's dominant nature is mainly seen in the placement of Saturn with Police Officers, Politicians, and Sports Coaches.

This is a strong position for Saturn, even though Leo is its sign of detriment.

Sun	Moon	Mercury	Venus
AIDS Dancers Lawyers Lesbians Only Child Sex Offenders	Activists Inheritance Only Child Technical	Drug Abuse Models Psychiatrists Stroke	Philosophers Sex Abuse Wealthy
Mars	**Jupiter**	**Saturn**	**TOTAL**
Astrologers Dancers Sex Workers Sports	IQ (high) Police	Astrologers Homosexuals Lesbians Medical Musicians Police Politicians Songwriters Sports Coaches	Astrologers Fashion Designers Lawyers Lesbians Models Politicians

Virgo ♍ "The Perfectionist"
Axis of Humility

Quadrant: Relations	**Element: Earth**	**Mode: Mutable**
Temperament: Introvert	**Gender: Masculine**	**Perspective: Objective**

Virgos are conservative, reserved, polite and soft spoken. They are methodical and practical in the way they lead their lives and have a logical, detail-oriented mindset. They are strong-willed and pursue their goals with determination. They appreciate things of beauty, are very tidy by nature, and like to keep their surroundings organized and clean.

Virgos can be emotionally cold, critical of others, and hesitant to commit themselves to new relationships. A Virgo individual expects a lot from others and from themselves. They have strong guilt complexes, since they distrust emotion and sexuality. This causes much difficulty for them. Their need for control and fear of chaos and injury often lead them to the very experiences they seek to avoid.

As a result, Virgo is extremely frequent with people who have suffered difficulties in their lives, such as AIDS, Cancer, Drug Abuse, and Heart Attack victims. The sex-related groups are also very strong in Virgo as it helps ease their anxieties (which completely contradicts its traditional name of the Virgin).

Overall, Virgo is the most difficult sign position, especially for Venus, Mars, and Saturn.

Sun	Moon	Mercury	Venus		Mars
Cancer Transvestites	Actors Alcoholics Business Owner Drug Abuse Out-of-Body	Lawyers Models Sex Offenders Sex Workers	AIDS Astronauts Child Abuse Drug Abuse Journalists Medical	Priests Psychiatrists Sex Abuse Sex Offenders Sex Workers	Culinary Depression Mental Illness Out-of-Body Prisoners Sex Workers Suicide

Jupiter	Saturn			TOTAL	
AIDS Fashion Life (<29 yrs) Medical Only Child Out-of-Body Sex Offenders Sports	AIDS Astrologers Cancer Child Abuse Children (none) Lesbians Marriage (none)	Medical IQ (high) Military Murder Victims Mystics Out-of-Body		Actors AIDS Cancer Child Abuse Drug Abuse Heart Attack Medical Military	Only Child Out-of-Body Priests Prisoners Sex Abuse Sex Offenders Sex Workers

Libra ♎ "The Protagonist"
Axis of Independence

Quadrant: Groups	**Element: Air**	**Mode: Cardinal**
Temperament: Extrovert	**Gender: Feminine**	**Perspective: Objective**

Librans are thoughtful, insightful, and sociable people with the ability to charm their many friends and associates. The most distinct characteristic of Librans is their direct presentation. With their astute minds and social intelligence, they are able to promote causes and rally support for their ideas.

Librans are active and open-minded by nature and are highly inclusive. They have a true distaste for injustice or any form of ugliness. Though Librans are peace-loving by nature, they are quick to stand up for others. Librans can often be found in self-help movements that promote higher standards of well-being and goodness in all.

The Sun and Saturn in Libra is strong for groups of people that are thoughtful in nature and need to express their ideas. Libra is in exaltation for Saturn, but in fall for the Sun. Jupiter appears to be the most difficult planet in Libra.

Sun	Moon	Mercury	Venus
Activists Child Performers Sex Workers Songwriters Spiritual Teachers	Alzheimer's Child Prodigy Culinary Nazis Professors Sex Abuse Sex Workers Teachers (K-12)	Sex Workers	Alcoholics
Mars	**Jupiter**	**Saturn**	**TOTAL**
Murder Victims Salespeople	Mental Handicap Murderers	Artists Child Performers Comedians Medical IQ (high) Mystics Spiritual Teachers	Activists IQ (high) Sex Workers

Scorpio ♏ "The Deep Diver"
Axis of Magnetism

Quadrant: Groups   **Element:** Water   **Mode:** Fixed

Temperament: Introvert   **Gender:** Feminine   **Perspective:** Subjective

Scorpios are known for their intense, magnetic, and powerful natures. They believe life is meant to be lived to the fullest and show incredible dedication to their goals. Scorpios have outstanding imaginations and intuitive powers. A mysterious aura surrounds them and people always find something exciting and fascinating about them. They may look reserved, but are always aware of everything that goes on around them or concerns them. Scorpio's deep insight into everything makes them difficult to be understood.

Scorpios are totally committed in relationships, but once threatened or betrayed, they can display highly manipulative, possessive behavior. When their deep attachments are threatened, they are willing to do anything necessary to hold on to a cherished person, goal or idea. It is, therefore, extremely important for Scorpios to maintain their integrity.

Scorpio's profound, deep-thinking nature is shown with the strength of groups such as Astrologers, Mystics, Psychiatrists, and Spiritual Teachers. Their tendency towards difficult life situations is seen with AIDS patients, Child Abuse Victims, Criminals, Mental Illness, Prisoners, and Transvestites.

Jupiter's best position is in Scorpio, with many people achieving notoriety.

Sun	Moon	Mercury	Venus	Mars
Computers	Criminals	Astrologers	Child Abuse	Drug Abuse
Heart Attack	Lawyers	Criminals	Child Performers	Medical
Police	Murderers	Heart Attack	Comedians	Murder Victims
Priests	Only Child	Homosexuals	Computers	Philosophers
	Psychiatrists	Mental Illness	Life (<29 yrs)	
	Same Job (10+)		Mystics	
			Prisoners	
			Spiritual Teachers	
			Transvestites	

Jupiter	Saturn		TOTAL	
Birth Defect	Activists	Education (low)	AIDS	Life (<29 yrs)
Child Performers	AIDS	Journalists	Astrologers	Mental Illness
Homosexuals	Business Owners	Mental Illness	Child Abuse	Mystics
Models	Child Abuse	Spiritual Teachers	Child Performers	Police
Professors	Children (4+)	Stroke	Comedians	Prisoners
Racers	Comedians	Transvestites	Computers	Psychiatrists
Sports	Corporate Tycoons	Wealthy	Criminals	Royalty
Wealthy			Heart Attack	Spiritual Teachers
			Homosexuals	Transvestites

SAGITTARIUS ♐ "THE SEEKER"
AXIS OF COMMUNICATION

QUADRANT: GROUPS	**ELEMENT: FIRE**	**MODE: MUTABLE**
TEMPERAMENT: EXTROVERT	**GENDER: MASCULINE**	**PERSPECTIVE: SUBJECTIVE**

Sagittarians are optimistic, full of energy, and have versatile, adventurous minds. They are quick learners and profound philosophical thinkers, gifted with foresight and good judgment. They enjoy traveling and exploring new places and are continually searching for new experiences. They love challenges in life and live by their own standards.

They are social beings who love to share ideas and entertain others, but are very forthright, sometimes to the point of being rude. Sagittarians are susceptible to anger, but easily get back to their cheerful selves. They are often found pursuing goals in new projects, which can be demanding for other people around them. It can be difficult to be in relationships with them, since they seek trust from others while requiring freedom for themselves.

The Sagittarian urge to understand life is seen with the frequency of Scientists in this sign. The Moon in Sagittarius is especially strong for technical groups, while Jupiter is strong for creative thinkers.

Mars is strong for people with mental problems, while Saturn is common for socially defiant people.

Sun	Moon		Mercury	Venus	
Instrumentalists	Adopted Astronauts Depression Fashion Marriage (15+)	Math-Physics IQ (high) Scientists Songwriters Technical	Comedians Computers Prisoners Scientists	Artists Journalists Stroke	

Mars		Jupiter	Saturn		TOTAL
Alcoholics Alzheimer's Astrologers Children (4+) Culinary	Infant Mortality Literature Mental Handicap Mental Illness Twins	Artists Dancers Lawyers Lesbians Out-of-Body Singers Transvestites Twins	AIDS Cancer Children (none) Composers Criminals Depression	Infant Mortality Mental Illness Murderers Prisoners Scientists	Composers Scientists Stroke Twins

Capricorn ♑ "The Mountain Climber"
Axis of Responsibility

Quadrant: Universal	**Element:** Earth	**Mode:** Cardinal
Temperament: Introvert	**Gender:** Masculine	**Perspective:** Objective

Capricorns are hard-working, trustworthy, and reliable people. They take life seriously, are self-contained, and self-satisfied. Mostly cautious, confident, strong-willed, reasonable and hard working, Capricorns are as steady as a rock. They love being at the top of the ladder and being in commanding, respectable and rewarding positions of authority. Rational, logical and clearheaded, Capricorns have great concentration skills which help them in achieving success in life.

However, Capricorns tend to be pessimistic, suspicious, resentful, and stubborn individuals. Though surrounded by friends, they are lonely at heart. They can spread gloom and tension in a minute and are quite capable of depressing everyone around them. They often feel burdened by responsibilities, which others, they feel, leave neglected.

Capricorn is strong for analytical groups such as Math-Physics, Scientists, and Technical. Fighters and the Military reveal the hard-working, tough, durable side of this sign.

Mercury is very creative and philosophical in Capricorn. Mars, exalted in Capricorn, is strong with highly intelligent groups of people. The Moon, in its detriment, is common with people who experience much conflict with others.

Sun		Moon	Mercury		Venus
Adopted	Scientists	Military	Composers	Musicians	Homosexuals
Composers	Singers	Nazis	Fighters	Philosophers	Inheritance
Dancers	Songwriters	Police	Homosexuals	Scientists	Math-Physics
Fighters	Suicide	Prisoners	Inheritance	Songwriters	Only Child
Medical	Technical	Wealthy	Life (80+ yrs)	Technical	Suicide
Musicians	Twins		Military	Twins	Twins
Racers					

Mars	Jupiter	Saturn		TOTAL	
Birth Defect	Composers	Adopted		Adopted	Math-Physics
Comedians	Out-of-Body	Alzheimer's		Birth Defect	Military
Computers	Psychiatrists	Astronauts		Comedians	Musicians
Directors		Culinary		Composers	Scientists
Math-Physics		Priests		Fighters	Technical
Professors		Transvestites		Homosexuals	Twins
Spiritual Teacher				Inheritance	Writers
Technical					

Aquarius ♒ "The Humanist"
Axis of Charisma

QUADRANT: UNIVERSAL	ELEMENT: AIR	MODE: FIXED
TEMPERAMENT: EXTROVERT	GENDER: FEMININE	PERSPECTIVE: OBJECTIVE

Aquarians are intelligent, clear and logical people. They have high intellectual ability and are usually pathfinders for society. They have strong convictions and do not easily accept when they are wrong. When engaged in serious conversation they are edgy, enthusiastic and lively. Aquarians are in favor of reforms, change, and advancement of the human race, while hating dishonesty and treachery.

Since they love to read, write and contemplate the world, they need time to themselves. Such is their peace-loving nature that they would do their best to cooperate and compromise with everyone around them. For this reason, they are highly appreciated and tend to have long, close relationships.

Aquarius is strong with groups that are especially talented in relating to others such as Actors, Comedians, Marriage 15+ yrs, and Salespeople.

The Moon, Mercury, and Saturn show very positive results in Aquarius, with many creative, intelligent people. The Sun in Aquarius gives highly successful groups of people, but also those who cause or are subjected to violence.

Sun		Moon	Mercury	Venus	
Activists	Murder Victims	Composers	Artists	Actors	Medical
Child Abuse	Nazis	Computers	Astronauts	Alcoholics	Police
Comedians	Only Child	Musicians	Comedians	Children (none)	Racers
Homosexuals	Philosophers	Psychiatrists	Literature	Comedians	Singers
Inheritance	Politicians	Salespeople	Math-Physics	Marriage (15+)	Sports
Literature	Presidents	Same Job (10+)	Military	Marriage (none)	Suicide
Marriage (15+)	Suicide	Sex Workers	Politicians		Wealthy
Military	Wealthy		Psychics		
			Suicide		
			Writers		

Mars	Jupiter	Saturn		TOTAL	
Composers	Alzheimer's	Actors		Actors	IQ (high)
Journalists	Culinary	Child Prodigy		Artists	Mental Handicap
Life (80+ yrs)	Education (low)	Mental Handicap		Comedians	Murder Victims
Only Child	Marriage (none)	Models		Culinary	Nazi
Prisoners	Stroke	Royalty		Education (low)	Only Child
Salespeople				Life (<29 yrs)	Racers
				Literature	Salespeople
				Marriage (15+)	Suicide

Pisces ♓ "The Artisan"
Axis of Humility

QUADRANT: UNIVERSAL	ELEMENT: WATER	MODE: MUTABLE
TEMPERAMENT: INTROVERT	GENDER: FEMININE	PERSPECTIVE: SUBJECTIVE

Pisceans are friendly and pleasant by nature, with a true sense of kindness and compassion. They are highly non-judgmental and understanding people. They enjoy time alone where they use their highly intuitive and imaginative minds for spiritual and artistic pursuits. They have a great adaptive quality and can change with the times.

Pisceans lack decisiveness and can get diverted from their purpose easily. They can be impractical dreamers and are disorganized in things large and small, causing distress to people near them. They often get involved in religious pursuits, enjoying the ideas and leadership of others. They do not attach great value to money and would rather spend on others than increase their own level of comfort and luxuries. This leaves them vulnerable to people less considerate than themselves.

Pisces' reflective and creative nature is most demonstrated by the frequency of Artists and brilliant thinkers in this sign. Mercury in Pisces shows itself to be a very good position for intelligent thought, in spite of being its fallen position.

Pisces is the last sign of the zodiac and represents the final stage of life. People with beneficial planets in this sign tend to live long lives, since they enjoy reclusive living.

Sun		Moon		Mercury	
Artists	Math-Physics	Child Abuse	Actors	Nobel Prize	Psychiatrists
Astronauts	IQ (high)	Children (none)	Alcoholics	Only Child	Racers
Business Owner	Mental Handicap	Fighters	Artists	Poets	Scientists
Cancer	Musicians	Instrumentalists	Cancer	Politicians	Sports
Dancers	Psychics	Murder Victim	Infant Mortality	Presidents	Sports Coaches
Lawyers	Sex Offender	Novelists	Life (80+ yrs)	Priests	Technical
Life (80+ yrs)	Sports Coaches	Poets	Marriage (none)	Professors	Wealthy
			Math-Physics		

Venus	Mars	Jupiter	Saturn	TOTAL	
Alcoholics	Child Prodigy	Scientists	Composers	Artists	Math-Physics
Artists	Education (low)		Criminals	Astronauts	Mental Handicap
Cancer	Life (<29 yrs)		Directors	Cancer	Musicians
Life (80+)	Literature		Mental Handicap	Child Prodigy	Poets
Literature	Politicians		Same Job (10+)	Composers	Politicians
Musicians	Psychiatrists		Spiritual Teacher	Education (low)	Psychics
Twins	Suicide		Sports	Life (80+ yrs)	Sports
	Writers			Marriage (none)	Wealthy
					Writers

THE 9ᵀᴴ HARMONIC

The path of the Sun, known as the ecliptic, can be divided in many different ways. The Babylonians divided the zodiac into 12 signs, the Egyptians divided it into 36 decans, and the Vedic astrologers of India use 27 lunar mansions (nakshatras).

Many astrologers also find that important astrological information can be found by using harmonics, which are created by dividing the zodiac into sections, each with 12 smaller zodiac signs. A harmonic has the same pattern as the zodiac, but is smaller to the exact ratio of the harmonic being used. (This is fractal geometry applied to the zodiac.) Here is an example of a simple pattern repeating itself up to 3 times:

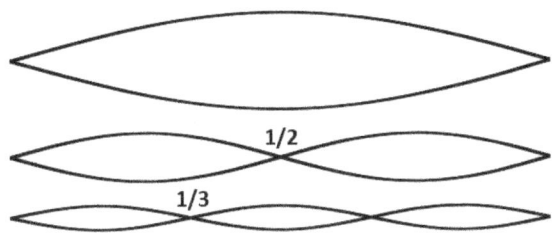

For years, astrologers have been seeking to gain knowledge of the meaning of planetary positions in various harmonics. The theory is that each harmonic has its own interpretation and can be used to describe specific aspects of life. For instance, in Vedic astrology, the 10ᵗʰ harmonic is used to describe a person's vocation and professional success. The strengths and weaknesses of this chart would be interpreted to describe your career. The 12ᵗʰ harmonic is taken to describe one's parents.

The most widely used method is to divide the zodiac into 9 equal sections, with 108 signs in total (9x12). This is called the 9ᵗʰ harmonic of the zodiac. In this harmonic, there are 9 places on the ecliptic where each sign can be found. These signs are 1/9th their original size, but add to 30 degrees in total. A chart can be created where planets in each sign are combined together into a 12 sign diagram, looking just like a normal birth chart.

In Vedic astrology, the 9ᵗʰ harmonic is called the Navamsha chart and represents relationships. It is considered the inner counterpart to the 1ˢᵗ harmonic or natal chart. The 1ˢᵗ harmonic is the tree, the 9ᵗʰ harmonic is the fruit. The 1ˢᵗ harmonic is the outer or visible life, while the 9ᵗʰ harmonic shows the soul of a person.

Planetary positions in the 9ᵗʰ harmonic have great importance in our birth charts, adding tremendous depth to the analysis. For instance, a person with their Sun in the Scorpio navamsha of Gemini will be far more introspective and intense than a typical Gemini. Also, conjunctions of planets in the 9ᵗʰ harmonic chart shine right through as major forces in our personalities, which are usually hidden from sight in the normal birth chart.

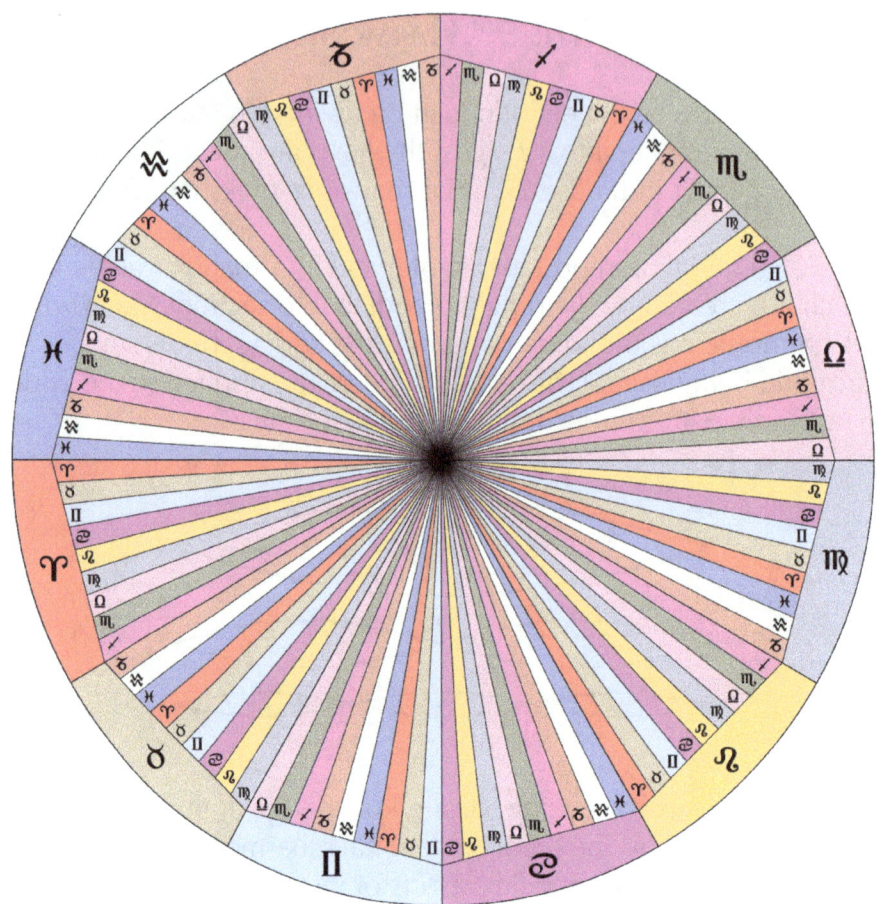

Any astrological analysis not including the 9th harmonic chart will only give a partial picture of a person. The 9th harmonic chart will often be the more dynamic chart and dominate the character of the individual. In these cases, the 1st harmonic chart will not be as relevant to the character of the person.

In my analysis of the 9th harmonic of the Tropical (seasonal) zodiac, just as many key findings were seen as in the 1st harmonic study and it was found that the two harmonics even have a strong correlation with each other. Twenty-three of the 93 study groups were statistically strong in the same sign(s) in both the 1st & 9th harmonics. If there were no correlation between the influences of planetary positions in the two harmonics, then only about 5 study groups would have been strong in the same sign, which is what was found when I studied the Constellation (Sidereal) zodiac. This is almost 5 times greater than expected, with a probability of 1 in 14,000 of being a random accident.

Even more importantly, the signs in which these study groups were strong could have been anticipated through knowledge of the zodiac signs. For instance, Child Prodigies were strong in both harmonics in nurturing, home-loving Cancer, while Military

Aries	Taurus	Gemini	Cancer	Leo	Virgo
Cancer	Alcoholics	Astrologers	Alzheimer's	Children 4+	AIDS
Heart Attack	Comedians	Business Owner	Child Prodigy	Fashion	Alcoholics
Politicians	Directors	Mental Illness	Corporate	Designers	Child Performer
Sports Coaches	Heart Attack		Culinary		Children (none)
	Nobel Prize		Fashion		Criminals
	Only Child		Presidents		Inheritance
	Psychiatrists		Psychiatrists		Marriage (none)
	Racers		Singers		Military
	Sex Abuse				Out-of-Body
	Suicide				

Libra	Scorpio	Sagittarius	Capricorn	Aquarius	Pisces
Astronauts	Child Performer	Corporate	Activists	Actors	Royalty
Stroke	Computers	Instrumentalists	Child Abuse	Children 4+	
	Fighters	Marriage 15+	Composers	Literature	
	Homosexuals	Salespeople	Computers	Marriage 15+	
	Math-Physics	Scientists	Drug Abuse	Novelists	
	Mental Illness	Wealthy	Homosexuals	Only Child	
	Murderers		Military	Prisoners	
	Prisoners		Models	Writers	
	Psychiatrists		Out-of-Body		
	Technical		Politicians		

personnel were strong in both harmonics in the hard-working, disciplined signs Virgo and Capricorn. Psychiatrists were strong in both harmonics in probing, analytical Scorpio, while Scientists were strong in both harmonics in exploratory, knowledge-seeking Sagittarius. Also, Actors were strong in the charismatic and intellectual sign Aquarius.

(For the summary, only the groups with overall sign strength are listed. This was done by totaling the amount of planetary positions for each study group in all seven planetary bodies (Sun through Saturn) combined. The blue highlights show the findings that are statistically strong in both the 1st harmonic and 9th harmonic studies.)

Top Findings (by probability of being an accident)

#	Category	Planet	Sign	Value
1.	Business Owners	Mercury	Libra	28,280
2.	Psychics	Saturn	Sagittarius	23,215
3.	Obesity	Mars	Cancer	15,193
4.	Criminals	Jupiter	Virgo	13,772
5.	Alzheimer's	Mercury	Cancer	12,894
6.	Astrologers	Venus	Cancer	9,744
7.	Scientists	Sun	Taurus	7,722
8.	Military	Mercury	Capricorn	4,473
9.	Homosexuals	Venus	Capricorn	4,321
10.	Directors	Mercury	Taurus	3,394
11.	Racers	Jupiter	Aries	3,323
12.	Child Abuse	Saturn	Capricorn	2,728
13.	Spiritual Teachers	Jupiter	Leo	2,634
14.	Children (none)	Saturn	Virgo	2,518
15.	Police	Saturn	Leo	2,449
16.	Medical	Mars	Pisces	2,315
17.	Mental Illness	Sun	Gemini	2,240
18.	Comedians	Mars	Taurus	1,977
19.	Math-Physics	Saturn	Taurus	1,704
20.	Priests	Sun	Leo	1,571
21.	Psychiatrists	Jupiter	Cancer	1,441
22.	Presidents	Mercury	Cancer	1,430
23.	Astrologers	Sun	Libra	1,394
24.	Culinary	Mercury	Cancer	1,348
25.	Sex Abuse	Mercury	Leo	1,260
26.	Child Prodigy	Mars	Libra	1,228
27.	Life (<29 yrs)	Jupiter	Virgo	992
28.	Sex Workers	Mars	Scorpio	942
29.	Salespeople	Saturn	Capricorn	904
30.	Transvestites	Venus	Scorpio	898
31.	Composers	Venus	Capricorn	894
32.	Royalty	Jupiter	Libra	809
33.	Widowed	Venus	Scorpio	806
34.	Transvestites	Saturn	Virgo	763
35.	Nobel Prize	Venus	Taurus	742
36.	Only Child	Sun	Aquarius	725
37.	IQ (high)	Moon	Leo	675
38.	Criminals	Mercury	Virgo	600
39.	Stroke	Jupiter	Libra	593
40.	Sex Abuse	Saturn	Capricorn	566

NAVAMSHAS

Navamsha is the name for the 1/9th portions of each zodiac sign in Vedic astrology. Navam means nine in the Tamil language spoken in Southern India.

For each sign, a compact description of its navamshas is shown below as well as a listing of the groups with statistical strength in each navamsha. These listings help us to understand which sections contribute the most for each study group. The planetary effects vary in different portions of each sign.

For instance, we see that Astrologers are strong in 1st harmonic Scorpio. But in reality, Astrologers are not strong in every part of Scorpio. They are actually strongest in the Scorpio and Sagittarius navamshas of Scorpio. Similarly, Musicians are not strong in every part of Pisces. They are strongest in the Leo and Pisces navamshas of Pisces. This adds great detail to the evaluation of planetary positions in the zodiac.

This study also shows us that the effects of sign-navamsha combinations can vary greatly. In Aries, the Gemini navamsha was strong for those with Psychics. However, in Taurus, the Gemini navamsha was strong for those with Mental Illnesses. The mental aptitude for which Gemini is known clearly expresses itself much better in Aries than Taurus.

There were many cases where reversing the sign and navamsha yielded the same result::

Group	Sign-Navamsha	Sign-Navamsha
Fashion Designers	Taurus-Leo	Leo-Taurus
Composers	Gemini-Capricorn	Capricorn-Gemini
Lesbians	Cancer-Leo	Leo-Cancer
Sex Workers	Cancer-Leo	Leo-Cancer
Military	Virgo-Capricorn	Capricorn-Virgo

In other cases, reversing the sign and navamsha gave reversed effects:

Group	Sign-Navamsha
Murderers	Taurus-Leo
Suicide	Leo-Taurus
Criminals	Scorpio-Aquarius
Police Officers	Aquarius-Scorpio
Psychiatrists	Cancer-Scorpio
Mental Illness	Scorpio-Cancer
Nazis	Scorpio-Aquarius
Murder Victims	Aquarius-Scorpio

Aries

Navamshas in Aries

Aries: Math-Physics, Politicians, Salespeople, Technical

These are forceful, independent, hardworking people with definite goals. They are, however, very self-centered, impatient and insensitive to others.

Taurus: Birth Defect

These are highly determined, money-oriented, sexually magnetic people. However, they are self-centered, stubborn, and unwilling to listen to others.

Gemini: Psychics

These are very alert, intelligent, adventurous people. However, they can be unreliable, impatient, and scattered.

Cancer: Child Prodigy, Engineers, Models, Same Job (10+ yrs)

These people are caring, romantic, and highly imaginative. However, they are also volatile, over-sensitive, and depressive.

Leo: Salespeople

These are confident, creative, romantic people who inspire confidence. They are egotistical though, demanding, and lack self-awareness.

Virgo: Mental Handicap, Sports

These are highly organized, hard-working, dutiful people. However, they can be extremely critical, pessimistic, and prone to feelings of guilt.

Libra: Depression

These are independent, pioneering, self-willed people who have big dreams. They are self-absorbed though, insensitive, and rebellious.

Scorpio: Math-Physics
These are very aggressive, intense, ambitious people. They are also devious, dangerous, and overly sexual.

Sagittarius: Homosexuals

These are adventurous philosophers who carve out their own merry paths. They can also be presumptuous, immature, and ungrounded.

Taurus

Navamshas in Taurus

Capricorn: Infant Mortality, Mental Handicap, Out-of-Body

These are very conservative, reliable, practical people. However, they dislike change and are stubborn and bossy.

Aquarius: Children (4+), Spiritual Teachers, Stroke

These are very attractive, diplomatic, and sensual lovers of luxury. However, they can be vain, materialistic, and superficial.

Pisces: Instrumentalists, Musicians, Royalty, Singers

These are very patient, self-contained, artistic people. However, they are lazy, withdrawn, and illogical.

Aries: Computer Programmers

These people are confident, practical achievers with high charisma. However, they are also self-indulgent, insensitive, and have a strong temper.

Taurus: Alzheimer's, Murderers, Nobel Prize, Priests, Psychiatrists

These are very physical, sensual, practical people who cannot be pushed. They are stubborn, unimaginative, intolerant, and materialistic.

Gemini: Child Abuse, Mental Illness

These people are charming, persuasive, with a good business sense. They are also materialistic, untrustworthy, and argumentative.

Cancer: Fashion, Inheritance, Marriage (none), Models, Sex Workers, Sports

These people are patient, caring, lovers of nature and family. However, they are passive, lazy, sensitive, and narrow-minded.

Leo: Directors, Fashion Designers, Murderers

These people are strong, charismatic, attractive enjoyers of life. However, they are also proud, inflexible, and elitist.

Virgo: Child Abuse, Same Job (10+ yrs)

These are practical, hard-working, realistic and capable people. They are prone to physical problems though, are pessimistic, and overly rational.

Gemini

Navamshas in Gemini

Libra: Stroke, Teachers (K-12)

These are intelligent, witty, and friendly idea people. However, they can be overly intellectual and talkative.

Scorpio: Actors, Fighters, Prisoners, Salespeople, Sports, Suicide

These people are deep thinkers capable of discovering hidden truths. They are perceptive, humorous, and passionate. They can be cynical and hurtful, though, and have sordid thoughts.

Sagittarius: Corporate Tycoons, Engineers, Instrumentalists, Marriage (15+ yrs), Wealthy

These are highly inquisitive, talkative, and exciting people. They are restless and indiscreet, though, and can lack emotional substance.

Capricorn: Birth Defect, Child Abuse, Drug Abuse, Medical, Politicians, Widowed

These clear thinking people are good planners, being practical and reliable. They can be overly reasonable and cautious.

Aquarius: Children (none)

These people are social butterflies who are highly tactful and likable. They can also be superficial and unwilling to face difficulties.

Pisces: Marriage (15+ yrs), Sex Offenders, Stroke

These are deeply reflective, imaginative people who love reading and all forms of artistic expression. They can be delusional, though, and are impractical thinkers.

Aries: Lesbians

These people are pioneering thinkers with a lot of energy and self-confidence. They are also argumentative people, who are impatient and insensitive.

Taurus: Infant Mortality, Sports

These people have good practical, physical intelligence and are very sensual people. They can also be closed-minded and self-absorbed.

Gemini: Engineers

These are entertaining, intelligent, problem-solvers. However, they can also be unreliable, and superficial.

Cancer

Navamshas in Cancer

Cancer: Corporate Tycoons, Fashion Designers, Songwriters

These are emotional, kind-hearted, imaginative people. However, they can also be shy, cautious, and moody.

Leo: Fashion Designers, Lesbians, Psychics, Sex Workers, Teachers (K-12)

These people are affectionate, devoted, and attractive people. They are also clannish and can be oversensitive to criticism.

Virgo: None

These are attentive, dutiful, and helpful people, though are timid, introverted, and fussy.

Libra: Astronauts, Corporate Tycoons, Salespeople, Spiritual Teachers, Twins

These people are visionary, kind-hearted humanitarians. They tend to isolate themselves, though, and become burdened by the suffering of others.

Scorpio: Astronauts, Engineers, Professors, Psychiatrists

These are highly passionate people with great, investigative minds. They can be overly emotional, secretive and self-defensive.

Sagittarius: Comedians, Instrumentalists, Lesbians, Murderers

These are idealistic, imaginative people with much emotional enthusiasm. They tend to be volatile and unrealistic.

Capricorn: None

These people are responsible, reliable, and family-oriented. They are overly cautious and serious.

Aquarius: Corporate Tycoons

These are charming, likable, social and caring people. However, they can become over-dependent on others, indecisive, and defensive.

Pisces: Sex Abuse Victims

These are sweet, intuitive, imaginative people. They can be withdrawn, unrealistic, insecure and self-doubting as well.

LEO

Navamshas in Leo

Aries: Police Officers, Sports, Sports Coaches

These people are confident, assertive, and proud leaders. They can also be arrogant and abrasive.

Taurus: Dancers, Fashion Designers, Literature, Military, Suicide

These are strong, sensual, magnanimous people who are also productive and reliable. They can also be stubborn and overly self-assured.

Gemini: Astrologers, Fashion Designers, Sex Offenders

These are popular, creative people who make friends easily. They can be restless, though, and need constant entertainment.

Cancer: Culinary, Education (low), Lesbians, Sex Workers, Singers

These are generous, loving, emotional people who inspire great loyalty. They are also self-absorbed, vulnerable and easily hurt.

Leo: AIDS, Astronauts, Heart Attack, Lawyers, Math-Physics

These people are magnetic, proud, and creative. But, they are also arrogant and stubborn.

Virgo: Directors, AIDS, Business Owners, Child Abuse, Murderers, Songwriters

These are dedicated, loyal people who put others first. They are also prone to self-pity and a lack of confidence.

Libra: Activists, Lawyers, Singers

These people are open-minded, intellectual orators with great confidence in the power of their own minds. They can be condescending, though, and unwilling to listen to others.

Scorpio: Lesbians, Models

These are magnetic, intense, loyal individuals with great ambition. They are also possessive, fanatical, and volatile.

Sagittarius: Presidents, Same Job (10+ yrs)

These people are adventurous, honorable, philosophical leaders who seek personal growth. They can also be blunt and restless.

Virgo

Navamshas in Virgo

Capricorn: Military, Only Child

These are disciplined, hard-working people who are realistic and humble. They are also overly serious and critical.

Aquarius: Alzheimer's, Child Abuse, Culinary, Only Child, Prisoners, Sex Offenders

These people are polite, moral people who seek to help others. They can also be fussy and critical of relationship partners.

Pisces: None

These people are modest, helpful, and understanding. However, they are also timid and worrisome.

Aries: Cancer, Dancers, Only Child, Out-of-Body

These intelligent people have highly productive analytical minds. They can also be know-it-alls and be critical of others.

Taurus: Astrologers, Child Abuse, Out-of-Body, Sex Abuse, Widowed

These are dependable, handy, reserved people who enjoy nature. They are narrow-minded, stubborn, and predictable, though.

Gemini: Actors, AIDS, Alzheimer's, Culinary, Fashion Designers, Lesbians

These people are highly analytical, logical, problem-solvers who love reading and learning. They are easily distracted, though, and can get lost in fearful thinking.

Cancer: None

These are conscientious, supportive, and sensitive people. They can be shy, fearful, and get stuck in the past.

Leo: None

These people are humble, productive achievers with high moral standards. They can also be intellectual snobs who lack self-confidence.

Virgo: AIDS, Children (none), Marriage (none), Out-of-Body, Transvestites

These are hard-working perfectionists with great analytical ability. However, they can be critical, shy, and guilt-ridden.

Libra

Navamshas in Libra

Libra: None

These are highly intellectual, socially aware, and independent people. However, they can also be opinionated and intrusive.

Scorpio: None

These people have deeply penetrating, insightful minds and strong principles. They are also argumentative and cynical.

Sagittarius: None

These people have innovative ideas, grand philosophies and are excellent communicators. They also can have difficulty with day-to-day realities and finding practical outlets for ideas.

Capricorn: Sports Coaches

These are people with practical intelligence, foresight, and organizational ability. They can be uptight and unemotional, though.

Aquarius: Life (<29 yrs), Salespeople, Songwriters

These are attractive, unique individuals who are dignified, idealistic, and open-minded. They are also naive, lack stamina, and are averse to struggles.

Pisces: Astronauts, Comedians

These are creative, imaginative people who have a deep understanding of life. They are escapists, withdrawn, and difficult to understand.

Aries: Twins

These are honest, freedom-loving people with lots of ideas and initiative. They have difficulty working with others, though, and are intolerant of weakness.

Taurus: Child Prodigy, Computers, Depression, Infant Mortality, Lesbians, Models

These people are realistic thinkers with philanthropic goals. They also enjoy solitude and have a taste for the beautiful and unique. They can also be self-indulgent and willful.

Gemini: None

These are curious, intelligent, communicative, and friendly people. However, they can be unreliable and emotionally detached.

Scorpio

Navamshas in Scorpio

Cancer: Child Performers, Mental Illness, Suicide

These are romantic, committed people who are perceptive and intuitive. They are also over-emotional and tend to suffer depression.

Leo: Life (<29 yrs), Murder Victims

These are powerful, magnetic people with strong powers of persuasion. They can also be ruthless and intolerant of obstacles.

Virgo: Child Performers

These people have probing, analytical minds and they pursue goals with hard work and dedication. They are also judgmental and hurtful.

Libra: Royalty

These are observant, honest, self-aware people. However, they are skeptical of others and prone to negativity.

Scorpio: Astrologers, Lawyers, Mystics, Psychiatrists, Spiritual, Transvestites, Widowed

These are magnetic, probing, perceptive people. However, they are also possessive, secretive, and cynical.

Sagittarius: Activists, AIDS, Astrologers, Composers, Instrumentalists, Police

These are inquisitive, philosophical people with great integrity. They can also be uncompromising and stern judges of people.

Capricorn: Activists, Computer Programmers, Police

These people have a strong sense of purpose, are demanding and tenacious. They can also be dark, brooding, and cynical.

Aquarius: Actors, Comedians, Criminals, Nazis, Performers, Prisoners, Professors

These are magnetic, sociable people who are both self-controlled and graceful. They can be jealous, manipulative, and over-dependent on others, though.

Pisces: Transvestites

These people have a rich inner world, are mysterious and can be healing for others. However, they are secretive, prone to addictive behavior, and can be extremely subjective.

SAGITTARIUS

Navamshas in Sagittarius

Aries: Heart Attack, Stroke

These are bold, adventurous people with big hearts and big dreams. They are also impractical and impatient.

Taurus: Adopted, Life (<29 yrs)

These people are generous, extravagant, self-assured, lovers of life. They are also gluttonous and unwilling to accept responsibilities.

Gemini: None

These are very friendly, talkative, intelligent people who are life-long learners. They can also be immature and unreliable.

Cancer: Twins

These are caring, imaginative, creative people who see the world as their home. However, they are also unrealistic and gullible.

Leo: IQ (high), Twins

These people are exuberant, philosophical thinkers who inspire confidence in others. They also are arrogant, impatient, and bossy.

Virgo: Children (4+), Children (none), Literature, Poets, Scientists

These people are cautiously optimistic and work hard to understand life and categorize knowledge. They are also anxious, critical, and tend to be narrow-minded.

Libra: Sex Abuse Victims

These are open-minded, intellectual, well-spoken philanthropists. However, they can be impractical and overly abstract.

Scorpio: Journalists, Literature, Nobel Prize, Novelists, Poets, Technical, Writers

These people are seekers of deep understanding and are psychologically astute. They are both magnetic and courageous, but can be over-zealous and condemning of others.

Sagittarius: Psychics

These are entertaining, free-spirited, philosophical explorers. They are self-righteous and blunt, though, and inclined to exaggerate.

CAPRICORN

Navamshas in Capricorn

Capricorn: Composers, Homosexuals, Poets, Same Job (10+ yrs)

These are conservative, responsible, authoritative people. However, they are over-serious and unemotional.

Aquarius: None

These are elegant social climbers who are committed to others. However, they are uptight and prone to negativity.

Pisces: Child Abuse, Teachers (K-12), Writers

These people are idealists who are willing to work for the benefit of others. They are cautious people, though, who seek solitude and are easily depressed.

Aries: Priests, Scientists, Stroke, Technical

These are ambitious leaders who seek to control their environment. They are also insensitive and self-absorbed.

Taurus: Fighters, Journalists, Life (80+ yrs), Math, Scientists, Technical

These are stable, sensible people who are organized and patient. They can also be inflexible, routine, and materialistic.

Gemini: Business Owners, Composers, Homosexuals, Songwriters, Sports Coaches

These are serious thinkers who seek respect for their ideas, often through writing or teaching. They are also overly rational and pompous.

Cancer: Dancers, Psychiatrists

These are very responsible people who take on the burdens of family and friends. They can also give tough love and can have difficulty relaxing.

Leo: None

These are status seekers who are authoritative, loyal, and proud. They are also dictatorial and have an inferiority complex.

Virgo: Astronauts, Culinary, Inheritance, Math-Physics, Military

These are organized, technical people who are reliable and hard-working. However, they are critical and over-serious.

Aquarius

Navamshas in Aquarius

Libra: Business Owners

These are intelligent, sociable, and diplomatic people, but can be vain and haughty.

Scorpio: Alzheimer's, Culinary, Fighters, Murder Victims, Police, Sex Workers

These are highly magnetic, attractive people who are very self-controlled yet desire intense personal relationships. They are manipulative and distrustful of others.

Sagittarius: Astrologers

These are popular, enthusiastic people who love to share knowledge. They lack introspection and need excessive freedom from others.

Capricorn: Life (<29 yrs), Literature

These are polite, responsible people who are good at managing others. They are also unemotional and strongly self-controlled.

Aquarius: Artists, Children (4+), Novelists, Only Child, Salespeople, Technical

These are uniquely attractive individuals who are visionary and have a strong humanitarian bent. However, they are overly idealistic and they distrust emotions.

Pisces: Alzheimer's, Birth Defect, Education (low), Mental Handicap, Suicide

These are humanitarians, sensitive people, great helpers and romantic dreamers. However, they are also indecisive and escapist.

Aries: Comedians, Sports Coaches, Teachers (K-12)

These people are initiators of relationships who are very lively and intelligent. They are also overly competitive and can lack commitment.

Taurus: Actors, Comedians, Instrumentalists, Playwrights, Songwriters

These are very creative, harmonious people with a love for beauty and the arts. They are also stubborn, self-willed and vain.

Gemini: Life (<29 yrs), Same Job (10+ yrs), Scientists, Teachers (K-12)

These people are excellent communicators and diplomats who are friendly, clear-headed and able to understand all points of view. They may also lack depth and only skim the surface of many topics.

Pisces

Navamshas in Pisces

Cancer: Computer Programmers, Lawyers, Sports

These are introspective, imaginative, and are warm-hearted helpers of others. However, they are also withdrawn, over-sensitive, and moody.

Leo: Artists, Comedians, Education (low), Mental Handicap, Musicians, Writers

These are creative, romantic people who are able to bring out the best in others. They are self-doubting, though, and prone to idolizing others.

Virgo: Alcoholics, Astronauts, Criminals, Math-Physics, Nazis

These people are reserved and thoughtful, and are willing helpers who are excellent problem solvers. However, they are also critical and worriers, though, especially, about their health.

Libra: Child Prodigy, Math-Physics

These are imaginative, inventive thinkers with a deep sense of humanitarianism. They are also gullible and get caught up in the problems of others.

Scorpio: Artists, Cancer, Life (80+ yrs), Murder Victims, Scientists

These are intuitive, insightful, healing individuals with the courage to find truth. However, they are suspicious, manipulative, and overly emotional.

Sagittarius: Artists, Life (80+ yrs), Priests, Psychics

These are friendly, inspired, understanding people with strong religious/spiritual interests. They are also naive, easily deceived, and exaggerating.

Capricorn: Directors, Models, Politicians

These are helpful, dutiful people who take responsibility for the problems of the world. However, they are also depressive and moralistic.

Aquarius: Infant Mortality, Instrumentalists, Journalists, Marriage (none), Poets

These people are gentle, loving, compassionate helpers. However, they are easily taken advantage of and can be over-sensitive to criticism.

Pisces: Life (80+ yrs), Literature, Musicians, Spiritual Teachers

These are contemplative, idealistic, artistic people. However, they are withdrawn, delusional and disorganized.

LIBRA VS AQUARIUS

A question that I came across while doing this project was whether the current interpretations of Libra and Aquarius are accurate.

In the study of those who were Married for 15 years or more, Aquarius was the dominant sign. How could this be the case, if Aquarians are such independent, rebellious people? Well, they are not! That description better fits Libras (Scales of Justice), which was a dominant sign in the study of Activists, High IQ, Lawyers, and Teachers.

Many of the groups that are associated with attractiveness and relationships are statistically strong in Aquarius <u>and</u> weak in Libra (Comedians, Marriage 15+ years, Salespeople). This occurs in both the 1st harmonic and 9th harmonic studies.

Also, the groups that represent abundance and longevity are strong in Aquarius and weak in Libra (Children 4+, Marriage 15+ years, Long Life 80+, Same Job 10+). This is because the Fixed mode represents love and preservation, while the Cardinal mode represents initiative and power (the Mutable mode represents knowledge and wisdom). As a result, the three Air signs each have vastly different purposes for connecting with people:

1) Libra is a Cardinal sign, which seeks active manifestation of power through communication; 2) Aquarius is a Fixed sign, which seeks to share its love of knowledge through relationship; and 3) Gemini is a Mutable sign, which seeks to increase its awareness through communication.

Venus, the planet of love, beauty, and relationships, has many groups that are strong in the studies on Venus aspects and Venus on the Ascendant also strong in Aquarius (Actors, Artists, Children 4+, Culinary, and Salespeople). Actors, a group clearly focused on attractiveness, is strongest in Aquarius and with Venus on the Ascendant.

Finally, the Greek myth of Aquarius, the Water Bearer, was of a beautiful prince who Zeus wanted to be his lover. In order to have the boy nearby, Zeus gave him the role of water bearer. This is a Venusian theme. In 1877, the British astrologer Raphael described Aquarians as "Stable, good, kind-hearted, scientific, fond of learning and recreation, gentle, and even-tempered." These descriptions are a far cry from the current depiction of Aquarius as rebellious and eccentric. It is Librans who stand up for ideas and seek truth and justice, just as its Scales of Justice symbol would suggest. This change occurred in the 20th century largely due to Aquarius being associated with Uranus.

For all these reasons, I conclude that Libra and Aquarius should have their interpretations restored to their original meanings. Libra is direct and decisive, both intellectually and interpersonally, while Aquarius is more diplomatic and thoughtful.

Below is a table comparing the statistical results from the study of the signs Libra and Aquarius. Both the 1st and 9th harmonic findings are listed for each sign. The groups highlighted in blue are strong in Aquarius and Venus. No groups are strong in Libra and Venus! The groups highlighted in red are weak in Libra and strong in Aquarius or Venus. The groups that fit the Venusian model of an attractive, sociable type are strong in Aquarius and weak in Libra, while only Activists are strong in Libra and weak in Aquarius.

	Libra			Aquarius	
	1st Harmonic	9th Harmonic		1st Harmonic	9th Harmonic
Strong:	Activists	Astronauts	**Strong:**	Actors	Actors
	IQ (high)	Stroke		Artists	Children (4+)
	Sex Workers			Comedians	Literature
Weak:	Comedians	AIDS		Culinary	Marriage (15+ yrs)
	Engineers	Comedians		Education (low)	Novelists
	Fashion	Homosexuals		Life (<29 yrs)	Only Child
	Fighters	Infant Mortality		Literature	Prisoners
	Life (80+ yrs)	Lesbians		Marriage (15+ yrs)	Writers
	Marriage (15+ yrs)	Models		IQ (high)	
	Musicians	Only Child		Mental Handicap	
	Psychiatrists	Poets		Murder Victims	
	Racers	Police		Nazis	
	Same Job (10+)	Salespeople		Only Child	
	Scientists	Same Job (10+)		Racers	
	Singers	Suicide		Salespeople	
	Sports	Writers		Suicide	
	Stroke		**Weak:**	Astrologers	Activists
	Suicide			Depression	Alcoholics
	Technical			Infant Mortality	Racers
				Lesbians	Sex Abuse
				Sex Abuse	

Venus Aspects	Venus on Asc
AIDS	Actors
Child Abuse	Artists
Child Performer	Birth Defect
Child Prodigy	Child Abuse
Children (4+)	Culinary
Education (low)	Lesbians
Out-of-Body	Models
Psychics	Salespeople
	Sex Abuse

As an example, here is the chart of Marilyn Monroe – perhaps the greatest sex symbol the USA has ever known. She is Air-Water dominant, which is the Feminine-Intuitive type. But, she also has a Moon-Jupiter conjunction in Aquarius. Current astrology would interpret this to indicate having great imagination, intellectual ability, and an independent nature. It makes much more sense if these planets were interpreted in the manner currently attributed to Libra. The interpretation would change to her having incredibly feminine beauty and refined people skills!

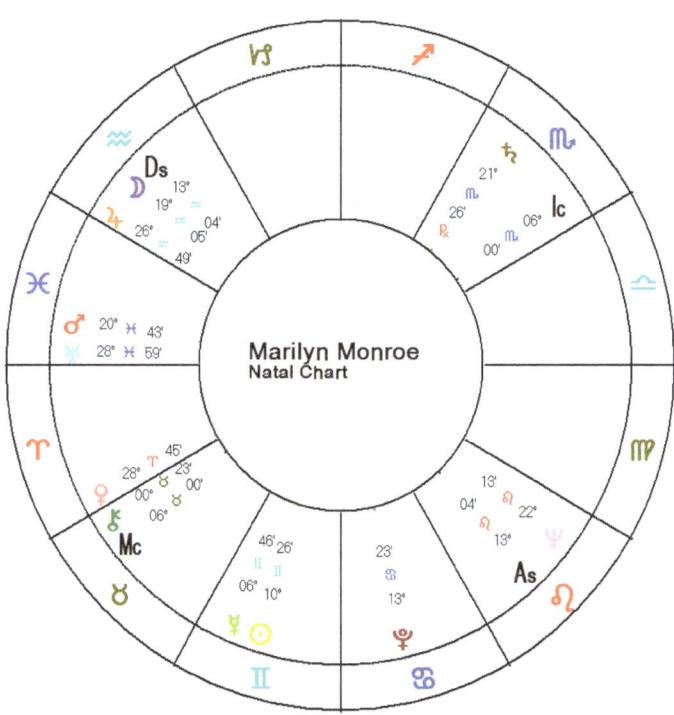

Hugh Hefner, the owner of Playboy magazine and the notorious playboy mansion, has Mars conjunct Jupiter in Aquarius and a Moon-Venus conjunction in Pisces. No planets are in Libra, yet Aquarius and Venus are very strong.

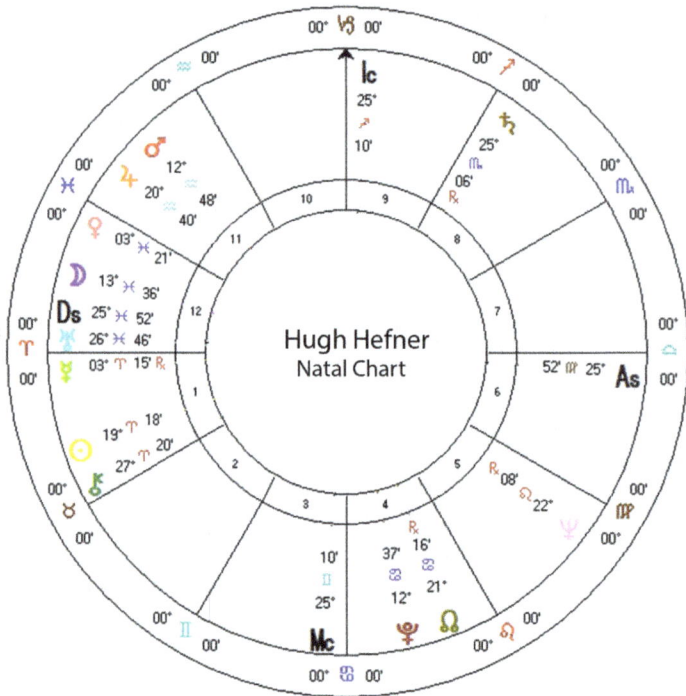

Nicola Tesla, the great scientist and inventor, has the Moon conjunct Mars in Libra, yet no planets in Aquarius. (He also has a Mercury-Jupiter conjunction in Gemini in the 9th harmonic chart.)

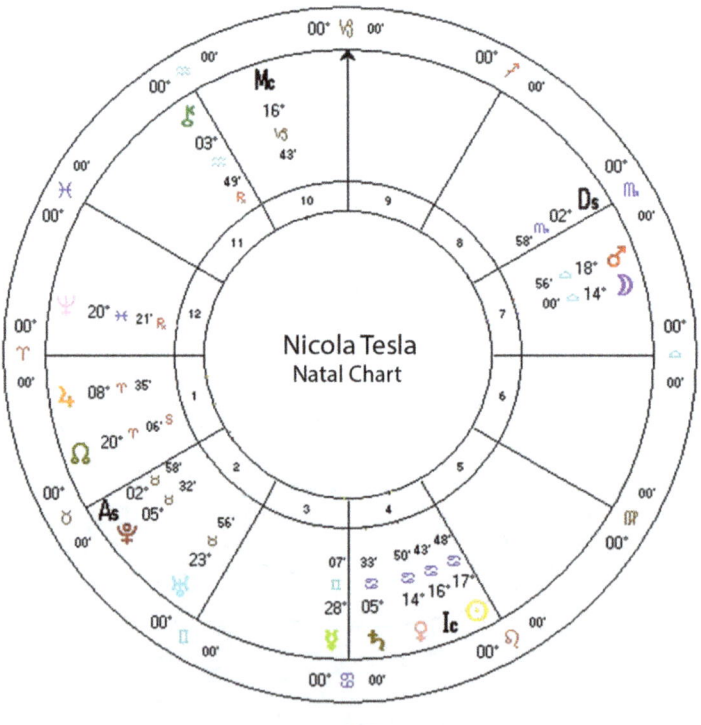

The Diurnal Cycle

The earliest form of astrology was probably based on the diurnal cycle of the Sun in our sky, since it has such noticeable effects on human life. It is based on the Earth's rotation on its axis, called primary motion. We wake up when the Sun is rising at the Eastern horizon (the Ascendant) and end each day of work with the setting Sun at the Western horizon (the Descendant). Noon is when the Sun is at its highest point in the sky (the Midheaven), and the heart of the night, midnight, is when the Sun is at its opposite point (the Imum Coeli - Latin for "the bottom of the sky"). The rising and setting points on the ecliptic gives us the East-West axis (though only due East or West at the Equinoxes). The ecliptic's highest point above the horizon and lowest point below the horizon gives us the South-North axis (which is reversed in the Southern hemisphere).

The rising, noontime, setting, and midnight positions of the Sun give us these four major axis points. The idea that these four points are powerful positions for the Sun was then extended to the rest of the visible planetary bodies. A representation of these axis points is that of two intersecting lines with the circle, representing the Sun, traveling clockwise, suggesting similarities with the cross of the Christian faith. (The cross could also be taken to represent the Sun's annual path through the zodiac. Both interpretations symbolize the self's journey through time and space.)

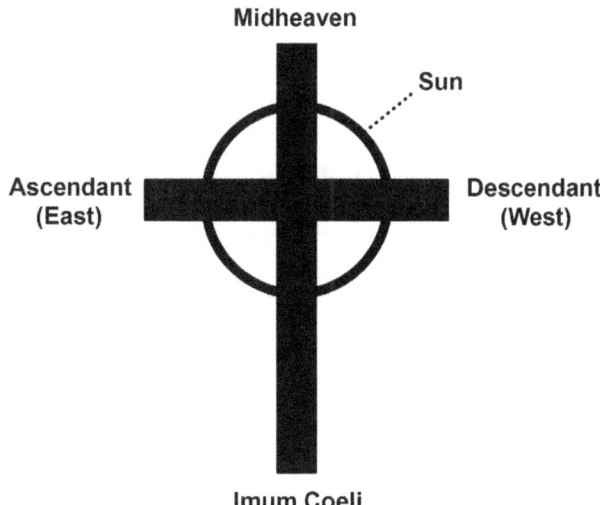

These four axis points have very specific and distinct meanings. Planets on the Ascendant are taken to have a definite influence on how a person is seen and expresses their personality. It represents the moment of waking up from sleep and initiating plans for the day. Planets on the Midheaven influence one's public efforts and career status. It represents the time of aspiring towards goals at work or school.

Planets on the Descendant indicate the type of people you tend to encounter in your life. It represents the moment of leaving work or school and relaxing with friends and family.

Planets on the Imum Coeli indicate the nature of your private life and represents the moment of going to sleep.

The Midheaven-Imum Coeli axis represents the basic foundations and responsibilities of one's life: home (IC) and public life (MC). The Ascendant-Descendant axis represents the creative and social aspects of one's life: self-expression (Asc) and relationships (Dsc). It is much like the base of a tree and its branches, which intertwine with the branches of other trees.

In studying the diurnal cycle of the planets, I tested the effect of the zodiac sign on the Ascendant. The Rising sign is thought to have tremendous influence on the way we express ourselves in public, our persona. However, the results of this study led me to question its effectiveness. I also tested the effect planets have at the axis points, in the spaces between the axis points (houses), as well as above or below the horizon.

People who are born above the horizon are thought to exist in the open, visible world with their predominant activities taking place in the daytime (highlighted in blue). People who are born below the horizon are thought to exist beneath view, functioning in private with their predominant activities taking place in the night.

In the study, most, if not all, of the groups of people born at night were people who do most of their productive work or activities in private (highlighted in red). The groups of people born in the daytime were more mixed between people with a public and private orientation.

Horizon

Above	Below
Business Owners	AIDS
Child Performers	Cancer
Drug Abuse	Homosexuals
Lawyers	Life (80+ yrs)
Lesbians	Literature
Marriage (15+ yrs)	Nobel Prize
Mental Illness	Obesity
Military	Poets
Mystics	Psychiatrists
Novelists	Scientists
Politicians	Sex Workers
Royalty	Songwriters
Widowed	Transvestites
	Writers

The Four Axis Points

Every planetary body crosses the four axis points throughout each day. The Sun always crosses the Eastern horizon on the ecliptic. However, the other planetary bodies do not move strictly on the ecliptic. They move above and below the ecliptic as well. This movement is called latitude, while movement along the ecliptic is called longitude.

Since all planets besides the Sun have a latitude movement, they usually do not cross the Eastern horizon while on the ecliptic. Nor do the planets always cross the horizon at the same time as its zodiacal degree, since the ecliptic is not usually perpendicular to the horizon. The planets mostly cross the Eastern horizon before or after its zodiacal degree crosses the Eastern horizon. The same is true at the Western horizon. Each planet has its own unique orbit around the Sun as well its own diurnal arc around the Earth.

The result is that planets which are measured to be conjunct the Ascendant or Descendant are often not on the horizon at all. A planet which is shown above the Ascendant or Descendant in a birth chart will often actually be below the horizon and vice versa.

Typically, planets are less than two zodiacal degrees away from the Ascendant or Descendant when they are on the horizon. However, the Moon is frequently five degrees away. Pluto, which can have a very high latitude, could be more than 30 zodiacal degrees away from the Ascendant or Descendant when it is actually on the horizon, especially for locations far from the Equator.

As a result, using the ecliptic to calculate when a planet is near the horizon creates imprecise results. When measuring a planet's distance from the horizon, each planet's arc must be calculated individually. The method used in this study was to divide each of the four quadrants of a planet's diurnal arc into 1/10 portions. A planet in a 1/10 portion of its diurnal arc next to an axis point would be considered to be on that axis point.

For instance, the Sun on the Ascendant would be 9 degrees above or below the horizon, which is 10% of 90 degrees (for a perfectly square axis cross). Any planet would be considered on its Ascendant if it were within 10% of its arc above or below the Eastern horizon. For the Midheaven, a planet would be within 10% of its upper arc to the East of the Midheaven or to the West, being the highest 5% of its total diurnal arc.

The Midheaven and Imum Coeli are the same for the Sun and all planetary bodies, since these points define the peak and low point of the diurnal cycle, being the midpoint of each day and night. In the Northern hemisphere, the Midheaven points directly due South and the Imum Coeli points due North.

In the study of the planets on the four axis points, most categories fit with the traditional meanings for the planets. The Sun, representing leadership, is strong for Politicians, Professors, and the Military on the Midheaven. The Moon, representing femininity, is strong for Models on the Eastern horizon. Mercury, representing the intellect, is strong for Mystics, Priests, and Psychics on the Eastern horizon. However, the Moon on the Western horizon is also an indication of loneliness, being strong with people who are unmarried and childless.

Venus, representing the arts, pleasure, and attractiveness, is strong for Actors, Artists, Culinary, and Models on the Eastern horizon. On the Imum Coeli, Venus is strong for contemplatives such as Poets and Spiritual Teachers since these people enjoy solitude. Mars, representing technical ability, is strong for Mathematicians-Physicists, Scientists, and Technical people on the Eastern horizon. Mars' more harmful side is seen on the Midheaven, being strong for Criminals and Drug Abusers.

Jupiter, representing positivity, exploration, and success, is strong for Actors, Nobel Prize winners, Presidents, and members of Royalty on the Eastern horizon. Saturn, which represents discipline and endurance, was surprisingly strong for Actors and Models on the Eastern horizon and strong for Astrologers, Mystics, and Psychics on the Midheaven.

Uranus, representing awakening, is strong for people with high IQs on the Eastern horizon, but did not show any strong pattern across the study groups. Neptune, representing spirituality, is strong for Astrologers, Mystics, and Psychiatrists on the Eastern horizon.

Pluto, representing power and the taboo, is strong for Fighters and Athletes on the Midheaven and Criminals and Murderers on the Imum Coeli. Chiron, representing wounds, is strong for Criminals and Suicides on the Western horizon and strong for people with AIDS, Birth Defects, and Murderers on the Imum Coeli.

Top Findings (by probability of being an accident)

#	Category	Planet	Angle	Value
1.	Murderers	Chiron	IC	37,291
2.	Lawyers	Moon	Dsc	16,265
3.	Inheritance	Sun	IC	9,640
4.	Obesity	Sun	IC	6,712
5.	Cancer	Mars	Asc	5,179
6.	Psychics	Jupiter	Asc	3,313
7.	Culinary	Venus	MC	2,847
8.	Child Abuse	Mars	IC	2,453
9.	Depression	Jupiter	Asc	2,136
10.	Child Abuse	Moon	Asc	1,656
11.	Widowed	Mercury	MC	1,400
12.	Marriage (15+ yrs)	Jupiter	Asc	1,217
13.	Models	Chiron	IC	1,093
14.	Salespeople	Saturn	Asc	963
15.	Life (<29 yrs)	Venus	MC	945
16.	Nobel Prize	Jupiter	Asc	841
17.	Children (4+)	Mercury	MC	802
18.	Cancer	Mercury	IC	788
19.	Models	Chiron	MC	714
20.	Child Abuse	Venus	Asc	639
21.	Criminals	Jupiter	IC	634
22.	Murder Victims	Saturn	MC	618
23.	Presidents	Uranus	Asc	611
24.	Widowed	Jupiter	MC	604
25.	Philosophers	Uranus	Dsc	492
26.	Murderers	Chiron	MC	474
27.	Scientists	Uranus	IC	431
28.	Life (80+ yrs)	Sun	IC	416
29.	Children (none)	Mars	Asc	409
30.	Adopted	Mercury	MC	402
31.	Murder Victims	Venus	MC	394
32.	Marriage (none)	Moon	Dsc	382
33.	Child Prodigy	Mars	Dsc	374
34.	Activists	Chiron	Dsc	358
35.	Depression	Mars	Asc	301
36.	Widowed	Mars	Dsc	291
37.	Suicide	Mars	Dsc	284
38.	Mystics	Mercury	Asc	274
39.	Salespeople	Mars	IC	274
40.	Presidents	Jupiter	Asc	248

	Sun	Moon	Mercury	Venus	Mars
Ascendant (East)	Poets Stroke	Child Abuse Models Obesity Wealthy	Alzheimer's Culinary Mystics Priests Psychics Sports Coaches Widowed	Actors Artists Birth Defect Child Abuse Culinary Lesbians Models Salespeople Sex Abuse	Business Cancer Children (none) Depression Fashion Heart Attack Math-Physics Scientists Technical
Midheaven	Drug Abuse Math-Physics Military Politicians Professors	Actors Spiritual Teacher	Adopted Astronauts Child Prodigy Children (4+) Fashion Racers Widowed	Adopted Alzheimer's Culinary Life (<29 yrs) Medical Military Murder Victims	Criminals Drug Abuse
Descendant (West)	Astrologers Astronauts Birth Defect Business Child Performers Medical Police Spiritual Teacher Twins	Children (none) Lawyers Life (80+ yrs) Marriage (none) Sports Coaches Transvestites	Corporate Drug Abuse Journalists Police Presidents	Activists Child Abuse Fighters Life (80+ yrs) Transvestites	Child Performers Child Prodigy Nazis Poets Suicide Widowed
Imum Coeli	Inheritance Life (80+ yrs) Literature Obesity Writers	Astrologers Comedians Composers Fighters Mystics Sports Stroke	Cancer Sex Workers	Alzheimer's Child Abuse Culinary Infant Mortality Inheritance Poets Spiritual Teacher	Business Child Abuse Corporate Literature Salespeople Sex Abuse

Jupiter	Saturn	Uranus	Neptune	Pluto	Chiron
Actors Depression Engineers Marriage (15+) Mental Illness Mystics Nobel Prize Presidents Psychics Royalty Sex Workers Stroke	Actors Children (none) Marriage (15+) Mental Illness Models Salespeople	Adopted Alcoholics Corporate Depression IQ (high) Medical Police Presidents Twins	Cancer Journalists Nazis Sex Offenders	Child Prodigy Comedians Depression Fashion Professors	Psychiatrists Singers
Actors AIDS Alzheimer's Astronauts Murder Victims Spiritual Teacher Widowed	Astrologers Child Abuse Mystics Poets Psychics Stroke Twins	Engineers Fashion Singers	Astrologers Infant Mortality Mystics Out-of-Body Police Presidents Psychiatrists Transvestites	Astronauts Fighters Mental Illness Psychics Sex Offenders Sports Sports Coaches	Actors Child Prodigy Engineers Journalists Lawyers Marriage (none) Models Murderers Psychics Royalty Sex Abuse Singers
AIDS Lesbians Salespeople	Lawyers Psychiatrists Sports Coaches	Philosophers Sex Offenders Sports	Children (none) Dancers Drug Abuse Inheritance Journalists Only Child	Fighters Inheritance IQ (high) Lawyers Marriage (15+) Prisoners Twins	Artists Criminals Suicide
Alcoholics Criminals Homosexuals Presidents Prisoners	Adopted AIDS Alcoholics Fighters Spiritual Teachers	Education (low) Murderers Out-of-Body Scientists	Alzheimer's Child Abuse Culinary Fighters Police	Corporate Criminals Murderers Presidents	AIDS Alcoholics Birth Defect Models Murderers Salespeople Sex Workers

Houses

Houses are divisions of space created by the daily rotation of Earth on its axis. Originally, the usage of the word house and sign were synonymous. Later, houses became a method for counting the number of signs from the Rising sign, the sign on the Ascendant, to the planetary sign positions. Today, there are many house systems, all starting from the Ascendant and following the path of the ecliptic.

The most commonly used house systems in the West divide the space surrounding the Earth using the horizon and Midheaven to create four quadrants, just like the four seasons. Each of these quadrants can then be trisected to create 12 sections called houses. It is thought that these 12 houses correspond in meaning to the 12 signs of the ecliptic.

Western astrology is founded upon two cycles: 1) the annual cycle, with its four seasons and 12 signs, and 2) the daily cycle, with its four phases and 12 houses. These cycles mirror each other: Spring and morning, Summer and afternoon, Autumn and evening, Winter and sleep.

The Sun and planets travel through the four cardinal points of the zodiac just as they travel through the four axis points of our sky:

The Spring Equinox corresponds with the Eastern horizon (sunrise).
The Summer Solstice corresponds with the Midheaven (high noon).
The Autumn Equinox corresponds with the Western horizon (sunset).
The Winter Solstice corresponds with the Imum Coeli (midnight).

The houses are ordered in the same direction as the ecliptic, but in the opposite order from which the planets move through the sky. If the daily and annual cycles are mirror images of each other, why don't the houses follow the same path as the Sun?

The Sun travels from the Ascendant to the Midheaven to the Descendant to the Imum Coeli each day. Should not the houses be numbered in the order of this path? How is it that the 12th house, which represents all things hidden, is where planets rise and display themselves at the beginning of each day? Should this space not be the first house? Also, why doesn't the Midheaven start the 4th house if the Summer Solstice is the start of the 4th sign, Cancer? Are they not both the pinnacle of the Sun's daily and annual cycles?

The answer to this question lies in the fact that astrology is person-centered and Earth-based, not heliocentric and Sun-based. Just as the direction of the zodiac signs is based on the annual movement of the Earth through the solar system, the direction of the houses is based upon the daily rotation of the Earth. The Eastern horizon spins through the signs in the same direction as the ecliptic, not backwards.

I studied many house systems in depth: Whole signs, Equal houses, Placidus, Bhava (Hindu) houses, Porphry houses, and Gauquelin's Diurnal Arc Sectors. I analyzed them ordered backwards and forwards and came to the conclusion that **Diurnal Arcs in traditional zodiacal order gave, by far, the most accurate results.**

This system takes the actual arc of each planet and divides it into 12 sections. It divides each semi-arc, above and below the horizon, into 6 houses by dividing the time each planet spends in each hemisphere by six. The location of a planet 1/6 of the time through the lower semi-arc is the 6th house cusp, 2/6 is the 5th house cusp, 3/6 is the 4th house cusp (IC), 4/6 is the 3rd house cusp, and 5/6 is the 2nd house cusp. The same is done with the upper semi-arc for each planet.

The Placidus house system uses this exact method, but only for the Sun's diurnal arc - the ecliptic. It then places every planet into the Sun's arc houses. However, since planets are usually not on the ecliptic, this method often incorrectly places planets that are above the horizon into the 1st & 6th houses and planets below the horizon into the 7th & 12th houses. Each planet has its own path and its own houses, DIURNAL ARC HOUSES.

Gauquelin Sectors

The French statistician, Michel Gauquelin, published a large statistical research project in the 1970s. He found that planets have the greatest effect in the sectors just after the axis points, which are considered by house systems to be the weakest sectors. According to his understanding of astrological theory, the houses 1, 4, 7, and 10 are the strongest. But, his study found that the sectors largely occupying the spaces known as houses 3, 6, 9, and 12 are actually the strongest.

To investigate his work, I obtained Gauquelin's database of 15 study groups and replicated his study. I compared his groups with my control group, using the 1% end of the normal curve as statistical strength (1 in 100). What I found was that of the 15 groups, 7 were statistically strongest in the Mutable houses - taken as whole, 5 were strongest in Fixed houses, and none were strongest in the Angular houses. It appeared that he was right.

When studying the individual findings in the specific houses, a different story unfolded: 14 in Mutable houses, 11 in Fixed houses, and 8 in Angular houses - still weighted towards Mutable houses.

However, 7 of the 14 findings in Mutable houses were in the 9th house! The 9th house is well known to be a house of great strength and benevolence! Gauquelin's assumption that the 9th house is a space of weakness was incorrect and distorted his conclusions.

The Fire houses (1, 5, & 9) were actually the most prominent. Of 33 findings, 15 were in Fire houses, 8 in Earth houses, 3 in Air Houses, and 7 in Water houses.

Gauquelin Study Groups (by probability of being an accident)

Group	Planet	House	Value
Actors	Mercury	10th	107
Business	Mercury	11th	133
Business	Jupiter	12th	4,097
Journalists	Moon	11th	104
Journalists	Sun	1st	610
Journalists	Mercury	1st	1,216
Medical	Mars	12th	145
Medical	Sun	2nd	2,293
Medical	Mercury	5th	145
Medical	Mars	9th	213
Medical	Saturn	9th	119
Mental Illness	Saturn	1st	30,013
Mental Illness	Sun	2nd	1,181
Mental Illness	Mercury	2nd	530
Mental Illness	Saturn	3rd	898
Mental Illness	Saturn	4th	131
Mental Illness	Saturn	5th	117
Military	Jupiter	12th	690
Military	Jupiter	9th	58,816
Murderers	Venus	1st	498
Murderers	Venus	6th	117
Musicians	Mars	10th	380
Musicians	Mars	4th	116
Painters	Mercury	2nd	837
Painters	Saturn	5th	431
Politicians	Sun	9th	448
Politicians	Jupiter	9th	313
Sports	Mars	12th	1,496
Sports	Sun	5th	316
Sports	Mars	9th	332
Writers	Moon	12th	4,136
Writers	Sun	2nd	249
Writers	Moon	9th	503

Michel Gauquelin staked his conclusions on the "Mars effect", being strong for Sports figures in the 9th and 12th houses. My studies have also shown planets to be beneficial in the 12th house. Considering that planets first rise and become visible in this house, we should redefine it as a house of creative expression, just like Pisces, not a house of hidden activities - which is better suited to the 4th house. The 4th house starts at the Imum Coeli - the lowest, most invisible point in the daily cycle.

In the 93 fact-based Rodden study groups, the main distribution of strong findings was in the Angular houses: 52 in Angular houses, 37 in Fixed houses, and 27 in Mutable houses. Nearly twice as many findings were in Angular houses than Mutable houses. The distribution by element was spread evenly: 32 in Fire houses, 25 in Earth houses, 25 in Air houses, and 35 in Water houses.

There are four quadrants of houses in an astrological chart, beginning with an Angular (Cardinal) house, followed by a Succedent (Fixed) house, and ending with a Cadent (Mutable) house.

The Angular houses begin at one of the 4 Axis points (Ascendant, Midheaven, Imum Coeli and Descendant) and are considered to have the greatest impact in the birth chart. Angular houses rule those critical things in our life, such as our appearance and how we behave (1st), our family life (4th), our married life or partnerships (7th), and career (10th).

Succedent (Fixed) houses follow (succeed) the Angular houses. These are productive houses in which matters take root and flourish, such as possessions (2nd) or children (5th). They have a stable, fixed, unchanging quality, deriving from their central position in each quadrant. The Succedent houses appertain to the Angular houses. In this way, the second house, which succeeds the first house of the self, signifies things that belong to the person. Similarly, the eighth house, which follows the seventh house of the partner, represents the belongings of the partner.

Cadent (Mutable) houses are the last houses of each quadrant. The Greeks called the Cadent houses apoklima, which literally means "falling" or "decline," because these houses were seen to be falling away from the strength of the Angular houses. Cadent houses are places of reflection, therefore, usually considered by astrologers as less fertile and productive places than either Angular or Succedent houses, with the notable exception of the ninth house.

The Fire Houses (1, 5, 9) are called the Dharma houses, which means "action" because dharma relates to our right actions done in the pursuit of our passions. They represent where we can best express ourselves and follow our true purpose.

The Earth Houses (2, 6, 10) are called the Artha houses, meaning "wealth". They describe what we value, work for, as well as our resources.

The Air Houses (3, 7, 11) are called the Kama houses, meaning "desire". They show what we desire to learn, and how we interact and communicate with others.

The Water Houses (4, 8, 12) are called the Moksha houses, meaning "liberation." They represent the areas in life where we liberate ourselves from the bonds of the material world, and delve deeper into the inner dimensions of our psyche and soul.

Top Findings (by probability of being an accident)

1. Child Abuse	Mars	4th	4,512,699
2. Obesity	Sun	4th	39,301
3. Business Owners	Saturn	12th	24,539
4. Prisoners	Mars	7th	13,419
5. Nazis	Mars	7th	10,833
6. Spiritual Teachers	Mars	1st	6,170
7. Sports Coaches	Sun	1st	4,441
8. Criminals	Saturn	9th	2,699
9. Culinary	Venus	10th	2,360
10. Songwriters	Saturn	11th	2,056
11. Computers	Sun	4th	2,052
12. Songwriters	Venus	5th	1,611
13. Lesbians	Mercury	11th	1,497
14. Computers	Venus	1st	1,284
15. Children (4+)	Mercury	9th	1,209
16. Infant Mortality	Mercury	6th	1,133
17. Mental Illness	Mercury	11th	1,099
18. Musicians	Mars	10th	914
19. Models	Mars	8th	851
20. Alzheimer's	Mercury	1st	685
21. Prisoners	Saturn	12th	681
22. Same Job (10+ yrs)	Mars	5th	569
23. Fighters	Sun	6th	561
24. Poets	Saturn	8th	534
25. Life (<29 yrs)	Sun	10th	489
26. Literature	Mercury	4th	484
27. Spiritual Teachers	Sun	1st	450
28. Journalists	Sun	6th	438
29. Models	Moon	7th	436
30. Composers	Jupiter	4th	419
31. Marriage (15+ yrs)	Jupiter	1st	395
32. Lawyers	Jupiter	11th	394
33. Mental Handicap	Mars	8th	391
34. Psychiatrists	Moon	12th	387
35. Drug Abuse	Mars	8th	379
36. Mental Illness	Sun	4th	368
37. Obesity	Moon	5th	345
38. Children (none)	Moon	7th	342
39. Only Child	Venus	7th	342
40. Psychiatrists	Sun	8th	341

1st House

Mode: Angular (Cardinal)   **Element: Fire**   **Horizon: Below**

The first house begins on the Ascendant with the Rising sign. It signifies a person's outward appearance, behavior, health, and vitality. The zodiac sign on the cusp of the 1st house helps to show how you present yourself to the world and how others see you.

The first house is associated with the sign Aries, thus describing the way you start something new, your physical energy, self-confidence, assertiveness, and aspirations. It can also show your overall outlook on life and ability to enjoy living.

Sex Workers and Sports Coaches are strong in this house, showing its connection with the Arian energy of aggression and sexuality.

Being that the first house is just below the horizon, representing the early morning hours, it is also a time of reflection and preparation. This is seen through the strength of Mystics, Psychiatrists, and Spiritual Teachers in this house. The Sun is especially strong in the first house for these types of people.

Sun	Moon	Mercury	Venus	Mars
Poets	Obesity	Alzheimer's	Cancer	Alcoholics
Priests		Culinary	Computers	Criminals
Singers		Instrumentalists	Culinary	Dancers
Spiritual Teachers		Marriage (15+)	Engineers	Marriage (15+)
Sports Coaches		Salespeople	Sex Workers	Presidents
Stroke				Spiritual Teachers
Technical				

Jupiter	Saturn		TOTAL	
Depression	AIDS		Alcoholics	Models
Marriage (15+)	Journalists		Alzheimer's	Mystics
Nobel Prize	Philosophers		Cancer	Psychiatrists
Psychics	Sports Coaches		Depression	Sex Workers
			Instrumentalists	Spiritual Teachers
			Marriage (15+)	Sports Coaches

2nd House

Mode: Succedent (Fixed)   **Element: Earth**   **Horizon: Below**

The 2nd house relates to your sense of stability, how you make money, and also your immediate environment. This is the house of money and material possessions. This includes your finances, personal belongings, and your relationship to the gifts you possess, both physically and spiritually.

Being associated with Taurus, it's a very physical and material realm that governs the five senses. The 2nd house is also associated with values, self-esteem, self-worth, income, and how at home we feel in our bodies and environments.

The groups of people showing longevity are strong in this house, especially for the Moon: Same Job (10+ yrs), Life Long (80+ yrs). The Moon is also strong in this house for creative, intelligent people such as Artists and Nobel Prize winners.

Sun	Moon	Mercury	Venus
Criminals	Artists	Adopted	Alcoholics
Lawyers	Astronauts	Literature	Cancer
Presidents	Child Prodigy	Mental Handicap	IQ (high)
Psychiatrists	Criminals	Presidents	Prisoners
Sex Abuse	Life (80+ yrs)	Sex Workers	Sex Abuse
	Marriage (15+ yrs)	Writers	Stroke
	Nobel Prize		
	Poets		
	Psychiatrists		
	Singers		
	Twins		
Mars	**Jupiter**	**Saturn**	**TOTAL**
Life (80+ yrs)	Same Job (10+ yrs)	Actors	Philosophers
Poets	Sports	Corporate Tycoons	Psychiatrists
Scientists			Same Job (10+ yrs)
Singers			Sex Abuse
Stroke			Sex Workers

3rd House

Mode: Cadent (Mutable)   **Element: Air**   **Horizon: Below**

Associated with Gemini, the third house governs your thinking process, style of communicating, and cognitive functioning. It signifies all aspects of communication: thinking, talking, your on-line persona, media, electronic devices, and the ways we give and receive messages.

It also determines the energy of your general exchanges, along with your immediate environment. This includes your siblings, cousins, neighbors and neighborhood, early education, and short trips.

The third house is strong for Salespeople, Scientists, and Writers, especially with the planets Mercury and Venus.

Sun	Moon	Mercury	Venus
Artists	Alzheimer's	Out-of-Body	Astronauts
Life (<29 yrs)	Culinary	Poets	Child Abuse
Same Job (10+ yrs)	Fighters	Salespeople	Marriage (none)
Scientists	Heart Attack	Scientists	Novelists
Stroke	Life (<29 yrs)	Writers	Poets
	Presidents		Scientists
	Singers		Sex Abuse
	Widowed		Writers

Mars	Jupiter	Saturn	TOTAL
Computers	Math-Physics	AIDS	Heart Attack
Instrumentalists	Nazis	Heart Attack	Life (<29 yrs)
Psychics	Presidents	Police	Literature
		Professors	Salespeople
		Teachers (K-12)	Scientists
		Wealthy	Transvestites

4th House

Mode: Angular (Cardinal) **Element: Water** **Horizon: Below**

The 4th house begins at the Imum Coeli, at the lowest point in the chart. It is known as the house of home and roots, and is the foundation of the whole birth chart. It concerns the home, security, mothers and lineage, children, and things in early life that served as a foundation for you. It also shows the way you nurture others and your approach to self-care.

The fourth house rules the home within you and the one you grew up in. Associated with Cancer, this house can show your roots, sense of security, and emotional foundation. This house also revolves around your connection to the past along with your ancestry line, as well as endings, such as a person's end of life.

The fourth house is strong for Composers and Literature writers, who spend much time working in private. The Sun is especially strong for these people. Jupiter is especially strong for Musicians.

However, the fourth house is the most difficult position of all the houses, being at the bottom of the chart. The Sun and Mars are particularly common positions for Child and Sex Abuse victims, Sex Offenders, and Murderers.

Sun		Moon	Mercury	
Children (none)	Poets	Child Abuse	Cancer	Literature
Computers	Scientists	Murderers	Child Prodigy	Sex Offenders
Literature	Sex Offenders	Priests	Comedians	Suicide
Mental Illness	Sex Workers	Racers	Infant Mortality	Teachers (K-12)
Murderers	Technical	Royalty	Inheritance	Technical
Nobel Prize	Writers	Twins	Life (80+ yrs)	
Obesity				

Venus	Mars	Jupiter	Saturn	TOTAL
Police	Cancer	Cancer	Business Owners	Cancer
	Child Abuse	Composers	Musicians	Child Abuse
	Marriage (none)	Criminals	Salespeople	Composers
	Murderers	Homosexuals	Sex Workers	Literature
	Novelists	Medical	Transvestites	Murderers
	Sex Abuse	Musicians		Obesity
	Sex Offenders	Singers		Sex Offenders
	Writers	Stroke		
		Suicide		

5TH HOUSE

MODE: SUCCEDENT (FIXED) **ELEMENT: FIRE** **HORIZON: BELOW**

Associated with Leo, the 5th house is all about self-expression, creativity, and celebration. This is a very fertile house in the sense that it rules the creation of all things. It shows how we express ourselves in artistic and dramatic ways, as well as the creation of children. This house governs fun and play, the joys of romance and love, and arts such as theater and music.

Ancient astrologers thought this was the house of good fortune and quite a positive and beneficial one. For this reason, they believed that Venus is strong here, which is seen with people who have 4 or more Children. But it is also a common position for unmarried people. The Moon is the strongest planet in this house, being common for intellectual thinkers and musical composers.

The outward planets Sun, Mars, and Jupiter are less successful in the fifth house, being common for many people with great interpersonal difficulties such as AIDS patients, Prisoners, and Widowers.

Sun	Moon	Mercury	Venus
Child Abuse	Composers	Adopted	Children (4+)
Child Performers	Computers	Alzheimer's	Homosexuals
Fighters	Musicians	Astrologers	Marriage (none)
Homosexuals	Obesity	Child Abuse	Priests
Poets	Philosophers	Culinary	Sex Workers
Suicide	Songwriters	Fighters	Songwriters
Widowed	Wealthy	Homosexuals	
		Out-of-Body	
		Songwriters	
		Twins	

Mars	Jupiter	Saturn	TOTAL
Culinary	AIDS	Child Prodigy	Children (4+)
Drug Abuse	Prisoners	Comedians	Comedians
Life (80+ yrs)	Sex Abuse	Infant Mortality	Culinary
Mental Handicap	Singers	Teachers (K-12)	Fighters
Obesity	Spiritual Teachers		Homosexuals
Same Job (10+ yrs)			Musicians
Transvestites			Out-of-Body
			Same Job (10+ yrs)
			Songwriters
			Transvestites

6TH HOUSE

MODE: CADENT (MUTABLE)   **ELEMENT: EARTH**   **HORIZON: BELOW**

Associated with Virgo, the 6th house is all about service and health. It shows our day-to-day life, as well as our level of mindfulness and problem-solving skills. This includes daily responsibilities, health habits, organization, daily routines and upkeep, fitness, self-care, diet, exercise, natural living, and being of service to people.

Ancient astrologers called the sixth house the house of bad fortune since it refers to sickness and servitude. The sixth is the house of servants, and so may also refer to our service to others, or that which we are obliged to do but do not necessarily want to do. It also reveals our duties and responsibilities, and the most routine aspects of work. Pets are also connected to this house since they are considered to be servants of people.

The sixth house is a position of difficulty for many groups, especially for people born with Birth Defects. However, there is no particular planetary position that is especially difficult. Mercury and Saturn are actually very productive in this house, being strong for Writers.

Sun	Moon	Mercury	Venus
AIDS	Lawyers	Birth Defect	Astrologers
Birth Defect	Marriage (none)	Fighters	Birth Defect
Fashion Designers	Sports Coaches	Journalists	Life (80+ yrs)
Fighters		Poets	Military
Journalists		Songwriters	Psychiatrists
Philosophers		Writers	Sex Offenders
			Teachers (K-12)
			Twins

Mars	Jupiter	Saturn	TOTAL
Directors	Children (none)	Engineers	Birth Defect
Wealthy	Teachers (K-12)	Lawyers	Fashion Designers
		Literature	Fighters
		Writers	Journalists
			Obesity
			Suicide
			Transvestites

7th House

Mode: Angular (Cardinal) **Element: Air** **Horizon: Above**

The 7th house begins at the Descendant, situated directly across from the first house of identity. Associated with Libra, It is all about relationships and other people that we come in contact with: marriages, business and domestic partnerships. It's where we join with others and get to know ourselves through our interactions and shows the type of mate the native is likely to be attracted to and all partnerships in general.

Creating a union and balance of two forces is the goal of this house and, therefore, symbolizes mediators and lawyers, as well as our open enemies.

For ancient astrologers, the seventh house was considered a fortunate one, which is seen for many groups of socially successful people such as Child Performers and Lawyers. The Moon and Mars are the most beneficial planets in the seventh house, while Saturn is the worst, being common for Murder Victims and Sex Abuse Victims.

Sun	Moon		Mercury	Venus
Instrumentalists	Activists	Life (80+ yrs)	Activists	Birth Defect
	Actors	Models	Birth Defect	Child Performers
	Children (none)	Mystics	Child Performers	Only Child
	Comedians	Priests	Drug Abuse	
	Fashion Designers	Spiritual Teachers	Philosophers	
	Lawyers	Teachers (K-12)	Police	
			Priests	

Mars		Jupiter	Saturn	TOTAL
Astronauts	Only Child	Child Abuse	Lawyers	Child Abuse
Dancers	Philosophers	Directors	Lesbians	Child Performers
Journalists	Politicians	Salespeople	IQ (high)	Children (none)
Lawyers	Presidents		Murder Victims	Directors
Lesbians	Prisoners		Priests	Lawyers
Nazis	Sex Offenders		Sex Abuse	Lesbians
	Stroke		Suicide	Life (80+ yrs)
				Mystics
				Only Child
				Priests
				Sports Coaches
				Stroke

8TH HOUSE

MODE: SUCCEDENT (FIXED)   **ELEMENT: WATER**   **HORIZON: ABOVE**

The 8th house, being associated with Scorpio, is one of the most psychological houses. It has to do with transformation through the birth-death-rebirth cycle, sex, and deep bonding. The 8th house is all about looking into the depths and getting to know ourselves and others intimately. It's the place of merging, where two become one, and is associated with reproduction and regeneration.

The 8th house is commonly referred to as the house of death. This house governs sexuality, an orgasm being a kind of miniature death. It also shows our relationship to other people's money: inheritances, a partner's money, debt, and taxes.

Ancient astrologers considered the eighth to be an unfortunate house. However, modern astrologers see elements of spirituality here, since the transformation of the soul is caused by psychological insight and healing. Psychiatrists and Philosophers commonly have planetary placements here.

Mercury is an exceptionally good placement here for analytical thinkers. Venus is common for people with complicated relationships such as Transvestites and Widowers. The Moon and Mars are the most difficult planets in this house for people prone to conflict.

Sun	Moon		Mercury	Venus
Birth Defect	Cancer	Life (<29 yrs)	Activists	Computers
Culinary	Computers	Murderers	Computers	Directors
Depression	Corporate Tycoons	Philosophers	Math-Physics	Homosexuals
Mental Illness	Drug Abuse	Prisoners	Mental Illness	Models
Nobel Prize	Fighters	Transvestites	Philosophers	Police
Philosophers	Lesbians		Professors	Technical
Prisoners			Technical	Transvestites
				Widowed
				Writers

Mars		Jupiter	Saturn	TOTAL
Astrologers	Mystics	Child Abuse	Fighters	Child Performers
Children (none)	Obesity	Child Performers	Sports	Computers
Drug Abuse	Salespeople	Criminals		Directors
Homosexuals	Sex Offenders	Life (80+ yrs)		Fighters
Instrumentalists	Sex Workers	Murderers		Homosexuals
		Only Child		Math-Physics
				Mental Illness
				Murderers
				Prisoners
				Psychiatrists
				Technical

9th House

Mode: Cadent (Mutable) **Element: Fire** **Horizon: Above**

The 9th house, associated with Sagittarius, represents expansive thinking and growth. This includes global travel, higher-level learning, publishing, and university teaching. It shows how we take risks and venture into areas of the world. While the opposite 3rd house is all about knowledge and information, the ninth house is about wisdom, philosophy, and religion.

The ninth house portrays our higher learning, including our spiritual understanding, belief systems, and personal philosophies. It also governs our higher cognition, dreams, and level of awareness. Ancient astrologers called this the house of the Sun God since the Sun was considered powerful in this position. Prominent Sports figures and Singers commonly have the Sun in the ninth house. The ninth house, as a whole, is the most beneficial. Child Prodigies, Children (4+), Marriage (15+), Politicians, Sports, and the Wealthy are all strong on this position.

Saturn is the one planet that does not fare well here, with many groups of problematic people strong in this position, such as Murderers, Criminals, and Drug Abusers.

Sun	Moon	Mercury	Venus
Children (4+)	Business Owners	Activists	Corporate Tycoons
Racers	Cancer	Alcoholics	Life (<29 yrs)
Singers	Medical	Child Performers	Murder Victims
Sports	Mental Illness	Children (4+)	
		Marriage (15+ yrs)	
		Politicians	
		Professors	
		Racers	
		Singers	

Mars	Jupiter	Saturn	TOTAL
Life (<29 yrs)	Adopted	Activists	Child Prodigy
	Fighters	Child Abuse	Children (4+)
	Journalists	Criminals	Depression
	Widowed	Depression	Marriage (15+ yrs)
		Drug Abuse	Politicians
		Heart Attack	Sports
		Murderers	Wealthy
		Only Child	Widowed
		Psychics	
		Wealthy	

10TH HOUSE

MODE: ANGULAR (CARDINAL)   **ELEMENT: EARTH**   **HORIZON: ABOVE**

The 10th house begins at the Midheaven, located at the very top of the birth chart. It represents the most visible and public areas of our life. This is the world's stage and how we're seen, our status and leadership. It's also about honors, achievements, fame, and public reputation.

Associated with Capricorn, the 10th house governs our career and shows our relationship to authority figures, including our father figures and the government. Traditions and institutional structures are represented by the 10th house, as well as traditions and legacies we leave in the world.

The Moon is strong in the tenth house with Nobel Prize winners and Professors. However, Mercury is common here for Murder victims and people who died young. Overall, the tenth house is a significantly worse position than the ninth house, being common for people struggling with Mental Health issues, Alcoholics, and Drug Abusers.

Sun	Moon	Mercury	Venus	Mars
Heart Attack	Astronauts	Life (<29 yrs)	Adopted	Activists
Life (<29 yrs)	Child Performers	Murder Victims	Alzheimer's	Directors
Politicians	Mental Handicap	Sports	Culinary	Infant Mortality
Professors	Nobel Prize	Widowed	Drug Abuse	Instrumentalists
	Out-of-Body		Sports	Math-Physics
	Professors		Teachers (K-12)	Musicians
			Transvestites	Singers
				Technical

Jupiter	Saturn		TOTAL	
AIDS	Children (none)		Adopted	Math-Physics
Composers	Computers		Alcoholics	Mental Illness
Military	Culinary		Child Performers	Murder Victims
Prisoners	Journalists		Culinary	Sports
Racers	Murder Victims		Drug Abuse	Widowed
Royalty	Novelists		Instrumentalists	
Widowed	Presidents			
	Stroke			

11TH HOUSE

MODE: SUCCEDENT (FIXED)　　　**ELEMENT: AIR**　　　**HORIZON: ABOVE**

The 11th house represents all that we associate with as Aquarian such as originality, inventions, astrology, sci-fi, and things unexpected or future-oriented. The 11th house is all about envisioning future possibilities, often with a group of like-minded people. It also shows our interactions with associations, groups of people working together, and friendships.

Ancient astrologers called the eleventh house the house of Good Spirit or Good Divinity and was considered a very beneficial place in the chart. It is a house of hopes, aspirations and expectations, a place for the fulfilling of desires. Worldly eminence and material abundance are denoted by this house.

The most common house for Astrologers is here, in the 11th house, since astrology is still an unchartered territory. Twins have many planets frequently in this house, showing that this house also represents comradery.

Sun	Moon	Mercury	Venus
Astrologers Cancer Child Abuse Mental Illness Twins	Sex Workers	Lesbians Medical Mental Illness Only Child Twins	Lawyers Mental Illness Stroke
Mars	**Jupiter**	**Saturn**	**TOTAL**
Children (4+)	Alcoholics Lawyers Priests	Fighters Philosophers Songwriters Sports Twins	Alzheimer's Astrologers Lawyers Lesbians Mental Illness Only Child Technical Twins

12th House

Mode: Cadent (Mutable) **Element: Water** **Horizon: Above**

The 12th house, associated with Pisces, is linked to dreams, spirituality, imagination, and creative arts such as film, dance, and poetry. It shows the parts of you that are hidden,- such as secrets, the subconscious mind, and past-life karma, as well as hidden places, such as jails and institutions.

Being the last house, it is associated with endings, the last stages of projects, old age, and surrender. Western astrologers have always regarded the twelfth house as a very unfortunate place and ancient astrologers called it the house of Evil Spirit, representing troubles, self-undoing, secret enemies, and imprisonment.

Many creative and mystical groups of people are common in the twelfth house. Artists and Astrologers have many planets frequent in this space. Contrary to tradition, there are no negative planetary positions in this house and many positive positions. The Moon, Mercury, Venus, and Jupiter all show very beneficial results here.

Sun	Moon	Mercury	Venus	Mars
Artists	Artists	Actors	Artists	Astrologers
Astrologers	Poets	Artists	Child Abuse	Cancer
Business Owners	Psychiatrists	Cancer	Dancers	Children (none)
Dancers	Singers	Depression	Lesbians	Comedians
Engineers	Transvestites	Novelists	Models	Engineers
Salespeople	Widowed	Obesity	Musicians	
	Writers	Spiritual Teachers	Sports Coaches	
		Sports Coaches		

Jupiter	Saturn			TOTAL	
Instrumentalists	Business Owners	Nazis	Actors	Mystics
Military	Children (none)	Presidents	Artists	Royalty
Psychics	Marriage (15+ yrs)	Prisoners	Astrologers	Salespeople
Royalty	Math-Physics	Sex Offenders	Business Owners	Sex Abuse
Salespeople	Military	Technical	Lesbians	Singers
Widowed			Models	

THE RISING SIGN

The Rising sign is the zodiac sign on the Eastern horizon, where the Ascendant is located. It is thought to show how a person expresses themselves outwardly, as their persona. The Rising sign displays an increasingly uneven distribution the farther North or South of the Equator people are born. This is because, in locations other than the Equator, the signs on the ecliptic cross the Eastern horizon at differing angles. The signs that take the longest to cross the horizon are Virgo-Libra, when Gemini-Cancer are at the Midheaven.

Below is a graph of the Rising signs of 10,000 random charts at 60°N, in places such as Alaska, Scandinavia, and Moscow. Eighty percent of the charts have their Ascendant in half of the signs, the signs where the declination of the Sun is decreasing (Cancer through Sagittarius). This is the time when the signs of Northern declination (Aries-Virgo) are on the Midheaven and the ecliptic is highest in the sky.

It is unlikely that human beings would have such a skewed distribution of personas. The findings amongst the study groups also do not show a very accurate portrayal of the zodiac signs and appear somewhat random.

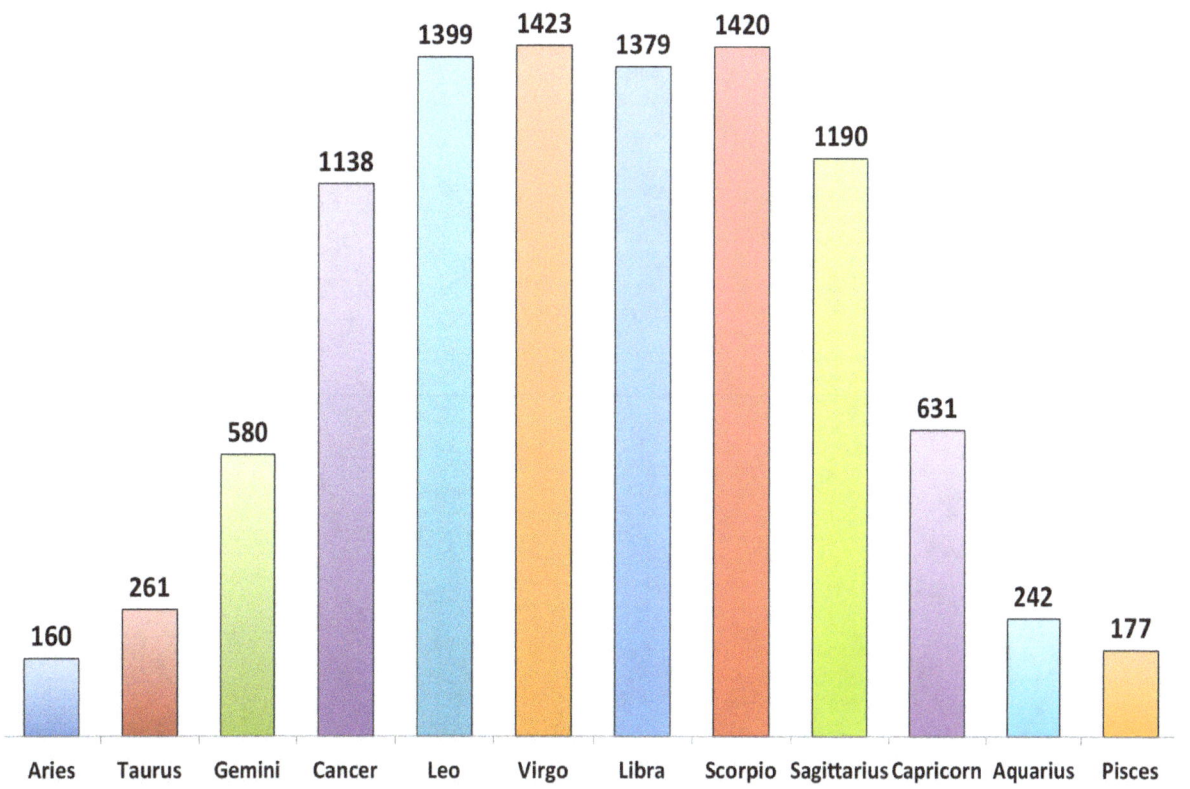

Aries	Taurus	Gemini	Cancer	Leo	Virgo
Astrologers	Artists	Heart Attack	Birth Defect	Engineers	Corporate
Child Abuse	Astrologers	Lesbians	Fashion	Royalty	Literature
Child Performer	Children (4+)	Marriage (none)	Infant Mortality	Sex Workers	Singers
Comedians	Computers	Math-Physics		Suicide	
Corporate	Drug Abuse	Musicians		Transvestites	
Heart Attack	Engineers	Twins			
Life (<29 yrs)	Inheritance				
Only Child	Poets				
Out-of-Body	Prisoners				
Sex Abuse	Salespeople				
Songwriters	Sex Abuse				
Transvestites	Teachers (K-12)				

Libra	Scorpio	Sagittarius	Capricorn	Aquarius	Pisces
Business Owners	Life (80+ yrs)	Adopted	Murder Victims	AIDS	Birth Defect
Prisoners	Mental Illness	Police	Royalty	Alcoholics	Directors
Widowed	Sports			Children (none)	IQ (high)
	Widowed			Comedians	Sex Offenders
				Education (low)	Sex Workers
				Lawyers	
				Novelists	
				Only Child	
				Philosophers	
				Priests	
				Prisoners	
				Sports Coaches	
				Stroke	
				Writers	

Top Findings (by probability of being an accident)

1. Royalty	Capricorn	3,405
2. Only Child	Aquarius	1,189
3. Birth Defect	Cancer	1,055
4. Sex Offenders	Pisces	775
5. Sex Abuse	Taurus	322
6. Lesbians	Gemini	280
7. Children (none)	Aquarius	258
8. Astrologers	Taurus	219
9. Heart Attack	Gemini	215
10. Obesity	Scorpio	203
11. Computer Programmers	Taurus	187
12. Business Owners	Libra	173
13. Artists	Taurus	136
14. Salespeople	Aries	132

Planetary Aspects

The original or archaic meaning of the word 'aspect' is to observe or to look at (from a specific perspective). In astrology, planets are judged to be in aspect when they are looking at each other from particular spatial angles and, thus, bring out specific characteristics of each planet. The new and full moons are the most observable form of aspects. These are the cyclical conjunctions and oppositions of the Sun and Moon. As viewed from the Earth, their positions in the zodiac, or ecliptic, are measured in longitude, with 0 degrees longitude being the vernal equinox, the first degree of Aries. Therefore, the positions of the Sun, Moon, and planets can be regarded as being in aspect with each other based on their positions in the zodiac.

The most powerful form of aspect that two planets can have is when they are visibly conjunct. This happens when any two of the Sun, Moon, or planets are at the same longitude (in the same zodiacal position) and the same latitude (number of degrees above or below the ecliptic). They show a profound expression in the person's character. The next most powerful aspect is when two planetary bodies are visibly opposite. This happens when they are at opposite longitudes and at opposite latitudes. They are largely experienced through inter-personal relations and objective conditions of the person. These visible and exact aspects produce eclipses when the Sun and Moon are together or apart, the visible conjunction being a solar eclipse and the visible opposition being the lunar eclipse.

Astrologers have also determined that there are many other meaningful aspects that planetary bodies can have, as measured by their longitude in the zodiac. These aspects are calculated by dividing the zodiac by a number, mainly 1, 2, 3, and 4. Since the zodiac takes a circular shape around us, we divide 360 degrees by this number. For instance, planetary bodies in opposition are 180 degrees apart, since it is calculated by dividing the circle in half. The trine is 120 degrees apart, since it is calculated by dividing the zodiac into thirds. Squares are 90 degrees apart, since they are calculated by dividing the zodiac into quarters. Each type of aspect is thought to add its own coloring to the two planets involved. For instance, a square between the Moon and Venus would cause conflict in intimacy, while a trine would cause intimacy to be more pleasant.

Many others aspects are also thought to have importance. The sextile divides the zodiac into sixths and occurs when planets are 60 degrees apart. The semi-sextile and inconjunct divide the zodiac into twelfths and occurs when planets are 30 degrees and 150 degrees apart, respectively.

Every planetary pair has a waxing and waning cycle. Waxing aspects are where the faster planet is less than 180 degrees past conjunction with the slower planet and are about initiating and growth. Waning aspects are where the faster planet is past opposition with the slower planet and are about fulfillment and resolution. As a result, the waxing square has a different effect than the waning square.

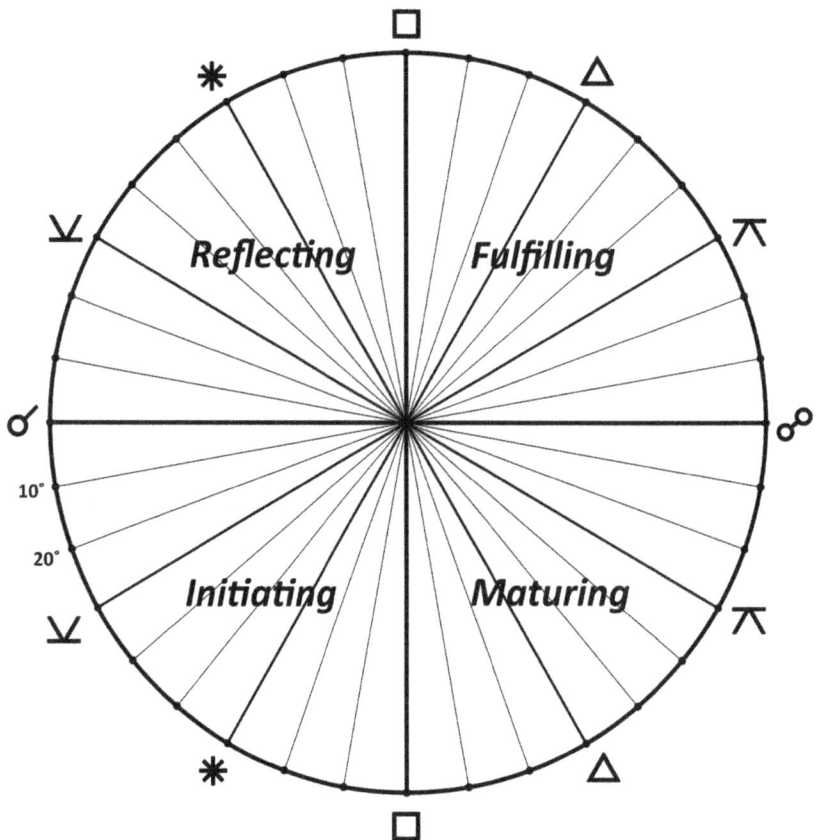

In this project, I studied the effects that these aspects show within the study groups in many different ways. I studied conjunctions alone as well as conjunctions combined with many other aspects. I also looked for the effect when two planetary bodies are parallel, that is, having the same declination (Appendix 3). The conjunction and declination studies showed good results. But the study showing the combined effects of conjunctions, oppositions, trines, squares and other minor aspects showed the best results.

The significance of the 9th harmonic, by sign, gave me a clue to finding the most significant aspect pattern for the planetary bodies of our solar system. Studying 9th harmonic charts revealed that planets were in a strong relationship with each other if they were together in the same sign and, therefore, a multiple of 40 degrees apart in the 1st harmonic. Planets that are in opposition and square in the 9th harmonic chart (10 degrees and 20 degrees apart in 1H chart) also show meaningful information. So, **I did a study on the 1/36th division of the zodiac, which is every 10 degrees**, by totaling the sum of all aspects for each planet combination. This is the first division that includes all of the major aspects as well as the nonile.

The following table shows the aspects included in the study as well the orbs used for each aspect. Larger aspect orbs were used for the more significant aspects.

Aspect	Symbol	Harmonic	Distance	Frequency	Orb
Conjunction	☌	1	0°	1	4°
Opposition	☍	2	180°	1	2°
Trine	△	3	120°	2	1.5°
Square	□	4	90°	2	1.5°
Sextile	✶	6	60°	2	1°
Semi-sextile	⋎	12	30°	2	1°
Inconjunct	⋏	12	150°	2	1°
Nonile	N	9	40°	6	1°
Semi-nonile	N_2	18	20°	6	2/3°
Quadri-nonile	N_4	36	10°	12	2/3°

The results largely showed what astrological theory would have expected.

The Sun was found to have a stronger influence when unaspected by other planets. The groups with an unaspected Sun are strong in leadership qualities: Directors, Politicians, Salespeople., Sports Coaches, and the Wealthy.

The Moon aspects showed mostly difficult results, being strong with groups having Alzheimer's, Mental Illnesses, and Obesity. Mercury aspects were frequent with many highly intelligent groups of people. Venus aspects were frequent for groups related to childhood. Mars aspects were frequent for Murderers and short-lived people, while infrequent for long-lived people.

Jupiter aspects were beneficent with the Sun and Mars, but difficult with the Moon and Venus. Saturn aspects were strong for Professors, but also for people with relationship difficulties like those who were Adopted, Unmarried and Childless.

The major exception to astrological theory were the results found for Neptune. The groups Business Owners, Corporate Tycoons, Lawyers, and the Military were all strong in Neptune. Neptune represents idealism, dreams, and spirituality, but it also expresses as materialism, deception, and a lack of respect for personal boundaries.

It is also interesting to note that prominent astrologers studying Neptune in the early 1900's such as Alan Leo and Sepharial observed that people with a strong influence of Neptune in their birth charts showed a strong likelihood to be involved in scandals.

Top Findings (by probability of being an accident)

1.	Professors	Moon-Saturn	14,730
2.	Homosexuals	Mercury-Venus	10,000
3.	Inheritance	Venus-Saturn	8,196
4.	Child Abuse	Venus-Jupiter	3,185
5.	Mental Illness	Moon-Neptune	1,417
6.	Models	Moon-Neptune	1,022
7.	Models	Mercury-Mars	995
8.	Culinary	Mars-Chiron	855
9.	Directors	Mars-Neptune	730
10.	Infant Mortality	Sun-Jupiter	713
11.	Business Owners	Moon-Pluto	710
12.	Activists	Mars-Uranus	709
13.	Mental Illness	Mercury-Neptune	639
14.	Culinary	Moon-Uranus	534
15.	Heart Attack	Moon-Saturn	504
16.	Wealthy	Mars-Pluto	465
17.	Philosophers	Mercury-Saturn	464
18.	Adopted	Moon-Jupiter	414
19.	Children (4+)	Venus-Saturn	398
20.	Corporate Tycoons	Sun-Neptune	300
21.	Education (low)	Mercury-Neptune	271
22.	Child Performer	Venus-Uranus	264
23.	Child Abuse	Moon-Jupiter	259
24.	Out-of-Body	Venus-Saturn	235
25.	Suicide	Sun-Pluto	210
26.	Child Prodigy	Sun-Venus	209
27.	Mental Illness	Moon-Venus	194
28.	Life (80+ yrs)	Mercury-Saturn	192
29.	Presidents	Mercury-Chiron	191
30.	Transvestites	Sun-Mars	189
31.	Psychics	Mercury-Venus	187
32.	Adopted	Venus-Pluto	182
33.	Medical	Moon-Jupiter	181
34.	Alzheimer's	Moon-Chiron	163
35.	Songwriters	Sun-Mercury	145
36.	Suicide	Mars-Uranus	135

Sun ☉ Aspects

Element: Fire (Yang)   **Perspective: Personal**   **Cognition: Will (Ego)**

The Sun is the masculine ego. It represents your identity and shows how you view yourself. Your outer life goals are linked to the Sun and it indicates how you can shine in your life. You will feel truly alive when living the traits indicated by your Sun. Therefore, it's likely that you will spend much of your life aiming to incorporate these traits into yourself. The older we get, the more we strive to express ourselves through our Sun's qualities.

The Sun appears to express itself more powerfully when unaspected since it naturally gives out its light when in contact with other planets. When unaspected, it retains its light and shines the brightest and gives people confidence in their talents and leadership ability. This is seen with Directors, Politicians, Salespeople, Sports Coaches, and the Wealthy. (This is the reverse of the traditional idea of planetary combustion, which states that the planets in conjunction with the Sun will be weakened - not the Sun.)

Songwriters are strongest with Sun aspects showing the creative influence the Sun has when influencing other planets. This is especially seen with Sun-Mercury and Sun-Jupiter aspects.

Moon	Mercury	Venus	Mars	Jupiter	TOTAL (strong)
Engineers	Children (4+)	Child Performers	Drug Abuse	Actors	Songwriters
Obesity	Songwriters	Child Prodigy	Presidents	Cancer	
Technical		Children (none)	Transvestites	Infant Mortality	
		Education (low)		Marriage (15+)	
		Out-of-Body		Murderers	
		Sex Offenders		Nobel Prize	
				Psychics	
				Royalty	
				Songwriters	

Saturn	Uranus	Neptune	Pluto	Chiron	TOTAL (weak)
Dancers	Astrologers	Business Owners	Lesbians	Depression	Directors
Sex Abuse	Singers	Corporate	Prisoners	Journalists	Inheritance
	Transvestites	Instrumentalists	Sex Workers	Sex Abuse	Only Child
		Songwriters	Sports	Spiritual Teacher	Politicians
		Teachers (K-12)	Suicide	Widowed	Salespeople
					Sex Workers
					Sports Coaches
					Wealthy

Moon ☽ Aspects

Element: Water (Yin) **Perspective: Personal** **Cognition: Emotion**

The Moon is the feminine ego. It represents your inner self and reveals your interior experience of being. It shows your habitual nature or subconscious patterns of thought and behavior. As such, it represents feelings and reflective thinking. While the Moon can show your natural strengths, it can also depict your weaknesses, which need to be overcome in order to be successful in life. The Moon depicts your personal roots, your home and family. It tells about your emotional nature, basic needs for comfort and is the key to how we experience intimacy in relationships.

The Moon's nurturing nature is seen with Culinary people. Its reputation for lunacy is justified by its frequency for people with Mental Illnesses. A surprising finding was that so many groups with Moon-Venus and Moon-Jupiter aspects have personal difficulties.

Sun	Mercury	Venus	Mars	Jupiter	TOTAL (strong)
Engineers	Comedians	Actors	Fighters	Adopted	Alzheimer's
Obesity	Military	Alzheimer's	Life (<29 yrs)	Child Abuse	Birth Defect
Technical		Child Abuse	Murderers	Homosexuals	Culinary
		Culinary	Presidents	Marriage (none)	Mental Illness
		Drug Abuse	Psychics	Medical	Transvestites
		Education (low)	Transvestites	Obesity	Widowed
		Mental Handicap	Wealthy	Royalty	
		Mental Illness	Widowed	Sex Abuse	
		Only Child		Sex Workers	
		Presidents			
		Transvestites			
Saturn	**Uranus**	**Neptune**	**Pluto**	**Chiron**	**TOTAL (weak)**
Adopted	Alzheimer's	Actors	Artists	Alzheimer's	Children (4+)
Comedians	Child Prodigy	Fighters	Business Owners	Culinary	Nobel Prize
Heart Attack	Comedians	Mental Handicap	Lesbians		Poets
Infant Mortality	Composers	Mental Illness			Psychiatrists
Life (80+ yrs)	Culinary	Models			
Priests	Fashion	Prisoners			
Professors	Heart Attack	Racers			
Scientists	Instrumentalists				
Sex Abuse	Philosophers				
Songwriters	Stroke				
Transvestites	Suicide				
	Widowed				

MERCURY ☿ ASPECTS

| ELEMENT: AIR | PERSPECTIVE: PERSONAL | COGNITION: LOGIC |

Mercury is a neutral planet. It represents the observer, awareness itself, that which perceives. Its symbol even looks like you! It is a circle (spirit) with a moon on its head (soul) on top of a cross (body).

Mercury tells us about your method of thinking, your learning habits and style of communication. It stands for the intellect, common sense, cold reasoning, commerce and business. Mercury's aspects also tell us about your academic schooling and learning experiences through life.

Engineers are highly influenced by Mercury, being strong with multiple aspects. Mercury-Mars aspects, which represents the union of thought, sexuality and aggression, are frequent for Murderers and Sex Offenders. Mercury-Uranus aspects are strong for Technical groups. Mercury-Chiron aspects are frequent for many types of analytical thinkers.

Sun	Moon	Venus	Mars	Jupiter	TOTAL (strong)
Children (4+)	Comedians	Birth Defect	IQ (high)	Artists	Birth Defect
Songwriters	Military	Computers	Literature	Birth Defect	Child Prodigy
		Corporate	Models	Composers	Education (low)
		Directors	Murderers	Education (low)	Psychics
		Homosexuals	Sex Offenders	Engineers	
		Mystics	Transvestites	Stroke	
		Psychics			
		Same Job (10+)			
		Sports Coaches			
		Teachers (K-12)			
		Widowed			
Saturn	**Uranus**	**Neptune**	**Pluto**	**Chiron**	**TOTAL (weak)**
Artists	Adopted	Education (low)	Adopted	Alcoholics	Alzheimer's
Education (low)	Composers	Engineers	Psychics	Child Prodigy	Business Owners
Life (80+ yrs)	Computers	Instrumentalists	Sex Offenders	Engineers	Child Abuse
Novelists	Engineers	Lawyers	Spiritual Teacher	Military	Criminals
Philosophers	Suicide	Marriage (15+)		Politicians	Culinary
	Technical	Mental Illness		Presidents	Stroke
		Murderers		Psychiatrists	
		Technical		Technical	

Venus ♀ Aspects

Element: Earth (Yin) **Perspective: Personal** **Cognition: Sensuality**

Venus, the feminine persona, represents pleasure of the senses and the harmonious co-operation of beings. It tells us about love and relationships and reveals the type of relationships you value, what you are seeking from your most intimate relationship, as well as how you like to show your love. Venus also depicts art and beauty, which is created by joining of senses into patterns. It indicates what you enjoy and value as well as your ability to attain them and, therefore, is linked with money and resources.

Venus aspects are strong with people who have both positive experiences as children and having children. Venus-Saturn aspects have the most influence on childhood experiences. Venus-Mars aspects incline people to be attractive. Venus-Neptune aspects are frequent for both Musicians and powerful groups of people. Venus-Chiron aspects are frequent for people with relationship wounds.

Sun	Moon	Mercury	Mars	Jupiter	TOTAL (strong)
Child Performers	Actors	Birth Defect	Politicians	Child Abuse	AIDS
Child Prodigy	Alzheimer's	Computers		Depression	Child Abuse
Children (none)	Child Abuse	Corporate		Infant Mortality	Child Performers
Education (low)	Culinary	Directors		Out-of-Body	Child Prodigy
Out-of-Body	Drug Abuse	Homosexuals			Children (4+)
Sex Offenders	Education (low)	Mystics			Education (low)
	Mental Handicap	Psychics			Out-of-Body
	Mental Illness	Same Job (10+)			Psychics
	Only Child	Sports Coaches			
	Presidents	Teachers (K-12)			
	Transvestites	Widowed			

Saturn	Uranus	Neptune	Pluto	Chiron	TOTAL (weak)
Adopted	Child Performers	Business Owners	Activists	AIDS	Literature
Birth Defect	Child Prodigy	Composers	Adopted	Lesbians	Murderers
Children (4+)	Marriage (15+)	Computers	AIDS	Priests	Obesity
Inheritance	Police	Corporate	Artists	Psychics	Psychiatrists
Marriage (15+)		Lawyers	Cancer		Twins
Mental Handicap		Military	Murder Victims		
Nazis		Musicians	Presidents		
Out-of-Body		Only Child	Spiritual Teacher		
Sports		Singers			
Wealthy					

Mars ♂ Aspects

Element: Fire (Yang)　　　**Perspective: Personal**　　　**Cognition: Desire (Id)**

Mars, the masculine persona, represents the urge towards fulfilling individual desires. It is the planet that tells us about your assertion and drive to dare to do something and grow as an individual. It tells how you function in a competitive, pressurized situation, your stamina, determination, and how you experience conflict and risk. It also shows how you behave sexually.

Mars aspects are frequent with people who have lived less than 29 years and infrequent with people who have lived more than 80 years. This is due to Martian attraction to competition and conflict. This is most noticeable with Moon-Mars and Mars-Uranus aspects. Also, noticeable is that people with positive, enduring relationships have little influence from Mars.

There is a technical side to Mars that is revealed with its aspects to Saturn and Pluto, which itself has an investigative and probing nature.

Sun	Moon	Mercury	Venus	Jupiter	TOTAL (strong)
Drug Abuse	Fighters	IQ (high)	Politicians	Astrologers	Culinary
Presidents	Life (<29 yrs)	Literature		Astronauts	Life (<29 yrs)
Transvestites	Murderers	Models		Marriage (15+)	Psychics
	Presidents	Murderers		Psychiatrists	Wealthy
	Psychics	Sex Offenders		Racers	
	Transvestites	Transvestites		Wealthy	
	Wealthy				
	Widowed				
Saturn	**Uranus**	**Neptune**	**Pluto**	**Chiron**	**TOTAL (weak)**
Cancer	Activists	Criminals	Adopted	Alzheimer's	Homosexuals
Child Abuse	Business	Directors	Math-Physics	Culinary	Life (80+ yrs)
Culinary	Criminals		Scientists	Fighters	Mental Illness
Math-Physics	Dancers		Wealthy	Only Child	Military
Professors	Murder Victims			Sex Abuse	Nobel Prize
Psychics	Suicide				Philosophers
Wealthy					Scientists
					Twins

Jupiter ♃ Aspects

Element: Air (Yang)   **Perspective: Collective**   **Cognition: Intellect**

Jupiter represents one's philosophy and pursuit of goals. It is the planet of growth, abundance, and wisdom and, therefore, can also be linked to excess. It highlights the areas in which we experience enthusiasm and confidence. Its position indicates where we are trusting in life, our ideals and values, and our search for and discovery of personal meaning.

Jupiter's effect of expanding our minds and success in life is seen with Sun-Jupiter aspects. Surprisingly, Jupiter in aspect with the feminine planets, Moon and Venus, are frequent for people who have childhood and relationship difficulties. It also has the effect of increasing our capacity for risk and danger, which is most frequently seen with Pluto aspects.

Sun	Moon	Mercury	Venus	Mars	TOTAL (strong)
Actors	Adopted	Artists	Child Abuse	Astrologers	Birth Defect
Cancer	Child Abuse	Birth Defect	Depression	Astronauts	Child Abuse
Infant Mortality	Homosexuals	Composers	Infant Mortality	Marriage 15+	Life (<29 yrs)
Marriage 15+	Marriage (none)	Education (low)	Out-of-Body	Psychiatrists	Obesity
Murderers	Medical	Engineers		Racers	Sex Abuse
Nobel Prize	Obesity	Stroke		Wealthy	Spiritual Teacher
Psychics	Royalty				Stroke
Royalty	Sex Abuse				
Songwriters	Sex Workers				

Saturn	Uranus	Neptune	Pluto	Chiron	TOTAL (weak)
Activists	Heart Attack	Actors	Business	Child Performer	Business Owner
Homosexuals	Life (<29 yrs)	Computers	Child Abuse	Education (low)	Corporate
Journalists		Directors	Criminals	Infant Mortality	Fashion
Spiritual Teacher		Fighters	Murder Victims	Lesbians	Police
		Murder Victims	Life (<29 yrs)	Singers	Transvestites
		Journalists	Sex Abuse	Sports	
		IQ (high)	Sex Offenders	Sports Coaches	
		Obesity	Sports		
		Out-of-Body	Sports Coaches		
		Songwriters	Stroke		
		Sports	Suicide		
		Sports Coaches	Writers		
		Technical			

SATURN ♄ ASPECTS

ELEMENT: EARTH (YIN) **PERSPECTIVE: COLLECTIVE** **COGNITION: SUPEREGO**

Saturn represents the control of behavior and thought. It tells us about authorities in our lives, such as parental figures, governmental laws, and cultural morals. It shows how we accept responsibilities and boundaries as well as the limitations imposed upon us. Therefore, Saturn can have a serious, depressing effect. This planet is also linked with our ambitions, and how we approach the practical and professional side of our lives.

Saturn as a symbol of authority is seen with Professors, and its relation to longevity is evident with the Long Life group. With Venus, Saturn is an indication of lasting relations and wealth. With Chiron, it has the effect of limiting our joy and vitality. People with Saturn-Mars and Saturn-Pluto aspects are often especially profound thinkers.

Sun	Moon	Mercury	Venus	Mars	TOTAL (strong)
Dancers	Adopted	Artists	Adopted	Cancer	Adopted
Sex Abuse	Comedians	Education (low)	Birth Defect	Child Abuse	Birth Defect
	Heart Attack	Life (80+ yrs)	Children (4+)	Culinary	Children (none)
	Infant Mortality	Novelists	Inheritance	Math-Physics	Heart Attack
	Life (80+ yrs)	Philosophers	Marriage (15+)	Professors	Homosexuals
	Priests		Mental Handicap	Psychics	Marriage (none)
	Professors		Nazis	Wealthy	Priests
	Scientists		Out-of-Body		Professors
	Sex Abuse		Sports		Racers
	Songwriters		Wealthy		
	Transvestites				

Jupiter	Uranus	Neptune	Pluto	Chiron	TOTAL (weak)
Activists	Alcoholics	AIDS	AIDS	Adopted	Astrologers
Homosexuals	Engineers	Astrologers	Alzheimer's	Birth Defect	Corporate
Journalists		Birth Defect	Child Prodigy	Homosexuals	Engineers
Spiritual Teacher		Business Owner	Culinary	Marriage (none)	Psychiatrists
		Children (4+)	Math-Physics	Mental Handicap	Salespeople
		Criminals	Medical	Salespeople	
		Drug Abuse	Philosophers	Songwriters	
		Engineers	Police	Sports	
		Literature	Psychiatrists		
		Models			
		Mystics			
		Obesity			
		Psychics			

Uranus ♅ Aspects

Element: Air (Yang)   **Perspective: Impersonal**   **Cognition: Intuition**

Uranus represents the process of awakening to truth. It shows us where we are original and unconventional, since it brings us a connection to our inner self. It can manifest as unusual and brilliant genius, willful rebelliousness, or shocking impulsiveness. It tells us the areas in which you break with convention, become innovative and experiment in new and exciting ways. Unlike any other planet, Uranus spins sideways on its axis.

Uranus manifests its enlightening influence differently based on the planet it aspects. With the Moon, it reveals creative intelligence. With Mercury and Chiron, it shows technical ability. With Venus, it manifests as talent during childhood.

Uranus is also famous for delivering sudden shocks and disruptions into our lives (rude awakenings). This most clearly manifests with Mars and Jupiter aspects.

Sun	Moon	Mercury	Venus	Mars	TOTAL (strong)
Astrologers	Alzheimer's	Adopted	Child Performers	Activists	Activists
Singers	Child Prodigy	Composers	Child Prodigy	Business	Composers
Transvestites	Comedians	Computers	Marriage (15+)	Criminals	Marriage (15+)
	Composers	Engineers	Police	Dancers	Suicide
	Culinary	Suicide		Murder Victims	Widowed
	Fashion	Technical		Suicide	
	Heart Attack				
	Instrumentalists				
	Philosophers				
	Stroke				
	Suicide				
	Widowed				
Jupiter	**Saturn**	**Neptune**	**Pluto**	**Chiron**	**TOTAL (weak)**
Heart Attack	Alcoholics	Child Prodigy	Composers	Adopted	Artists
Life (<29 yrs)	Engineers	Computers	Criminals	AIDS	Business Owners
			Sports	Astrologers	Literature
				Child Performers	Math-Physics
				Computers	Politicians
				Drug Abuse	
				Philosophers	
				Scientists	
				Technical	

Neptune Ψ Aspects

Element: Water (Yin)   **Perspective: Impersonal**   **Cognition: Imagination**

Neptune represents the dream and illusion of being an individual in an objective universe. It, therefore, blurs reality and fantasy and shows your tendency for imagination and escapism. Neptune often shows where we feel lost and without foundation, since it represents the forgetfulness of our true self. It also reveals how you seek personal glamour and glory. This can manifest as deliberate deceptiveness and insincerity.

Neptune is commonly thought of as a planet representing spirituality. But, these findings completely contradict that notion. Instead, Neptune reveals itself to be strong with groups which have great personal ambition, aggression, and selfish motives. These groups reveal Neptune's willingness to attain goals through deception.

Neptune's dual nature of being imaginative yet deceptive is seen most when in aspect with the Sun and Venus, with Corporate Tycoons and Business Owners. With the Moon, Neptune is a strong indication of mental problems. Neptune-Pluto aspects appear to be an indication of violence.

Sun	Moon	Mercury	Venus	Mars	TOTAL (strong)
Business Owners	Actors	Education (low)	Business Owners	Criminals	Corporate
Corporate	Fighters	Engineers	Composers	Directors	Instrumentalists
Instrumentalists	Mental Handicap	Instrumentalists	Computers		Lawyers
Songwriters	Mental Illness	Lawyers	Corporate		Mental Handicap
Teachers (K-12)	Models	Marriage (15+)	Lawyers		Mental Illness
	Prisoners	Mental Illness	Military		Military
	Racers	Murderers	Musicians		Murder Victims
		Technical	Only Child		
			Singers		

Jupiter	Saturn	Uranus	Pluto	Chiron	TOTAL (weak)
Actors	AIDS	Child Prodigy	Comedians	Actors	Artists
Computers	Astrologers	Computers	Lawyers	Adopted	Dancers
Directors	Birth Defect		Life (80+ years)	Priests	IQ (high)
Fighters	Business		Medical	Sports	Spiritual Teacher
Murder Victims	Children (4+)		Military	Wealthy	Twins
Journalists	Criminals		Nazis		
IQ (high)	Drug Abuse		Suicide		
Obesity	Engineers		Transvestites		
Out-of-Body	Literature				
Songwriters	Models				
Sports	Mystics				
Sports Coaches	Obesity				
Technical	Psychics				

Pluto ♇ Aspects

Element: Fire/Water **Perspective: Impersonal** **Cognition: Shadow**

Pluto represents the shadow aspect of our subconscious mind. Under its influence, powerful transformations can occur through uncovering hidden knowledge. It also indicates what we seek to influence and control. Pluto shows us in what areas of life you need to let go, and whether you're likely to find the process painful or joyful.

When Pluto is in aspect to planets, it often manifests deep thinking qualities. This is most apparent with Mercury, Mars, and Saturn. However, Pluto aspects can also be a strong indication of unhealthy relationships and life shortening circumstances. This is seen most frequently with Pluto aspects to the Sun, Venus, Jupiter, and Neptune.

Sun	Moon	Mercury	Venus	Mars	TOTAL (strong)
Lesbians	Artists	Adopted	Activists	Adopted	Adopted
Prisoners	Business	Psychics	Adopted	Math-Physics	Lesbians
Sex Workers	Lesbians	Sex Offenders	AIDS	Scientists	Spiritual Teacher
Sports		Spiritual Teacher	Artists	Wealthy	
Suicide			Cancer		
			Murder Victims		
			Presidents		
			Spiritual Teacher		

Jupiter	Saturn	Uranus	Neptune	Chiron	TOTAL (weak)
Business	AIDS	Composers	Comedians	Mental Illness	Only Child
Child Abuse	Alzheimer's	Criminals	Lawyers	Twins	Songwriters
Criminals	Child Prodigy	Sports	Life (80+ yrs)		
Murder Victims	Culinary		Medical		
Life (<29 yrs)	Math-Physics		Military		
Sex Abuse	Medical		Nazis		
Sex Offenders	Philosophers		Suicide		
Sports	Police		Transvestites		
Sports Coaches	Psychiatrists				
Stroke					
Suicide					
Writers					

CHIRON ⚷ ASPECTS

ELEMENT: EARTH/AIR **PERSPECTIVE: COLLECTIVE** **COGNITION: TRAUMA**

Chiron, the Wounded Healer, represents our emotional and physical wounds. It reveals a part of us that is hurt, vulnerable, and insecure. Chiron helps us to heal our wounds, and then look beyond personal realms, to see the suffering of others, and to become teachers and healers.

Chiron's association with wounds is most apparent through its Saturn aspects. Venus aspects indicate difficulties in relationships as seen with AIDS patients and Priests. Chiron's healing qualities are expressed through its analytical nature, seen when in aspect to Mercury and Uranus.

Sun	Moon	Mercury	Venus	Mars	TOTAL (strong)
Depression	Alzheimer's	Alcoholics	AIDS	Alzheimer's	Culinary
Journalists	Culinary	Child Prodigy	Lesbians	Culinary	Lesbians
Sex Abuse		Engineers	Priests	Fighters	Prisoners
Spiritual Teacher		Military	Psychics	Only Child	Widowed
Widowed		Politicians		Sex Abuse	
		Presidents			
		Psychiatrists			
		Technical			

Jupiter	Saturn	Uranus	Neptune	Pluto	TOTAL (weak)
Child Performer	Adopted	Adopted	Actors	Mental Illness	Novelists
Education (low)	Birth Defect	AIDS	Adopted	Twins	Murder Victims
Infant Mortality	Homosexuals	Astrologers	Priests		Nobel Prize
Lesbians	Marriage (none)	Child Performers	Sports		Sex Workers
Singers	Mental Handicap	Computers	Wealthy		Sports Coaches
Sports	Salespeople	Drug Abuse			Writers
Sports Coaches	Songwriters	Philosophers			
	Sports	Scientists			
		Technical			

Case Studies

Here are a number of case studies that demonstrate the use of the 1st and 9th harmonics, as well as a few predictive techniques implementing the 9th harmonic.

Bill Gates and Steve Jobs

These are the 1st harmonic natal birth charts of Bill Gates and Steve Jobs, two of the most important figures in the computer technology revolution. They both have the Moon in the Aries-Gemini navamsha (which is strong in the Brilliant Minds group), the Sun in Water signs, and Mercury in Air signs. This tells us that they are intuitive leaders with good thinking and communication skills. The charts show some similarity but nothing remarkable.

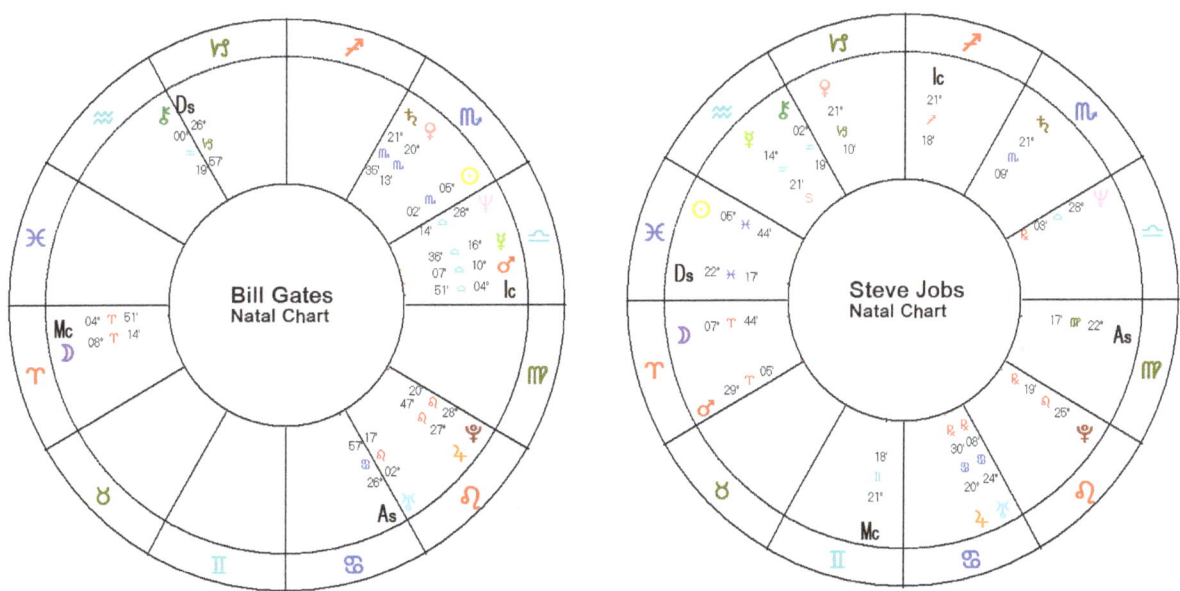

In their 9th harmonic charts, an amazing correspondence reveals itself. They both have the Sun in Leo, and Moon in Gemini conjunct Neptune. They both have Mercury in Aquarius and Saturn in Capricorn involved in a conjunction with inner planets. They have Venus in opposite signs, Jupiter and Mars are transposed and they both have Chiron in Libra.

These charts represent the inner workings of Bill Gates and Steve Jobs and an argument could be made that they are twins on a spiritual level. Both men have tremendous confidence in their intelligence and a great work ethic demonstrated by the planets in Capricorn. The Moon-Neptune conjunction in Gemini indicates their great need to achieve their ideals through intelligence.

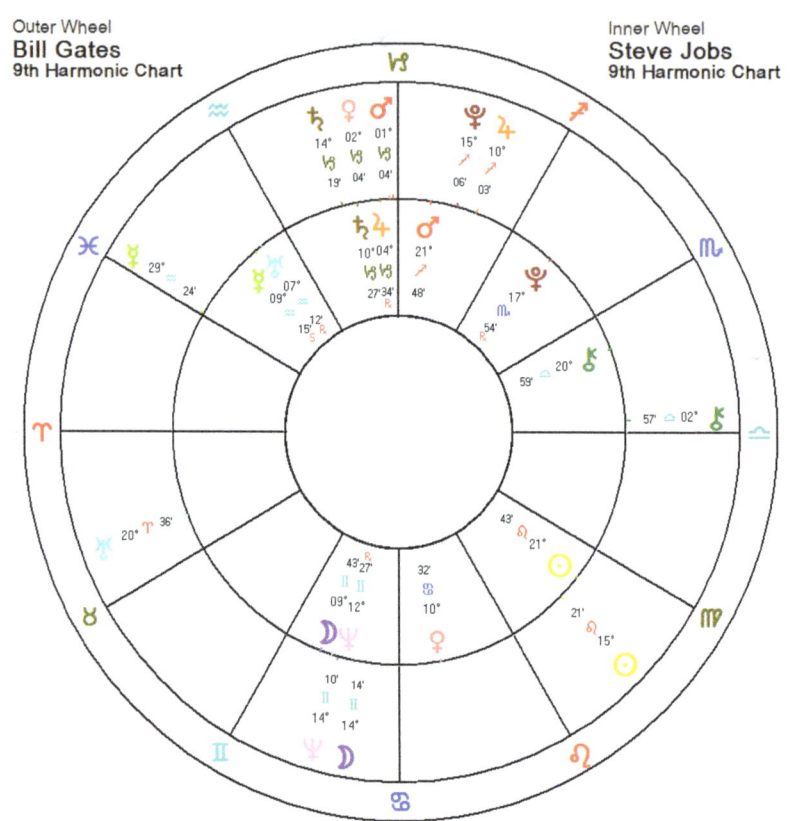

Mars and Athletes

It is commonly thought that Mars represents aggression and is therefore the critical element for a successful athlete. While this did not show up as an overall trend for the 1,214 person Sports study group (though Aries was strong), the charts of the three most prominent athletes of our time, Pele, Michael Jordan, and Wayne Gretzky show clear signs of Mars. All three of them have Mars in the 1st degree of Cancer or Capricorn in the 1st or 9th harmonic charts, indicating that these parts of the zodiac have great importance. This is not surprising since they are the points at which the Sun changes direction by declination and signal the beginning of Summer and Winter.

Michael Jordan has Mars in the 1st degree of Cancer in his 9th harmonic chart. In this chart, he also has an extremely tight Sun-Moon conjunction in Gemini (Air Jordan). Wayne Gretzky has Mars in the 1st degree of Cancer in the 1st harmonic. Pele has Mars in the 1st degree of Capricorn in the 9th harmonic chart. (This also indicates that the degree positions in the 9th harmonic chart are significant.)

It is interesting, that Pele is the only one of them with Mars in Capricorn, the low point of the solar cycle. It is likely that Mars is behaving as though it is in Cancer since he was born in the Southern hemisphere, where the seasons are reversed.

OJ Simpson

Looking at OJ Simpson's 1st harmonic chart, there is very little evidence of an athlete or of someone with violent tendencies. The Sun, Mercury, Venus stellium in Cancer and his Pisces Moon would lead most astrologers to interpret him as a sensitive, caring introverted type.

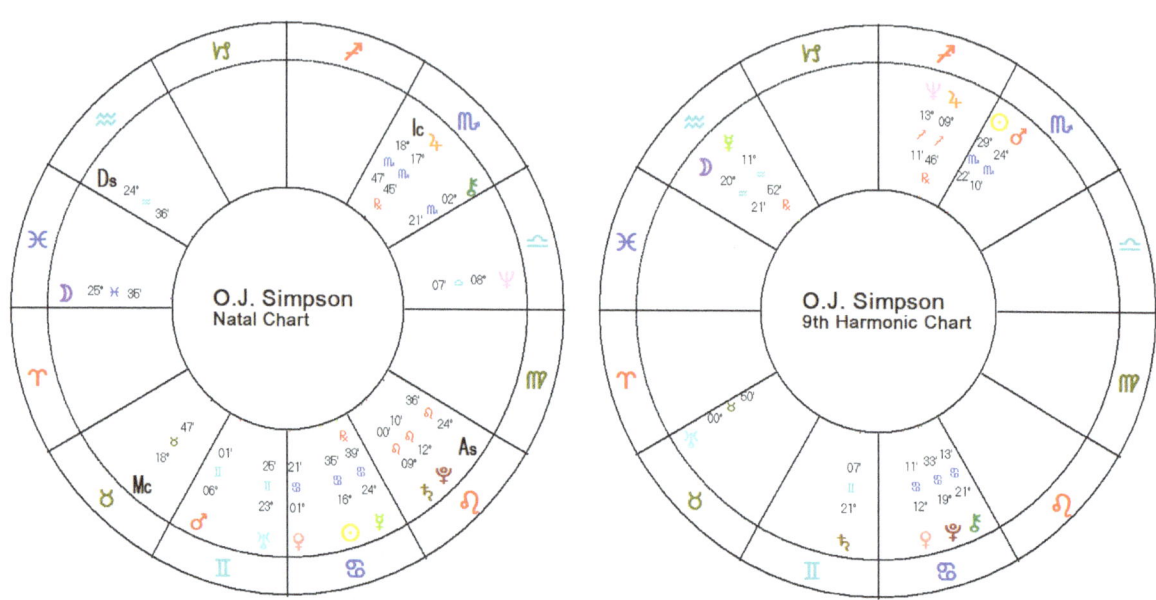

However, a quick glance at his 9th harmonic chart reveals an entirely different picture. He has the Sun and Mars together in Scorpio, revealing an intense, aggressive nature. Mercury and the Moon are together in Aquarius showing his charisma and facility at communicating with the public. He also has Venus, Pluto and Chiron together in Cancer, which is highly symbolic of co-dependent, hurtful relationships.

Using the 9th harmonic chart also enables us to see events in a person's life that would have gone unnoticed using the 1st harmonic alone. On June 12, 1994, Nicole Brown Simpson was murdered and OJ Simpson was found fleeing in a car. This was highly suspicious behavior for a man found innocent by our courts. When we look at transits to his 1st harmonic chart, Chiron and Saturn were square by transit to his natal Mars. These are small indications of violence but hardly life-altering indicators.

However, looking at transits to his 9th harmonic chart reveals a whole different story. Pluto was conjunct with his 9th harmonic Sun (and Mars). There is no bigger transit than this in astrological symbolism. It speaks of having one's life significantly transformed in turbulent circumstances. The Pluto on Mars transit indicates that he was feeling very energized and aggressive at the time as well.

RELATIONSHIP COMPATIBILITY

When looking at the compatibility of two persons' charts, we can directly compare the charts by viewing the interrelations of the planets (synastry) or we can create a new chart using the midpoints of the planets in both charts (composite chart).

When comparing OJ's chart with Nicole's chart, the most obvious indication of violence is that they both have Mars within 10 degrees of each other's Sun. In addition, Nicole's Chiron is on OJ's Descendant indicating that Nicole is a relationship partner who can be woundful or be wounded by OJ. She also has Pluto on her own Descendant, showing a tendency to attract powerful, controlling people into her life.

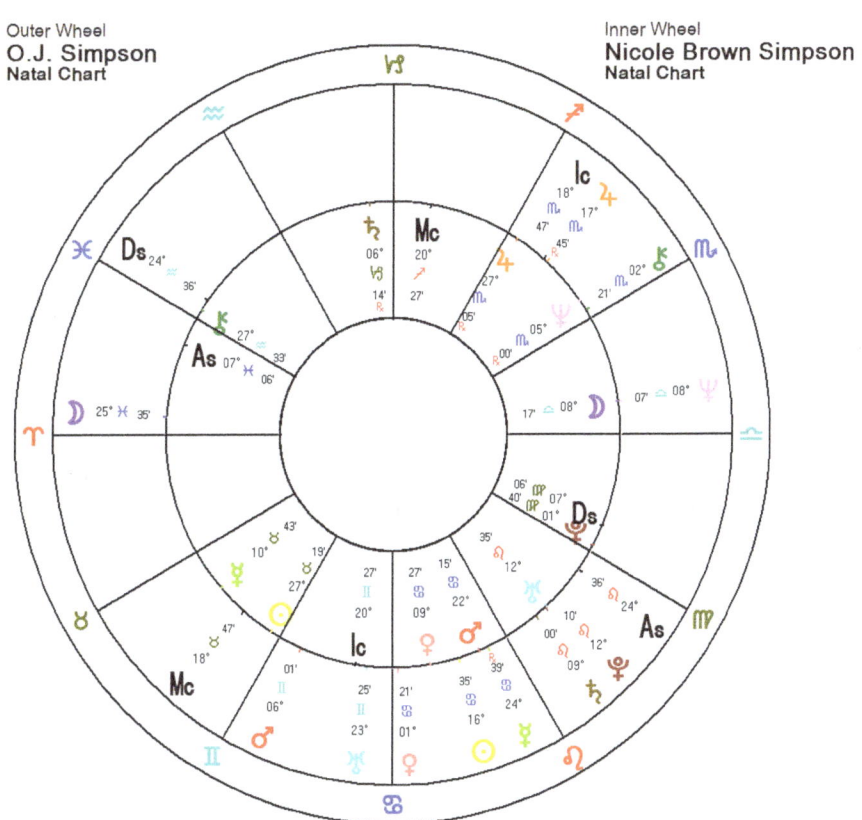

When looking at the composite chart, we see that the Sun and Mars are conjunct. The Sun is also trine Saturn and sextile Pluto - all significant indicators of tension. The Moon is in the powerful Capricorn-Capricorn navamsha indicating deep emotional repression and the inability to give and receive nurturing. The Moon is also conjunct Chiron, and opposite Mars indicating the possibility of hurtful experiences within the relationship.

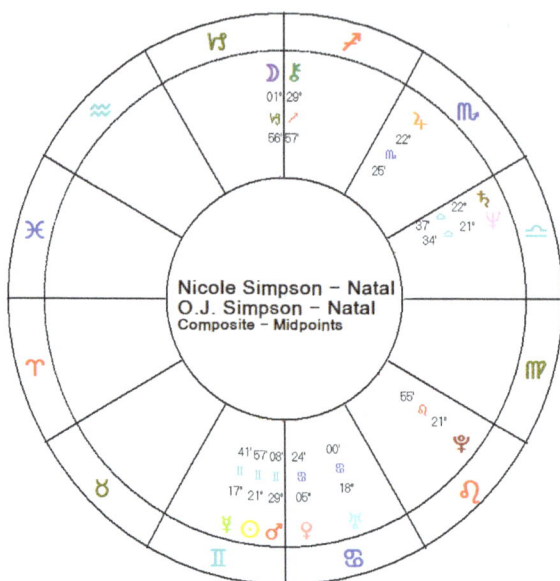

There are two major techniques for observing the transitions that a relationship may undergo through time. The first is to view the transits to the composite chart and the second technique is to calculate a secondary progressed composite chart for the time in question.

In the case of OJ and Nicole Simpson, there were no major transits at the time of the murder, but the secondary progressed composite chart contains some very interesting findings. Venus has moved into an exact conjunction with Pluto indicating a likelihood for intense passion, transformation and danger in love. The other major finding is that the composite Moon has returned to its natal position in Capricorn with Chiron (in the powerful Capricorn-Capricorn navamsha) showing deeply hurt feelings.

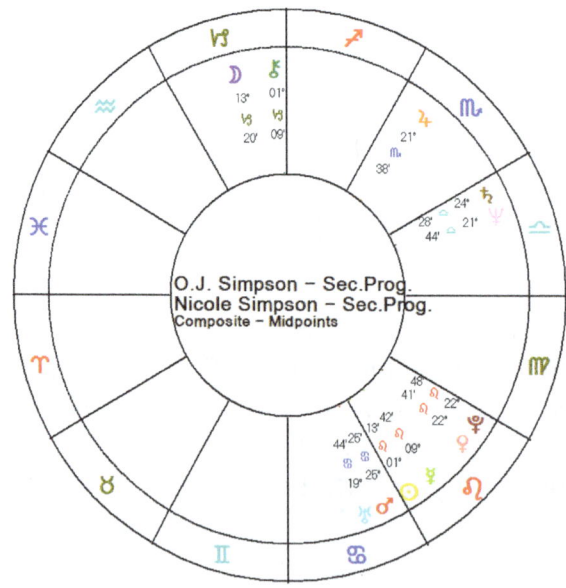

ADOLF HITLER

For Adolf Hitler, I will apply all of the techniques studied in order to show how to fully analyze a birth chart. The outer chart is the 1st harmonic, the inner chart is the 9th harmonic. The declination chart is on the left and fixed stars conjunctions are on the lower right.

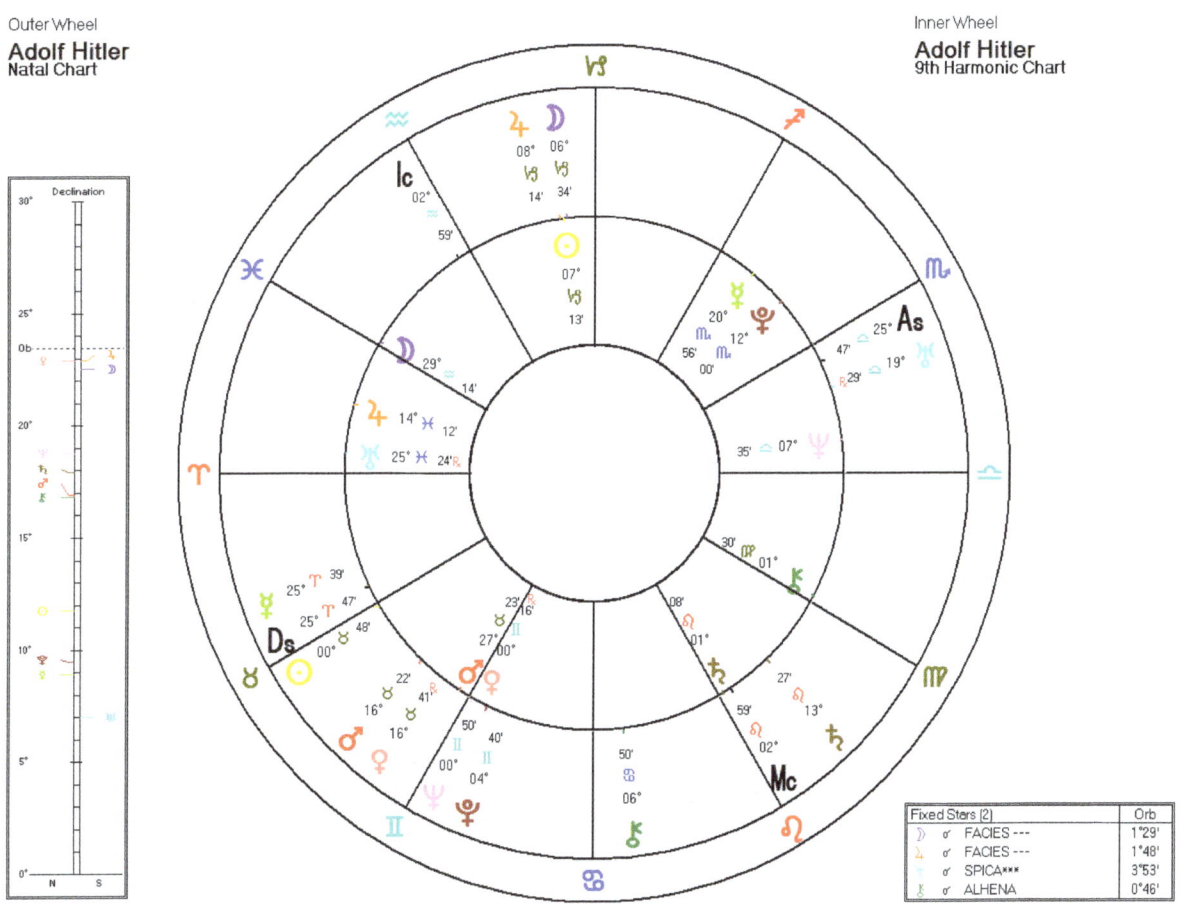

The first thing to do is look for sign, element, mode, and quadrant strength in the 1st and 9th harmonic charts. Hitler has many planets in Taurus and Capricorn and has a heavy dominance in the 1st quadrant. Hitler's strength in the Earth element shows his practicality, toughness, and stubborn determination to make his goal become reality. The dominance of planets in the 1st quadrant shows his focus on personal goals as well as a lack of awareness of the needs of others. This quadrant is very strong amongst athletes.

The next step is to look for conjunctions between planets, especially amongst the Sun to Saturn in both the 1st and 9th harmonic charts. In the case of Hitler, he has the Moon conjunct Jupiter in Capricorn, showing his fantastic executive leadership ability, yet a deeply repressed capacity for nurturing or empathy. This conjunction is common for people who were abused or adandoned in childhood.

The Moon and Jupiter are also in conjunction with the fixed star Facies, the Archer God of War. People with these conjunctions often cause violence or turmoil or have themselves been harshly treated and can have abrasive religious or philosophical views.

His Moon was also in the Sagittarius Lunar phase (240-270 degrees from the Sun) showing his ambition to spread his Aryan philosophy.

He has Mars conjunct Venus in Taurus, revealing a strong magnetism, sensuality, and determination. Mars and Venus are also the focal points of his Sun-Neptune and Sun-Pluto midpoints. This reveals his use of personal magnetism for purposes of deception and power.

In the 9th harmonic chart, Mercury is conjunct Pluto in the Aries-Scorpio navamsha. This reveals the key to Hitler's aggressive and sinister inner thoughts and shows penetrating and suspicious thinking, especially towards others.

Chiron is also conjunct with the Moon's North Node, indicating that his life purpose revolved around wounds inflicted on people.

The next step is to look for planets conjunct with the 4 Axis points, especially the Ascendant or Midheaven (within 6-10 degrees). He has Uranus on the Ascendant indicating that he seeks opportunity to awaken others through sudden change. His Sun and Mercury were also conjunct on the Descendant, showing his profound effect on others.

Then look at the planetary placements in the houses. Since I use Diurnal Arc houses, each planet has its own house sectors - found in many programs as Gauquelin Sectors. Sectors 1-3 are the 12th house and the last three sectors, 34-36, are the 1st house since Gauquelin counted them backwards (clockwise). Hitler had most of his planets in the Air houses, 3rd and 7th, showing his ability to communicate with and relate to people.

Any planet in the same sign in both the 1st and 9th harmonic also has a strong influence with the individual. This is especially true for Cancer and Capricorn since they are at the apex and trough of the solar cycle. In the case of Hitler, he has Mars in Taurus-Taurus (physical aggression) and Saturn in Leo-Leo (confident authority). His Mars in double Taurus is in aspect to Chiron (50 degrees) and parallel by declination, showing his capacity for aggression and physical violence. Mars in double Taurus is strong in the study of Murderers.

For more specific information, look at each planet by sign position for both harmonics and look at all aspects. For example, Hitler has his Sun in Taurus-Capricorn in tight 30-degree aspect with Neptune. This describes a very practical, hard-working person who is willing to use deception to fulfill his dreams.

(Finally, I would like to mention the 27 lunar mansions, the nakshatras. The ancient sages of India wrote about them in the Vedic scriptures and are likely the oldest form of astrology, predating Babylonian and Egyptian astrology. Each nakshatra covers the distance traveled by the Moon in a day. This is the lunar zodiac, viewed at night in reference to the stars. Whereas, the 12 tropical signs are the solar zodiac, measured by daylight in reference to the seasons.

Due to the lack of precision or agreement using the stars as a reference, my research did not yield meaningful results. However, I do often find them to provide very accurate descriptions of the person. Using the most common calculation (Lahiri), Hitler was born with the Moon in Purva Ashada, which means "the undefeated" and is called "the invincible star." It is considered a "fierce" nakshatra and often shows powerful leaders.)

Part II

The Study Groups

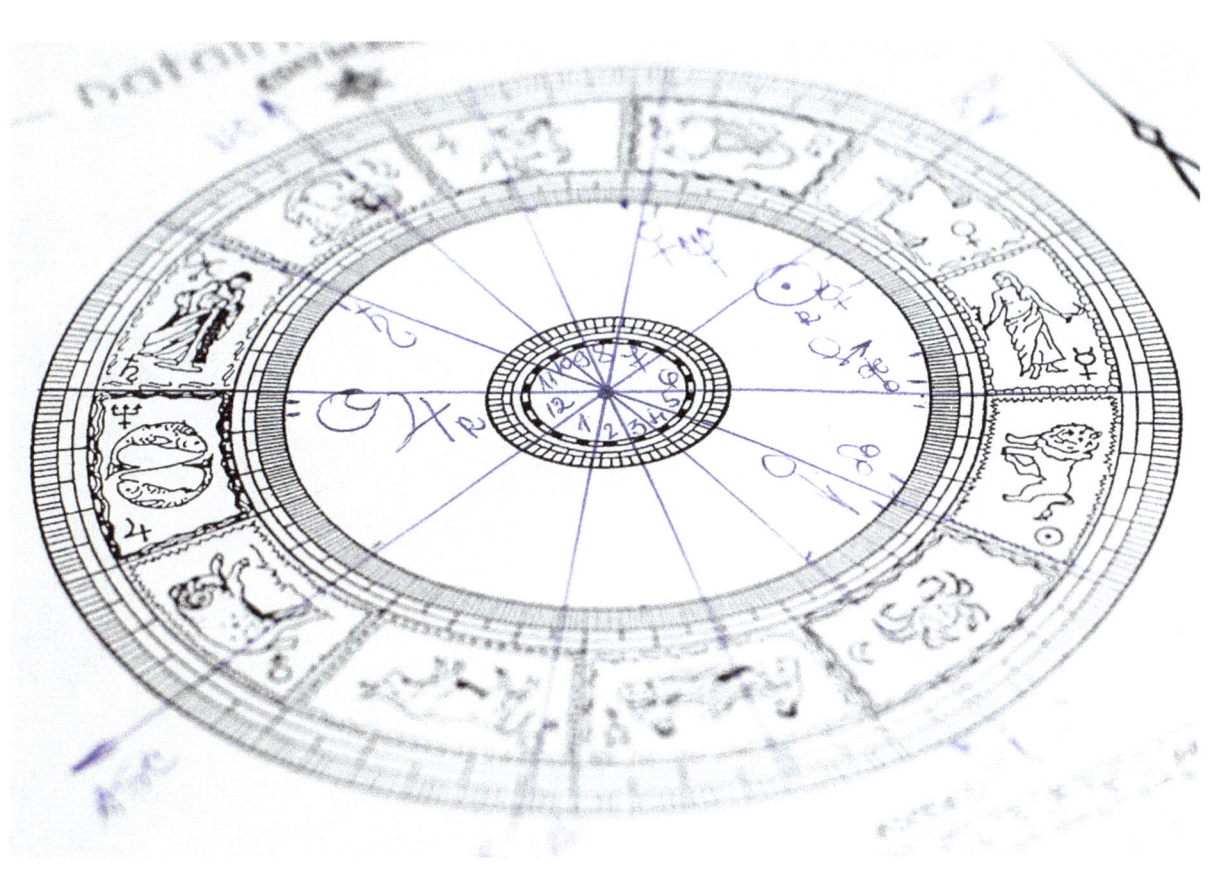

STUDY GROUPS

The following section contains an in-depth look at each of the 122 study groups. There are 93 Rodden groups that are based on factual biographical information, 14 Rodden groups based on subjective personality traits, and 15 Gauquelin groups.

For each group, I show the major themes as well as the most notable individual findings. The descriptions of the findings are short summaries of possible interpretations that are intended to instigate thought and let the reader come to their own conclusions.

Thereafter, I summarized all significant findings in a sign table that lists the strong 1st harmonic and 9th harmonic sign positions, houses, Rising sign, Moon's Node, and Lunar Phase - which is divided into 12 sections (Appendix 1). The number of people in each group study is listed in the upper left corner of this table.

A second planetary table lists the frequent 36th harmonic aspects, parallels of declination, the 4 Axis points, Sun/Moon midpoints, directional strength, and Moon's Node conjunctions.

Also shown are all the numerical findings (raw data) for each planet's sign positions in the 1st and 9th harmonic, houses, Rising sign, and Moon's Nodes. Findings in the upper 5% (95% or higher) of the normal curve, as compared with the control group, are highlighted in blue and those in the lower 5% are highlighted in red.

In each of the groups, there is a strong theme by sign and/or planet (highlighted by color). Out of the 93 fact-based groups studied, 30 of them are significantly strong or weak in the same sign in both the 1st and 9th harmonics.

Often the dominant sign and planet shows the commonly understood relationship between them. For example, people with careers in the Culinary industry are strong with both Cancer and the Moon. Also, Criminals are strong with both Scorpio and Mars.

The 14 personality trait study groups are listed after the fact-based groups. Many of these groups show interesting results, such as how Extroverts are dominant in the extroverted signs (Fire and Air).

Of the 15 Gauquelin groups, 11 of them show the same findings by sign, house, or planet as the comparative Rodden study group. For instance, both Rodden's and Gauquelin's group of Artists are strongest in Pisces.

Summary of Group Findings

Childhood	**Sign**	**House**	**Planet**
Adopted	Capricorn		Saturn/Pluto
Birth Defect	Cancer	6th	Saturn
Child Abuse	Virgo/Scorpio	4th	Moon/Venus
Child Performers	Libra/Scorpio	7th	Sun/Venus
Child Prodigy	Cancer		Mercury/Venus
Only Child	Aquarius		Venus
Twins	Sagittarius	11th	

Family Relations			
Children (4+)	Aries/Taurus/Aquar	9th	Mercury/Venus
Children (none)	Virgo		Saturn
Marriage (15+ yrs)	Sagittarius/Aquarius	1st	Jupiter
Marriage (never)	Virgo		Moon
Widowed			Moon/Chiron

Sexuality			
AIDS	Virgo		Venus
Homosexuals	Scorpio/Capricorn	5th	Jupiter
Lesbians	Leo	7th	Pluto/Chiron
Sex Abuse	Taurus	2nd	Saturn/Chiron
Sex Drive (high)	Virgo	5th	Venus
Sex Offenders	Virgo	4th	
Sex Symbols	Scorpio	8th	Venus
Sex Workers	Virgo/Libra		Jupiter
Transvestites	Scorpio		Moon/Mars

Life Span			
Infant Mortality	Cancer	6th	Moon
Long Life (80+ yrs)	Taurus/Pisces	2nd	Moon
Short Life (<29 yrs)	Scorpio	3rd	

Artistry	Sign	House	Planet
Artists	Pisces	12th	Venus
Culinary	Cancer	5th/10th	Moon/Venus
Fashion Designers	Cancer/Pisces		
Directors	Taurus	7th/8th	
Models		8th	
Musicians	Capricorn/Pisces	5th	
Composers	Capricorn		Moon/Merc/Uranus
Instrumentalists	Sagittarius		Neptune
Singers	Taurus/Gemini		
Songwriters		5th	Sun
Performers			
Actors	Aquarius		Venus
Comedians	Taurus/Scorpio/Aqua	8th	Moon
Child Performers	Libra/Scorpio	7th	Sun/Venus
Dancers	Aries/Pisces		
Sports			
Coaches	Aries	1st	
Players	Aries/Gemini	9th/10th	
Racers	Aries/Taurus		
Political			
Activists	Libra	7th/9th	Uranus/Pluto
Lawyers		7th/11th	Neptune
Nazis	Aries		Mars/Chiron
Politicians	Aries/Pisces		Mars
Presidents	Aquarius		Jupiter

Educational	**Sign**	**House**	**Planet**
Education (low)	Pisces		Mercury
Philosophers	Scorpio		
Nobel Prize	Taurus		Jupiter
Professors	Scorpio		Saturn
Psychologists	Cancer/Scorpio	2nd/8th	Chiron
Teachers (K-12)	Cancer/Libra		
Technical	Capricorn		Mercury
Astronauts	Libra/Aquarius		
Computers	Scorpio/Capricorn		Mercury/Uranus
Engineers			Mercury/Jup/Uranus
Math-Physics	Aries/Capricorn		Mars/Saturn
Medical	Virgo		Pluto
Scientists	Sagittarius/Capricorn	3rd	Moon
Writers	Aquarius		
Journalists		6th	
Literature	Aquarius	4th	Mercury
Novelists	Aquarius		Saturn
Poets	Pisces		
Mystics	Scorpio		Mercury
Astrologers	Gemini	12th	
Priests	Taurus/Pisces	7th/8th	Saturn
Psychics	Pisces	8th	Mercury
Spiritual Teachers	Scorpio/Pisces	1st	Jupiter

Cognition	Sign	House	Planet
Child Prodigy	Cancer		Mercury/Venus
IQ (high)	Libra		
Mental Handicap	Pisces	8th	Mercury
Mental Illness	Scorpio	11th	Moon/Mercury/Nept

Financial			
Business Owners	Gemini	12th	Pluto
Corporate Tycoons	Cancer		Neptune
Criminals	Scorpio		Mars
Inheritance	Capricorn	8th	Venus
Royalty			Jupiter
Salespeople	Aries		Saturn
Same Job (10+)	Taurus/Leo	2nd/5th	
Wealthy	Capricorn		Mars/Jupiter

Violence			
Child Abuse	Virgo/Scorpio	4th	Moon/Venus
Fighters	Scorpio/Capricorn	6th/8th	Moon
Military	Virgo/Capricorn		Neptune/Chiron
Murderers	Taurus/Scorpio	4th	
Murder Victims		10th	
Nazis	Aries		Chiron
Police	Leo/Scorpio	8th	Neptune
Prisoners	Scorpio/Aquarius		Mars
Sex Abuse	Taurus	2nd	Saturn/Chiron
Sex Offenders	Virgo	4th	
Suicide	Aquarius		Mars/Uranus/Chiron

Afflictions	**Sign**	**House**	**Planet**
AIDS	Virgo		Venus
Alcoholics	Taurus		
Alzheimer's	Cancer		Moon
Birth Defect	Cancer	6th	Saturn
Cancer	Virgo/Pisces	4th	
Depression	Cancer		
Drug Abuse	Scorpio		Sun/Mars
Heart Attack	Aries/Scorpio	3rd	Saturn
Infant Mortality	Cancer	6th	
Mental Handicap	Pisces	8th	Mercury/Venus
Mental Illness	Scorpio	11th	Moon/Merc/Neptune
Obesity			Moon/Jupiter
Out-of-Body (NDE)	Virgo/Capricorn		Venus
Sex Abuse	Taurus	2nd	Saturn/Chiron
Stroke			Chiron

Personality Traits			
Active	Aries	5th/11th	
Aggressive-Brash	Aries/Libra		
Brilliant Mind	Aquarius/Pisces	3rd/11th	
Eccentric		6th	Mercury
Emotional	Leo	5th	Venus/Mars
Extrovert	Aries/Aquarius		Mercury
Gracious-Sociable	Aquarius	11th	
Hard Worker	Virgo/Capric/Aquar		Saturn
Humorous	Taurus/Cancer/Leo		Venus
Introvert	Capricorn	12th	Saturn
Pioneer	Capricorn/Aquarius		
Principled	Taurus/Leo/Capricorn		
Sex Drive (high)	Virgo	5th	Venus
Sex Symbol	Scorpio	8th	

Gauquelin*	**Sign**	**House**	**Planet**
Actors	Sagittarius		Venus
Alcoholics	Taurus/Aquar/Pisces		45
Business	Virgo/Pisces	11th	Mars
Journalists	Capricorn	1st	Uranus
Medical	Pisces	12th	Neptune
Mental Illness	Aquarius/Pisces	2nd	Moon/Mercury
Military	Taurus/Capricorn	9th/12th	Pluto
Murderers	Taurus/Pisces	2nd	Neptune
Musicians	Taurus/Aquar/Pisces		Uranus/Pluto
Painters	Taurus/Pisces	2nd/5th	Pluto/Chiron
Politicians	Libra		Jupiter
Priests	Libra/Pisces	1st	Neptune/Chiron
Scientists	Sagittarius/Capricorn	7th	Sun/Neptune
Sports	Aries/Taurus/Aquar	5th/9th	Uranus
Writers	Cancer/Pisces	2nd/6th/10th	Neptune

*Gauquelin findings that are replicated by Rodden findings are highlighted in green.

ACTIVISTS

Theme: Libra (7th), Uranus/Pluto

Libra:
Activists participate in the awakening and transformation of society (Libra). The Libran intellect and humanitarianism combine to raise awareness of important causes.

Sun in Libra (1 in 136) & Sun in Aquarius (1 in 225):
Activists are highly mentally active and communicative (Air signs).

Mercury in Scorpio (1 in 398 - 9H):
They are interested in conspiracy theories and corruption. In Scorpio, Mercury produces a deep and penetrating thinker who can dig to the core of any issue.

7th & 9th Houses:
These houses are both concerned with influencing the values and beliefs of others.

Mercury in Air Houses (1 in 332):
Mercury is highly thoughtful and communicative in Air houses.

Mars-Uranus aspects (1 in 709) & Mars-Uranus parallels:
Mars and Uranus interact to create sudden change, such as awakenings.

Chiron on Descendant (1 in 358):
Activists are concerned with healing the wounds of other people.

Signs

Size: 383	1st Harmonic	9th Harmonic	Houses	Rising Sign
Moon	Gemini, Leo	Sagittarius	7th	Aries
Sun	Libra, Aquarius	Libra	8th	
Mercury		Scorpio	7th, 9th	Lunar Phase
Venus				
Mars		Leo	10th	☉/☽ Midpoint
Jupiter				Leo
Saturn	Scorpio	Aries, Taurus, Virgo	9th	Moon's Nodes
TOTAL	Libra	Capricorn		9th
(weak)	Capricorn	Aquarius		

1H	♈	♉	♊	♋	♌	♍	♎	♏	♐	♑	♒	♓
Moon	28	36	42	30	41	29	40	29	30	27	26	25
Sun	25	29	27	38	32	30	45	32	27	22	45	31
Mercury	26	19	31	29	32	38	36	41	30	33	35	33
Venus	29	22	27	42	26	39	29	32	32	31	40	34
Mars	29	30	40	33	40	29	38	42	28	31	21	22
Jupiter	24	28	35	28	33	33	35	36	40	27	30	34
Saturn	38	37	19	20	37	37	36	44	33	21	27	34
TOTAL	199	201	221	220	241	235	259	256	220	192	224	213
Rising	25	23	26	37	46	50	34	41	43	23	23	12
N. Node	27	29	25	29	36	32	39	34	33	37	38	24

9H	♈	♉	♊	♋	♌	♍	♎	♏	♐	♑	♒	♓
Moon	32	30	29	30	30	33	31	37	42	39	23	27
Sun	25	37	29	30	27	29	42	32	30	36	33	33
Mercury	28	32	37	27	27	32	26	47	29	35	36	27
Venus	29	39	30	32	27	31	25	30	35	39	33	33
Mars	34	28	23	29	42	34	37	32	32	31	30	31
Jupiter	28	37	35	34	28	35	31	31	31	35	20	38
Saturn	44	45	24	36	22	44	25	33	41	33	19	17
TOTAL	220	248	207	218	203	238	217	242	240	248	194	206

Houses	1st	2nd	3rd	4th	5th	6th	7th	8th	9th	10th	11th	12th
Moon	28	36	27	37	30	33	46	21	33	33	26	33
Sun	34	38	28	40	21	32	27	38	28	36	35	26
Mercury	35	24	42	28	28	31	39	21	38	30	40	27
Venus	29	35	33	34	28	28	27	31	33	35	26	44
Mars	36	26	37	19	36	20	32	31	28	42	39	37
Jupiter	40	29	28	28	33	39	36	25	31	26	40	28
Saturn	32	33	33	33	24	32	23	37	41	30	33	32
TOTAL	234	221	228	219	200	215	230	204	232	232	239	227
N. Node	28	38	27	32	29	35	31	33	44	38	23	25

Planets

Aspects	Parallels	4 Axis	Direction	☉/☽ Midpoint
Uranus	Mars-Uranus	Venus (Dsc)	Neptune (Rx)	
Venus-Pluto		Chiron (Dsc)	Pluto (Rx)	
Mars-Uranus				
Jupiter-Saturn				Moon's Nodes
				Pluto (S)

Actors

Theme: Aquarius, Venus

Aquarius (1 in 184):
Aquarius is in the 4th quadrant of the zodiac. These signs are concerned with humanity giving them the ability to connect with large groups of people. Many Aquarians have mass-appeal and are a particularly loquacious group of people who are skilled in the art of relating to people.

Earth & Air signs (1 in 760):
These two elements combine to create intellectual ability. Actors are able to detach themselves personally in order to perform the roles of other characters.

Sagittarius is weak:
Sagittarius is strong with fact-based scientists. Actors are more interested in exploring the subjective realities of people. Interestingly, the Gauquelin group of actors is strongest in the sign of Sagittarius. Perhaps, American actors of today are different from French actors of the 19th century.

Venus:
Venus relates to beauty and artistry. Actors have the Moon connected with Venus, giving them a human touch. Mercury with Venus shows their giftedness with communication and diplomacy. Venus retrograde is common for people with high attractiveness.

Signs

Size: 888	1st Harmonic	9th Harmonic	Houses	Rising Sign
Moon	Virgo		7th	
Sun				
Mercury	Pisces		12th	Lunar Phase
Venus	Aquarius	Taurus		Libra
Mars			8th	☉/☽ Midpoint
Jupiter	Taurus	Cancer		
Saturn	Aquarius		2nd	Moon's Nodes
TOTAL	Virgo, Aquarius	Aquarius	12th	Cancer
(weak)	Leo, Sagittarius	Sagittarius	3rd	

1H	♈	♉	♊	♋	♌	♍	♎	♏	♐	♑	♒	♓
Moon	63	84	76	76	76	91	78	65	57	71	83	68
Sun	84	76	79	69	68	73	79	74	64	75	79	68
Mercury	70	73	53	64	67	80	79	93	73	77	70	89
Venus	77	70	84	84	52	85	65	79	69	66	95	62
Mars	62	71	84	78	73	92	78	83	62	75	69	61
Jupiter	69	82	61	60	75	87	84	86	86	68	71	59
Saturn	58	73	51	67	68	77	86	84	77	76	100	71
TOTAL	483	529	488	498	479	585	549	564	488	508	567	478
Rising	47	44	74	95	92	103	110	73	90	68	51	41
N. Node	73	72	77	92	78	69	77	66	68	66	70	80

9H	♈	♉	♊	♋	♌	♍	♎	♏	♐	♑	♒	♓
Moon	78	84	85	64	79	77	58	78	57	65	83	80
Sun	74	78	70	76	73	75	73	75	72	73	67	82
Mercury	68	58	67	84	77	79	80	76	77	61	83	78
Venus	68	87	77	75	56	57	74	71	79	86	83	75
Mars	80	84	68	62	71	81	68	88	52	85	86	63
Jupiter	84	73	68	94	87	69	76	67	57	67	76	70
Saturn	75	74	86	71	58	78	62	80	74	72	79	79
TOTAL	527	538	521	526	501	516	491	535	468	509	557	527

Houses	1st	2nd	3rd	4th	5th	6th	7th	8th	9th	10th	11th	12th
Moon	80	86	67	78	72	75	89	61	76	78	51	75
Sun	89	74	74	55	66	76	67	69	58	81	82	97
Mercury	83	81	60	74	72	60	75	63	72	76	69	103
Venus	89	75	71	61	70	65	73	78	61	71	86	88
Mars	82	85	62	74	69	68	70	87	70	65	76	80
Jupiter	72	77	74	68	81	60	68	65	82	86	83	72
Saturn	72	94	70	76	81	61	61	71	73	67	79	83
TOTAL	567	572	478	486	511	465	503	494	492	524	526	598
N. Node	63	75	70	85	84	67	73	69	78	73	69	82

Planets

Aspects	Parallels	4 Axis	Direction	☉/☽ Midpoint
Moon-Venus	Mercury-Venus	Venus (Asc)	Venus (Rx)	
Moon-Neptune		Jupiter (Asc)		
Sun-Jupiter		Saturn (Asc)		
		Moon (MC)		**Moon's Nodes**
		Jupiter (MC)		Sun (S)
		Chiron (MC)		Mercury (S)
				Mars (S)

Adopted/Abandoned Children

Theme: Capricorn (10th), Saturn/Pluto

Capricorn:
Capricorns experience much solitude. They tend to be serious, driven people. Children who are abandoned often learn to become self-contained and self-reliant.

Moon-Jupiter aspects (1 in 414):
The Moon represents our early home environment. Jupiter influences us to expand and grow. People with Moon-Jupiter aspects often live very far from their birth places.

Moon conjunct Moon's South Node (1 in 859,445):
The South Node represents early life conditions. The Moon with this node indicates issues revolving around the home and mother.

Saturn:
Saturn with the Moon and Venus indicate seriousness and caution in intimate relationships.

Pluto:
Pluto with Mercury, Venus, and Mars indicate hidden difficulties regarding relationships. They tend to be private with others about their lives.

Signs

Size: 172	1st Harmonic	9th Harmonic	Houses	Rising Sign
Moon	Sagittarius			Aries
Sun	Capricorn			Sagittarius
Mercury		Libra, Pisces	2nd, 5th	**Lunar Phase**
Venus	Gemini		10th	
Mars				☉/☽ Midpoint
Jupiter		Libra	9th	
Saturn	Capricorn			**Moon's Nodes**
TOTAL	Capricorn		10th	1st
(weak)	Taurus		8th	

1H	♈	♉	♊	♋	♌	♍	♎	♏	♐	♑	♒	♓
Moon	12	11	10	16	13	15	17	12	21	16	13	16
Sun	15	11	18	16	17	14	12	12	16	20	12	9
Mercury	9	14	14	14	16	18	13	14	18	17	12	13
Venus	16	9	21	18	7	20	11	19	11	16	13	11
Mars	11	10	17	21	18	17	10	16	14	13	10	15
Jupiter	8	10	16	18	15	12	17	20	13	13	19	11
Saturn	16	14	11	8	14	7	13	16	16	27	16	14
TOTAL	87	79	107	111	100	103	93	109	109	122	95	89
Rising	14	7	9	19	14	17	15	21	24	13	13	6
N. Node	14	13	14	12	12	14	12	14	16	17	17	17

9H	♈	♉	♊	♋	♌	♍	♎	♏	♐	♑	♒	♓
Moon	14	15	13	16	11	17	12	15	14	15	12	18
Sun	19	20	7	13	13	18	14	13	11	17	16	11
Mercury	16	15	15	17	11	12	23	11	10	11	10	21
Venus	12	15	11	19	17	15	15	19	12	13	15	9
Mars	9	17	12	14	17	13	14	15	17	12	16	16
Jupiter	12	15	12	12	11	12	23	16	10	17	15	17
Saturn	19	12	19	19	10	19	6	9	18	18	11	12
TOTAL	101	109	89	110	90	106	107	98	92	103	95	104

Houses	1st	2nd	3rd	4th	5th	6th	7th	8th	9th	10th	11th	12th
Moon	15	10	17	19	12	18	11	12	15	14	14	15
Sun	18	17	18	11	13	15	8	6	13	21	20	12
Mercury	16	24	16	11	19	10	7	6	16	16	15	16
Venus	21	11	16	12	17	14	9	10	11	22	13	16
Mars	16	17	13	13	12	12	18	14	13	16	15	13
Jupiter	13	18	7	16	14	15	13	11	23	16	7	19
Saturn	11	11	16	20	15	9	16	12	14	13	16	19
TOTAL	110	108	103	102	102	93	82	71	105	118	100	110
N. Node	22	12	16	11	19	9	15	19	6	14	18	11

Planets

Aspects	Parallels	4 Axis	Direction	☉/☽ Midpoint
Saturn & Pluto	Sun-Venus	Uranus (Asc)		
Moon-Jupiter	Sun-Saturn	Chiron (Dsc)		
Moon-Saturn		Mercury (MC)		
Venus-Saturn		Venus (MC)		Moon's Nodes
Mercury-Uranus		Saturn (IC)		Moon (S)
Mercury-Pluto				
Venus-Pluto				
Mars-Pluto				

AIDS Patients

Theme: Virgo, Venus

Virgo:
Virgo is the sign most associated with health problems and negative life circumstances. Jupiter being strong in Virgo in both harmonics indicates a critical outlook towards personal growth and opportunities.

Pisces is weak:
Pisces, the opposite sign of Virgo, is associated with good health is the strongest sign for people who lived a long life (80+ years).

Venus:
People contract AIDS through sexual relationships. Venus is with Pluto and Chiron, indication wounds coming from hidden activities in relationships.

Signs

Size: 321	1st Harmonic	9th Harmonic	Houses	Rising Sign
Moon	Aries			Aquarius
Sun	Leo	Taurus	6th	
Mercury		Aquarius		Lunar Phase
Venus	Cancer, Virgo	Capricorn		Aries
Mars		Aries, Virgo		☉/☽ Midpoint
Jupiter	Virgo	Virgo, Capricorn	5th, 10th	Pisces
Saturn	Virgo, Scorpio, Sag	Aries, Capricorn	1st, 3rd	Moon's Nodes
TOTAL	Virgo, Scorpio	Virgo		Cancer
(weak)	Taurus, Libra, Pisces	Pisces		

1H	♈	♉	♊	♋	♌	♍	♎	♏	♐	♑	♒	♓
Moon	40	30	25	21	21	27	20	28	25	29	29	26
Sun	25	24	25	25	39	24	19	27	33	27	27	26
Mercury	29	18	26	22	30	26	25	34	23	31	31	26
Venus	26	17	24	44	21	39	18	27	21	28	34	22
Mars	20	26	25	29	38	28	33	30	31	22	14	25
Jupiter	29	21	28	31	20	45	19	36	33	25	18	16
Saturn	10	21	18	21	34	45	32	40	40	22	24	14
TOTAL	179	157	171	193	203	234	166	222	206	184	177	155
Rising	9	20	28	32	30	41	32	38	27	25	26	13
N. Node	28	34	21	35	25	20	27	29	25	20	28	29

9H	♈	♉	♊	♋	♌	♍	♎	♏	♐	♑	♒	♓
Moon	29	26	34	29	22	28	34	27	27	19	23	23
Sun	23	39	28	24	33	33	23	28	25	21	23	21
Mercury	28	24	21	30	17	26	32	27	29	24	36	27
Venus	24	21	29	22	29	28	29	22	33	35	23	26
Mars	35	13	31	30	34	35	31	29	18	21	20	24
Jupiter	27	27	24	33	34	35	11	22	29	37	21	21
Saturn	37	27	22	16	34	31	21	23	23	35	28	24
TOTAL	203	177	189	184	203	216	181	178	184	192	174	166

Houses	1st	2nd	3rd	4th	5th	6th	7th	8th	9th	10th	11th	12th
Moon	30	32	23	33	24	31	29	22	26	20	19	32
Sun	26	28	34	29	21	31	24	20	29	25	18	36
Mercury	25	30	33	29	24	28	27	21	25	23	26	30
Venus	30	29	27	29	23	27	29	28	21	31	23	24
Mars	19	27	16	30	34	26	27	26	31	26	29	30
Jupiter	24	26	22	24	37	20	31	20	30	36	32	19
Saturn	32	35	38	27	23	31	26	15	30	17	22	25
TOTAL	186	207	193	201	186	194	193	152	192	178	169	196
N. Node	26	29	18	23	26	24	34	27	30	28	33	23

Planets

Aspects	Parallels	4 Axis	Direction	☉/☽ Midpoint
Venus	Mars-Neptune	Jupiter (Dsc)	Saturn (D)	Pluto
Venus-Pluto		Chiron (Dsc)		
Venus-Chiron		Jupiter (MC)		
		Saturn (IC)		**Moon's Nodes**
		Chiron (IC)		Mercury (S)
				Mars (S)

ALCOHOLICS

Theme: Taurus (2nd)

Taurus:
The more negative characteristics of Taureans are expressed with alcoholics: over-indulgence, inertia, and laziness. Taurus is also the strongest sign in Gauquelin's alcoholic group.

Jupiter, Saturn, and Chiron on the Imum Coeli (1 in 826 - Jupiter):
This indicates that excessive and destructive behaviors may occur in private.

Signs

Size: 162	1st Harmonic	9th Harmonic	Houses	Rising Sign
Moon	Cancer, Virgo	Taurus		Aquarius
Sun		Taurus		
Mercury	Pisces	Aries	9th	Lunar Phase
Venus	Libra, Aquar, Pisces	Sagittarius	2nd	
Mars	Sagittarius	Cancer	1st	☉/☽ Midpoint
Jupiter	Gemini, Cancer		11th	Scorpio
Saturn				Moon's Nodes
TOTAL		Taurus, Virgo	1st, 10th	
(weak)	Taurus, Leo	Aquarius	6th	

1H	♈	♉	♊	♋	♌	♍	♎	♏	♐	♑	♒	♓
Moon	12	8	13	21	10	23	18	6	10	19	11	11
Sun	17	12	8	14	9	14	17	12	14	14	14	17
Mercury	14	13	7	11	10	18	13	17	14	14	11	20
Venus	15	10	10	10	13	13	18	11	11	9	23	19
Mars	13	8	12	13	9	21	14	17	19	9	16	11
Jupiter	10	9	21	20	12	15	17	16	9	15	8	10
Saturn	8	14	13	13	15	10	15	15	17	16	14	12
TOTAL	89	74	84	102	78	114	112	94	94	96	97	100
Rising	6	10	8	13	15	23	15	18	17	13	15	9
N. Node	14	11	11	16	14	13	11	11	19	12	16	14

9H	♈	♉	♊	♋	♌	♍	♎	♏	♐	♑	♒	♓
Moon	13	23	11	12	13	18	14	12	10	6	15	15
Sun	14	23	8	11	12	15	13	15	17	13	9	12
Mercury	20	10	9	13	19	19	11	18	6	16	8	13
Venus	14	17	13	13	7	15	15	10	23	12	10	13
Mars	9	12	16	20	11	17	10	16	15	13	11	12
Jupiter	9	16	15	16	10	13	13	19	14	15	11	11
Saturn	12	10	19	9	14	16	13	15	14	13	16	11
TOTAL	91	111	91	94	86	113	89	105	99	88	80	87

Houses	1st	2nd	3rd	4th	5th	6th	7th	8th	9th	10th	11th	12th
Moon	13	16	17	11	10	8	13	17	11	19	10	17
Sun	21	16	17	16	6	9	10	12	14	18	10	13
Mercury	18	19	15	13	11	9	10	8	19	17	11	12
Venus	14	24	13	15	7	13	11	10	13	17	10	15
Mars	20	11	14	12	12	9	12	13	14	18	10	17
Jupiter	14	10	12	13	17	9	10	16	18	7	22	14
Saturn	19	14	14	14	14	14	14	11	8	16	13	11
TOTAL	119	110	102	94	77	71	80	87	97	112	86	99
N. Node	14	17	17	15	18	16	10	14	10	8	11	12

Planets

Aspects	Parallels	4 Axis	Direction	☉/☽ Midpoint
Mercury-Chiron	Sun-Neptune	Uranus (Asc)		
		Jupiter (IC)		
		Saturn (IC)		
		Chiron (IC)		**Moon's Nodes**
				Mercury (N)

Alzheimer's Disease

Theme: Cancer, Moon

Moon:
The Moon is associated with dreams, memories, and the subconscious mind. Alzheimer's patients lose their memories and enter into a dreamlike reality with imagery from their subconscious. This confirms the Moon's association with lunacy.

Mercury in Cancer (1 in 12,894 - 9H)
Mercury in Cancer indicates a very subjective, personal style of thinking where facts and information are not as important as experiences. They are also more private with their thoughts.

Mercury in Fire Houses (1 in 6,115)
In the Fire houses, Mercury takes on a prominent role on their lives. Mercury is especially strong in the 1st house, showing that rational mind functioning (or its lack) is an attribute with which the person is identified.

Size: 141	1st Harmonic	9th Harmonic	Houses	Rising Sign
Moon	Cancer, Libra		3rd	
Sun				
Mercury		Cancer	1st, 5th	**Lunar Phase**
Venus			10th	Libra, Pisces
Mars	Sagittarius			☉/☽ Midpoint
Jupiter	Aquarius			Cancer
Saturn	Capricorn	Virgo, Pisces	8th	**Moon's Nodes**
TOTAL	Taurus	Cancer	1st, 11th	Leo
(weak)	Scorpio		7th	11th

Signs (header above 1st and 9th Harmonic columns)

1H	♈	♉	♊	♋	♌	♍	♎	♏	♐	♑	♒	♓
Moon	11	10	5	18	12	14	18	8	9	15	6	15
Sun	14	16	8	15	16	5	13	10	9	9	13	13
Mercury	16	14	11	9	11	12	12	9	10	13	14	10
Venus	11	14	11	10	12	13	11	7	10	13	18	11
Mars	10	14	9	11	14	20	10	3	16	12	13	9
Jupiter	14	11	10	15	6	13	12	10	13	8	17	12
Saturn	12	17	13	14	6	10	7	8	10	21	13	10
TOTAL	88	96	67	92	77	87	83	55	77	91	94	80
Rising	6	7	13	12	20	19	16	14	12	11	8	3
N. Node	16	9	13	9	23	11	14	10	7	6	12	11

9H	♈	♉	♊	♋	♌	♍	♎	♏	♐	♑	♒	♓
Moon	7	15	17	9	8	8	10	13	14	14	11	15
Sun	8	9	13	7	10	10	15	17	14	10	17	11
Mercury	10	12	14	24	16	12	10	11	6	11	8	7
Venus	16	11	8	15	11	8	9	16	15	7	14	11
Mars	11	12	14	17	11	10	9	13	11	8	11	14
Jupiter	16	11	10	13	11	6	9	11	16	8	15	15
Saturn	12	8	7	13	11	20	6	12	11	14	10	17
TOTAL	80	78	83	98	78	74	68	93	87	72	86	90

Houses	1st	2nd	3rd	4th	5th	6th	7th	8th	9th	10th	11th	12th
Moon	8	11	17	11	13	10	14	7	12	7	14	17
Sun	17	13	13	10	14	11	6	5	11	11	17	13
Mercury	24	11	15	8	17	5	5	8	12	11	17	8
Venus	18	17	13	10	10	9	6	6	8	21	15	8
Mars	12	11	10	8	15	11	10	16	9	11	17	11
Jupiter	15	9	14	11	7	13	10	11	12	14	11	14
Saturn	11	8	12	15	11	10	7	18	7	16	11	15
TOTAL	105	80	94	73	87	69	58	71	71	91	102	86
N. Node	11	11	8	12	15	7	9	11	15	16	21	5

Planets

Aspects	Parallels	4 Axis	Direction	☉/☽ Midpoint
Moon		Mercury (Asc)		Jupiter
Moon-Venus		Venus (MC)		Saturn
Moon-Uranus		Jupiter (MC)		
Moon-Chiron		Venus (IC)		Moon's Nodes
Mars-Chiron		Neptune (IC)		

ARTISTS

Theme: Pisces (12th), Venus

Pisces (1 in 41,048)
Pisces has long been associated with artistry. Their reflectiveness and sensitivity to their surroundings allows them to be productive during times of seclusion, which they enjoy.

12th House:
The 12th house corresponds with Pisces and shows where a person is most creative.

Virgo is weak:
The opposite sign of Pisces, Virgo focuses on logic and critical analysis. This is a classic right brain-left brain contrast.

Mars in Taurus (1 in 13,710):
Artists work with physical materials in order to create works of great beauty.

Mercury-Jupiter aspects and parallels:
Mercury and Jupiter together indicate excellent intellectual ability.

Signs

Size: 262	1st Harmonic	9th Harmonic	Houses	Rising Sign
Moon		Leo	2nd, 12th	Taurus
Sun	Pisces		3rd, 12th	
Mercury	Aquarius, Pisces		12th	Lunar Phase
Venus	Aries, Sagitt, Pisces	Sagittar, Aquarius	12th	Aquarius
Mars	Taurus			☉/☽ Midpoint
Jupiter	Sagittarius			Pisces
Saturn	Libra	Pisces		Moon's Nodes
TOTAL	♈ ♉ ♒ ♓		12th	
(weak)	♊ ♌ ♍ ♏	Virgo	7th	

1H	♈	♉	♊	♋	♌	♍	♎	♏	♐	♑	♒	♓
Moon	27	24	22	17	14	19	26	20	27	18	20	28
Sun	26	24	22	22	19	10	21	17	18	21	24	38
Mercury	27	25	17	18	15	18	17	20	16	23	35	31
Venus	34	21	18	18	19	18	13	12	33	17	24	35
Mars	13	36	18	18	29	20	24	25	18	22	21	18
Jupiter	25	16	16	21	18	24	23	22	33	21	24	19
Saturn	19	23	16	19	15	25	30	21	23	21	27	23
TOTAL	171	169	129	133	129	134	154	137	168	143	175	192
Rising	9	24	19	33	26	20	28	35	29	17	17	5
N. Node	22	9	25	28	23	28	18	29	20	17	26	17

9H	♈	♉	♊	♋	♌	♍	♎	♏	♐	♑	♒	♓
Moon	25	24	20	25	31	18	20	25	17	19	16	22
Sun	15	15	25	23	23	22	27	22	26	22	24	18
Mercury	26	17	25	20	17	22	24	22	19	20	26	24
Venus	19	21	18	27	26	17	13	24	33	20	28	16
Mars	17	21	29	24	22	22	25	14	26	20	22	20
Jupiter	20	27	26	27	22	21	19	25	19	23	16	17
Saturn	26	23	24	23	20	10	18	26	22	21	20	29
TOTAL	148	148	167	169	161	132	146	158	162	145	152	146

Houses	1st	2nd	3rd	4th	5th	6th	7th	8th	9th	10th	11th	12th
Moon	23	29	26	19	19	28	24	14	20	16	15	29
Sun	27	15	32	21	21	12	13	18	16	26	24	37
Mercury	25	26	18	25	22	13	14	17	18	26	24	34
Venus	26	26	20	24	21	19	17	10	19	22	26	32
Mars	29	23	22	23	27	13	17	28	6	22	26	26
Jupiter	23	22	26	11	23	23	15	22	25	26	27	19
Saturn	29	16	19	20	25	19	23	20	26	20	28	17
TOTAL	182	157	163	143	158	127	123	129	130	158	170	194
N. Node	23	18	22	20	25	21	18	23	21	25	22	24

Planets

Aspects	Parallels	4 Axis	Direction	☉/☽ Midpoint
Moon-Pluto	Venus-Uranus	Venus (Asc)	Chiron (D)	
Mercury-Jupiter	Mercury-Jupiter	Chiron (Dsc)		
Mercury-Saturn				
Venus-Pluto				Moon's Nodes
				Neptune (N)

ASTROLOGERS

Theme: Gemini, 12th House

Scorpio and Gemini:
Astrologers benefit from Scorpio's intuitively guided powers of insight and Gemini's ability to connect with people. Astrologers are comfortable discussing the secret areas of people's lives.

Venus in Cancer (1 in 266 - 1H, 1 in 9,744 - 9H):
Consulting astrologers frequently have close ties to clients. They enjoy being nurturing (Cancer) in relationships (Venus).

Aries and Taurus are weak:
These signs represents the individual will and self-absorption, which is lacking in counselors.

11th & 12th Houses:
These houses are concerned with the bigger issues in life. An emphasis in the 12th house shows a creative and spiritual nature.

Sun-Uranus aspects & Mercury-Uranus parallels (1 in 1,826):
Astrologers are unique individuals who are willing to communicate ideas outside of mainstream thinking. Their unique insights and intuition (Uranus) lead them into the hidden, occult, mystical aspects of life.

Saturn on the Midheaven (1 in 4,724):
Saturn is associated with maturity. Most astrologers become professionals in their later years, since it requires wisdom and financial security.

Size: 411	1st Harmonic	Signs 9th Harmonic	Houses	Rising Sign
Moon				Taurus
Sun		Libra	11th, 12th	
Mercury	Gemini, Scorpio		5th	Lunar Phase
Venus	Cancer	Cancer	6th	Aquarius
Mars	Leo, Sagittarius	Scorpio	12th	☉/☽ Midpoint
Jupiter		Sagittarius		Gemini, Aquar
Saturn	Cancer, Leo, Virgo	Gemini, Capricorn		Moon's Nodes
TOTAL	Leo, Scorpio	Gemini	11th, 12th	Gemini, 3rd
(weak)	Aries, Taurus	Aquarius		Pisces

1H	♈	♉	♊	♋	♌	♍	♎	♏	♐	♑	♒	♓
Moon	38	40	37	32	27	31	31	26	37	34	37	41
Sun	40	28	34	37	42	34	30	37	38	28	40	23
Mercury	26	23	42	33	32	29	36	51	38	35	34	32
Venus	32	33	26	54	26	43	27	41	34	26	42	27
Mars	24	28	37	35	53	35	35	37	43	26	26	32
Jupiter	21	33	27	32	40	37	37	46	39	36	37	26
Saturn	22	18	27	39	47	56	41	41	28	36	21	35
TOTAL	203	203	230	262	267	265	237	279	257	221	237	216
Rising	20	38	38	35	50	42	44	40	42	27	18	17
N. Node	43	32	49	35	22	33	24	26	39	34	31	43

9H	♈	♉	♊	♋	♌	♍	♎	♏	♐	♑	♒	♓
Moon	37	39	35	24	34	31	38	29	38	35	32	39
Sun	29	38	36	38	33	31	52	37	18	32	30	37
Mercury	36	35	39	24	31	31	40	24	40	41	31	39
Venus	26	28	40	55	39	30	23	37	39	35	31	28
Mars	39	31	32	33	31	37	29	48	37	27	30	37
Jupiter	31	35	39	44	32	37	30	27	48	30	25	33
Saturn	31	31	43	29	34	37	29	28	41	47	33	28
TOTAL	229	237	264	247	234	234	241	230	261	247	212	241

Houses	1st	2nd	3rd	4th	5th	6th	7th	8th	9th	10th	11th	12th
Moon	36	26	42	39	27	38	39	31	34	30	29	40
Sun	35	25	37	31	30	34	30	32	28	27	51	51
Mercury	33	33	39	29	40	31	27	25	31	30	46	47
Venus	33	38	42	19	29	41	28	27	35	36	42	41
Mars	32	31	27	40	41	31	38	32	22	30	40	46
Jupiter	39	31	24	31	35	35	39	30	31	40	35	41
Saturn	39	32	36	36	34	30	29	31	42	34	34	34
TOTAL	247	216	247	225	236	240	230	208	223	227	277	300
N. Node	29	28	45	26	35	35	33	31	34	34	39	42

Planets

Aspects	Parallels	4 Axis	Direction	☉/☽ Midpoint
Sun-Uranus	Mercury-Uranus	Sun (Dsc)	Neptune (D)	
Mars-Jupiter		Saturn (MC)		
		Neptune (MC)		
		Moon (IC)		**Moon's Nodes**
				Mars (S)

Astronauts

Theme: Libra/Aquarius

Libra & Aquarius:
Astronauts fly through the air (signs) up into outer-space. They are progressive people who seek to explore new frontiers.

Sun in Pisces (1 in 3,407):
Pisceans are dreamers who are comfortable being in isolation and enjoy working in solitude.

Mercury Retrograde (1 in 2,168):
Mercury retrograde indicates that a person prefers to spend time ruminating on their thoughts rather than communicating with others.

Signs

Size: 80	1st Harmonic	9th Harmonic	Houses	Rising Sign
Moon	Sagittarius		2nd, 10th	
Sun	Pisces	Aquarius		
Mercury	Aquarius	Virgo, Libra		Lunar Phase
Venus	Virgo		3rd	
Mars			7th	☉/☽ Midpoint
Jupiter	Gemini			Aquarius
Saturn	Capricorn	Aquarius		Moon's Nodes
TOTAL	Pisces	Libra	8th	Taurus, Libra (S)
(weak)				11th (S)

1H	♈	♉	♊	♋	♌	♍	♎	♏	♐	♑	♒	♓
Moon	5	5	7	8	7	3	7	6	11	7	7	7
Sun	5	9	4	8	6	7	9	8	1	2	6	15
Mercury	9	6	7	4	7	6	8	7	5	2	11	8
Venus	9	6	4	9	4	13	4	8	2	5	8	8
Mars	7	6	5	10	12	2	7	6	6	6	6	7
Jupiter	7	4	10	9	6	7	4	10	8	8	4	3
Saturn	5	1	0	2	8	4	3	6	9	23	9	10
TOTAL	47	37	37	50	50	42	42	51	42	53	51	58
Rising	3	6	6	4	12	6	12	11	7	7	3	3
N. Node	13	14	6	6	1	3	7	3	3	8	6	10

9H	♈	♉	♊	♋	♌	♍	♎	♏	♐	♑	♒	♓
Moon	8	4	10	7	5	9	8	5	6	6	6	6
Sun	7	4	2	6	6	6	8	6	7	9	11	8
Mercury	3	7	8	7	4	13	12	4	3	4	6	9
Venus	7	6	7	5	6	5	8	3	10	5	9	9
Mars	0	6	6	8	7	7	10	8	9	5	6	8
Jupiter	5	9	5	7	8	8	4	10	7	9	2	6
Saturn	10	1	4	5	6	8	7	7	7	7	12	6
TOTAL	40	37	42	45	42	56	57	43	49	45	52	52

Houses	1st	2nd	3rd	4th	5th	6th	7th	8th	9th	10th	11th	12th
Moon	3	11	6	7	5	6	10	5	7	11	5	4
Sun	9	6	7	6	2	8	5	7	8	6	11	5
Mercury	7	9	6	7	6	6	3	9	7	9	7	4
Venus	7	4	12	3	7	3	8	6	6	8	9	7
Mars	3	9	5	6	4	7	11	9	6	7	6	7
Jupiter	7	5	6	7	9	5	3	9	7	10	4	8
Saturn	4	10	7	7	8	4	6	10	5	5	4	10
TOTAL	40	54	49	43	41	39	46	55	46	56	46	45
N. Node	5	3	6	10	11	4	6	3	7	8	8	9

Planets

Aspects	Parallels	4 Axis	Direction	☉/☽ Midpoint
Mars-Jupiter	Venus-Uranus	Sun (Dsc)	Mercury (R)	
	Mars-Uranus	Mercury (MC)		
		Jupiter (MC)		
		Pluto (MC)		**Moon's Nodes**
				Chiron (N)

BIRTH DEFECT

Theme: Cancer, 6th House, Saturn

Cancer:
Cancerian lives revolve around their family, home, and early experiences. People with birth defects are more likely to be limited to their close circle of family relationships. Cancer Rising is the most prominent position (1 in 1,055).

6th House (1 in 1,072):
The 6th house is connected with Virgo and is associated with physical health issues.

Saturn aspects:
Saturn is the astrological symbol of caution, fear, and restriction. People with a strong Saturn influence are often forced to sacrifice personal freedom and to work through difficulties.

Moon's North Node in Pisces (1 in 3,669):
This position indicates that quietude, isolation, and reflection play an important role in a person's progress in life.

Signs

Size: 280	1st Harmonic	9th Harmonic	Houses	Rising Sign
Moon		Gemini	8th	Cancer
Sun		Cancer, Scorpio	6th	Pisces
Mercury		Taurus	6th, 7th	Lunar Phase
Venus	Gemini	Aries	6th, 7th	Aries
Mars		Capricorn		☉/☽ Midpoint
Jupiter	Scorpio	Pisces		Aquarius
Saturn	Taurus			Moon's Nodes
TOTAL		Cancer, Capricorn	6th	Cancer, Aquar
(weak)	Pisces	Sagittarius	3rd	Virgo (S), 6th

1H	♈	♉	♊	♋	♌	♍	♎	♏	♐	♑	♒	♓
Moon	24	22	28	19	25	24	23	22	21	23	30	19
Sun	26	18	22	28	19	28	23	17	23	28	26	22
Mercury	27	17	21	23	21	21	26	27	22	29	31	15
Venus	13	26	32	26	19	26	20	23	23	26	28	18
Mars	8	22	25	27	25	29	32	30	23	22	17	20
Jupiter	23	16	22	20	20	23	26	34	30	23	22	21
Saturn	24	40	22	17	26	25	25	24	16	21	24	16
TOTAL	145	161	172	160	155	176	175	177	158	172	178	131
Rising	13	9	22	44	34	23	27	25	30	18	16	19
N. Node	20	23	18	33	25	16	19	22	16	14	35	39

9H	♈	♉	♊	♋	♌	♍	♎	♏	♐	♑	♒	♓
Moon	24	25	31	23	24	25	28	16	16	28	22	18
Sun	23	21	26	32	24	20	12	35	22	27	22	16
Mercury	14	34	16	28	27	24	21	20	20	27	24	25
Venus	31	20	25	27	16	14	22	24	23	27	22	29
Mars	26	23	18	19	26	16	27	29	19	34	21	22
Jupiter	20	29	19	31	29	22	20	18	18	21	18	35
Saturn	21	21	27	24	19	31	25	21	21	27	23	20
TOTAL	159	173	162	184	165	152	155	163	139	191	152	165

Houses	1st	2nd	3rd	4th	5th	6th	7th	8th	9th	10th	11th	12th
Moon	25	22	21	22	21	22	25	31	26	25	17	23
Sun	27	19	12	21	23	28	23	26	18	22	32	29
Mercury	27	19	19	17	19	31	29	15	27	19	30	28
Venus	33	20	15	19	19	29	28	16	21	26	26	28
Mars	23	30	26	27	24	23	20	22	25	20	21	19
Jupiter	21	29	29	17	29	29	18	18	22	19	24	25
Saturn	20	20	18	19	23	28	28	27	26	26	25	20
TOTAL	176	159	140	142	158	190	171	155	165	157	175	172
N. Node	32	20	23	25	26	32	18	21	27	15	20	21

Planets

Aspects	Parallels	4 Axis	Direction	☉/☽ Midpoint
Moon & Mercury	Mars-Jupiter	Venus (Asc)		Venus
Jupiter & Saturn	Mars-Uranus	Sun (Dsc)		
Mercury-Venus	Venus-Jupiter	Chiron (IC)		
Mercury-Jupiter				**Moon's Nodes**
Venus-Saturn				Mercury (S)
Saturn-Neptune				Mars (N)
Saturn-Chiron				Mars (S)

Business Owners

Theme: Gemini, Pluto

Gemini:
Geminis are excellent networkers since they have great knowledge about many things and love talking with people. With their clever minds, they are also great negotiators.

Mercury in Libra (1 in 28,280 - 9H):
Businesspeople often have excellent communication skills.

Saturn in the 12th House (1 in 24,539):
Saturn in the 12th house indicates a discomfort with isolation.

Moon-Pluto aspects (1 in 710):
Businesspeople are often successful because they form close personal relations with others. The Moon with Pluto indicates involvement in family businesses, often due to powerful parental figures.

Signs

Size: 452	1st Harmonic	9th Harmonic	Houses	Rising Sign
Moon	Virgo		9th	Libra
Sun	Gemini	Gemini, Pisces	12th	
Mercury	Gemini	Libra		Lunar Phase
Venus	Taurus	Aries, Capricorn		
Mars	Gemini		8th	☉/☽ Midpoint
Jupiter		Gemini		
Saturn	Scorpio		4th, 12th	Moon's Nodes
TOTAL	Taurus, Gemini	Gemini	12th	Gemini (S)
(weak)	Sagittarius		2nd, 6th	10th

1H	♈	♉	♊	♋	♌	♍	♎	♏	♐	♑	♒	♓
Moon	46	39	27	29	44	52	42	41	33	34	28	37
Sun	39	39	55	44	32	38	36	33	27	40	24	45
Mercury	44	41	45	37	28	39	38	43	35	30	39	33
Venus	37	56	41	41	36	34	37	35	25	33	48	29
Mars	29	37	56	39	41	40	41	34	35	33	32	35
Jupiter	24	38	25	31	42	38	42	47	44	42	37	42
Saturn	45	35	39	39	35	37	34	51	40	30	30	37
TOTAL	264	285	288	260	258	278	270	284	239	242	238	258
Rising	17	33	33	39	55	46	63	43	51	37	21	14
N. Node	34	36	44	44	43	34	35	32	48	32	37	33

9H	♈	♉	♊	♋	♌	♍	♎	♏	♐	♑	♒	♓
Moon	32	45	45	43	44	39	30	37	42	24	32	39
Sun	27	29	48	32	36	42	44	41	35	28	38	52
Mercury	45	34	36	31	43	37	61	33	38	22	40	32
Venus	47	34	46	36	36	40	27	43	31	49	30	33
Mars	28	38	33	45	28	40	31	46	39	41	42	41
Jupiter	38	32	47	48	36	35	34	28	38	41	37	38
Saturn	40	42	43	35	33	42	37	38	42	38	34	28
TOTAL	257	254	298	270	256	275	264	266	265	243	253	263

Houses	1st	2nd	3rd	4th	5th	6th	7th	8th	9th	10th	11th	12th
Moon	34	40	32	40	46	40	44	23	47	31	35	40
Sun	42	25	41	39	33	26	33	38	35	41	40	59
Mercury	46	32	45	37	30	28	36	33	40	37	42	46
Venus	43	33	36	36	38	30	39	35	37	40	44	41
Mars	45	31	43	39	34	32	28	51	32	35	35	47
Jupiter	32	42	42	31	37	28	43	46	35	37	36	43
Saturn	40	29	26	49	34	31	36	45	26	32	43	61
TOTAL	282	232	265	271	252	215	259	271	252	253	275	337
N. Node	35	31	39	44	32	35	42	36	42	53	32	31

Planets

Aspects	Parallels	4 Axis	Direction	☉/☽ Midpoint
Moon-Pluto	Sun-Chiron	Mars (Asc)	Pluto (D)	Pluto
Sun-Neptune	Venus-Jupiter	Sun (Dsc)		
Venus-Neptune		Mars (IC)		
Mars-Uranus				Moon's Nodes
Jupiter-Pluto				

Cancer Patients

Theme: Pisces, 4th House

Virgo & Pisces:
These are the signs associated with health and wellness.

4th House:
The 4th house is at the bottom of the chart. A prominent 4th house indicates that a person has a substantial aspect of life hidden from others or from themselves, in their subconscious. Cancer is often a result of unresolved, buried frustration.

Mars on the Ascendant (1 in 5,179):
Mars on the Ascendant can indicate that a person's body is subjected to harm.

Neptune on the Ascendant (1 in 5,216):
Neptune on the Ascendant often shows that a person experiences losses within their life circumstances.

Signs

Size: 655	1st Harmonic	9th Harmonic	Houses	Rising Sign
Moon		Scorpio	9th	
Sun	Virgo, Pisces	Taurus, Libra	11th	
Mercury	Pisces	Libra	4th, 12th	Lunar Phase
Venus	Aries, Pisces	Capricorn	1st, 2nd	
Mars		Leo	4th, 12th	☉/☽ Midpoint
Jupiter	Taurus		4th	Pisces
Saturn	Virgo, Sagittarius		5th	Moon's Nodes
TOTAL	Virgo, Pisces	Aries	1st, 4th	Pisces (S), Libra
(weak)	Scorpio		6th, 7th, 8th	8th

1H	♈	♉	♊	♋	♌	♍	♎	♏	♐	♑	♒	♓
Moon	63	53	61	46	49	63	50	53	51	48	54	64
Sun	62	59	55	41	63	69	53	32	56	55	45	65
Mercury	50	52	44	46	51	66	57	48	61	54	54	72
Venus	69	61	39	64	52	58	42	62	45	39	62	62
Mars	49	44	58	68	69	60	53	62	52	53	45	42
Jupiter	62	63	40	60	52	63	61	48	63	48	49	46
Saturn	37	54	48	46	51	69	54	51	77	62	66	40
TOTAL	392	386	345	371	387	448	370	356	405	359	375	391
Rising	34	37	45	70	68	68	76	76	63	53	34	31
N. Node	53	55	57	63	39	74	67	48	47	40	59	53

9H	♈	♉	♊	♋	♌	♍	♎	♏	♐	♑	♒	♓
Moon	65	48	53	47	54	48	42	66	59	63	48	62
Sun	57	68	53	59	42	61	66	53	43	41	62	50
Mercury	53	61	40	52	60	58	69	49	56	46	57	54
Venus	56	50	51	61	45	57	51	58	51	68	57	50
Mars	54	53	50	58	66	53	54	64	44	51	57	51
Jupiter	60	59	61	54	53	55	50	51	62	49	53	48
Saturn	65	44	56	63	45	62	42	51	47	62	64	54
TOTAL	410	383	364	394	365	394	374	392	362	380	398	369

Houses	1st	2nd	3rd	4th	5th	6th	7th	8th	9th	10th	11th	12th
Moon	59	51	52	48	54	55	62	49	68	60	42	55
Sun	69	58	65	62	45	40	41	37	39	54	79	66
Mercury	69	67	64	68	43	40	50	28	42	48	59	77
Venus	78	75	60	50	55	36	49	36	37	51	69	59
Mars	62	60	61	68	50	42	41	47	53	40	58	73
Jupiter	65	50	59	66	61	51	46	55	48	55	52	47
Saturn	54	65	54	60	70	51	34	62	54	46	51	54
TOTAL	456	426	415	422	378	315	323	314	341	354	410	431
N. Node	49	50	57	56	56	48	62	65	46	59	44	63

Planets

Aspects	Parallels	4 Axis	Midpoints	Direction
Sun-Jupiter		Mars (Asc)	Pluto	Jupiter (D)
Venus-Pluto		Neptune (Asc)		
Mars-Saturn		Mercury (IC)		
				Moon's Nodes
				Chiron (S)

CHILD ABUSE VICTIMS

Theme: Virgo/Scorpio, 4th House, Moon/Venus

Virgo & Scorpio:
These are the two signs most associated with sexuality. Venus in Virgo (1 in 3,701) and in Scorpio (1 in 1,737) are especially strong. These positions indicate that a person is likely to form intense yet insecure relationships, with the possibility of trauma.

Mars in the 4th House (1 in 4,512,699) & Mars on the Imum Coeli (1 in 2,453):
This is the strongest finding in the entire study. It is indicative of violence and abusive sexual experiences occurring in the home.

Venus-Jupiter aspects (1 in 3,185), Venus on Ascendant, Descendant, and Imum Coeli:
Venus being so frequent on the 4 Axis points indicates that many of a person's most important life experiences revolve around relationships. Venus-Jupiter aspects describe these relationships as being frequent and abundant.

Moon on Ascendant (1 in 1,656), and Moon in Pisces (1 in 1,026):
These positions of the Moon show a person to be highly sensitive and emotional.

Aries is weak:
Child abuse victims lack the strength and confidence Arians are known for.

9th House is weak (1 in 11,705):
The 9th house is a place of positivity and good fortune.

Signs

Size: 266	1st Harmonic	9th Harmonic	Houses	Rising Sign
Moon	Pisces	Virgo	4th	
Sun	Aquarius		5th, 8th, 11th	
Mercury		Virgo	5th	Lunar Phase
Venus	Virgo, Scorpio	Sagittarius, Capr	3rd, 12th	Leo
Mars	Taurus	Leo, Aquarius	4th	☉/☽ Midpoint
Jupiter		Scorpio	7th, 8th	Leo
Saturn	Virgo, Scorpio	Capricorn	2nd, 9th	Moon's Nodes
TOTAL	Virgo, Scorpio	Capricorn	4th, 7th	Scorpio (S)
(weak)	Aries	Aries	6th, 9th	10th

1H	♈	♉	♊	♋	♌	♍	♎	♏	♐	♑	♒	♓
Moon	17	27	22	21	19	23	26	23	21	17	14	36
Sun	20	23	29	25	17	18	28	19	28	19	29	11
Mercury	16	23	24	23	10	28	22	27	30	24	27	12
Venus	23	22	18	25	15	41	10	37	21	16	18	20
Mars	18	28	26	18	26	29	25	20	14	25	24	13
Jupiter	21	21	22	22	27	22	20	26	29	20	21	15
Saturn	11	15	18	19	26	28	27	34	19	24	19	26
TOTAL	126	159	159	153	140	189	158	186	162	145	152	133
Rising	18	20	22	28	16	22	36	20	32	27	12	13
N. Node	26	34	23	24	26	23	15	19	21	26	12	17

9H	♈	♉	♊	♋	♌	♍	♎	♏	♐	♑	♒	♓
Moon	16	20	24	21	24	33	29	18	12	28	14	27
Sun	11	19	27	28	25	11	27	20	23	23	29	23
Mercury	14	22	23	22	20	30	25	21	19	21	22	27
Venus	27	21	16	15	28	18	15	19	30	32	22	23
Mars	15	18	23	20	29	24	11	25	28	17	32	24
Jupiter	17	25	21	25	16	23	28	33	22	22	20	14
Saturn	15	19	13	26	22	27	16	25	16	37	25	25
TOTAL	115	144	147	157	164	166	151	161	150	180	164	163

Houses	1st	2nd	3rd	4th	5th	6th	7th	8th	9th	10th	11th	12th
Moon	29	25	21	31	12	21	23	23	12	19	23	27
Sun	22	28	26	23	28	13	17	27	11	16	34	21
Mercury	22	22	31	23	28	16	26	16	7	24	27	24
Venus	25	21	31	22	20	18	25	22	12	16	17	37
Mars	17	18	27	44	15	21	21	21	17	22	22	21
Jupiter	19	27	18	20	24	15	30	29	13	23	21	27
Saturn	17	32	16	26	22	18	27	10	34	28	12	24
TOTAL	151	173	170	189	149	122	169	148	106	148	156	181
N. Node	26	22	16	16	20	15	29	19	26	31	17	29

Planets

Aspects	Parallels	4 Axis	Direction	☉/☽ Midpoint
Venus & Jupiter		Moon (Asc)	Chiron (D)	Venus
Moon-Venus	Moon-Uranus	Venus (Asc)		Pluto
Moon-Jupiter	Sun-Saturn	Venus (Dsc)		
Venus-Jupiter		Saturn (MC)		Moon's Nodes
Mars-Saturn		Venus (IC)		Mercury (S)
Jupiter-Pluto		Mars (IC)		
		Neptune (IC)		

Child Performers

Theme: Libra (7th), Scorpio, Sun/Venus

Scorpio:
Scorpio's personal magnetism and power allow child performers to rise faster than their peers. Venus in Scorpio (1 in 386) shows that the emphasis for child performers is on their attractiveness. Jupiter in Scorpio shows their influence on large groups of people.

7th House (1 in 546):
The 7th house shows the nature of our relationships with others. Child Performers have Mercury and Venus frequently in this position, showing their skill communicating and relating to people.

Venus aspects are frequent, especially Venus-Uranus aspects (1 in 264):
Venus, the planet of love, shows child performers to be very likable people from an early age. This is important for them, since they have not had time to develop other attributes through maturity and education. Venus with Uranus is indicative of a highly original and interesting presentation, which allows them to stand out in a crowd.

Signs

Size: 304	1st Harmonic	9th Harmonic	Houses	Rising Sign
Moon		Virgo	10th	
Sun	Libra		5th	
Mercury			7th, 9th	Lunar Phase
Venus	Scorpio		7th	Leo, Libra
Mars	Gemini			☉/☽ Midpoint
Jupiter	Scorpio	Scorpio		Virgo
Saturn	Libra			Moon's Nodes
TOTAL	Scorpio	Virgo, Scorpio	7th, 10th	
(weak)	Aries	Taurus, Leo		

1H	♈	♉	♊	♋	♌	♍	♎	♏	♐	♑	♒	♓
Moon	16	24	27	23	21	32	30	30	23	28	30	20
Sun	26	26	26	23	24	25	34	25	21	30	23	21
Mercury	19	29	15	21	25	25	33	31	30	25	26	25
Venus	26	17	25	29	15	33	22	39	19	23	34	22
Mars	18	25	36	28	34	28	22	25	27	21	18	22
Jupiter	20	25	21	28	24	23	23	38	27	20	24	31
Saturn	25	19	21	20	27	31	35	31	28	22	27	18
TOTAL	150	165	171	172	170	197	199	219	175	169	182	159
Rising	18	18	25	29	37	28	34	32	35	17	22	9
N. Node	25	23	28	27	32	29	22	21	19	25	29	24

9H	♈	♉	♊	♋	♌	♍	♎	♏	♐	♑	♒	♓
Moon	31	16	30	21	20	36	27	25	24	23	32	19
Sun	18	21	21	24	22	30	31	28	32	26	20	31
Mercury	28	22	24	27	27	30	20	25	27	22	23	29
Venus	23	25	30	18	26	23	20	27	29	25	29	29
Mars	23	20	25	29	18	31	18	33	28	29	27	23
Jupiter	23	22	23	33	21	21	27	37	22	26	21	28
Saturn	18	26	32	29	19	31	26	29	22	23	24	25
TOTAL	164	152	185	181	153	202	169	204	184	174	176	184

Houses	1st	2nd	3rd	4th	5th	6th	7th	8th	9th	10th	11th	12th
Moon	30	29	26	25	14	22	28	21	23	36	28	22
Sun	17	28	27	22	29	24	29	20	23	33	25	27
Mercury	20	25	25	25	29	25	29	18	32	25	20	31
Venus	29	22	23	22	27	24	32	22	25	25	27	26
Mars	25	30	20	27	23	26	29	29	27	25	20	23
Jupiter	23	32	30	21	18	22	26	14	30	30	31	27
Saturn	28	26	28	26	20	16	29	22	25	28	32	24
TOTAL	172	192	179	168	160	159	202	146	185	202	183	180
N. Node	28	27	28	24	26	24	22	26	25	21	28	25

Planets

Aspects	Parallels	4 Axis	Direction	☉/☽ Midpoint
Venus	Moon-Mars	Sun (Dsc)	Pluto (D)	Jupiter
Sun-Venus	Sun-Uranus	Mars (Dsc)		
Venus-Uranus				
				Moon's Nodes

Child Prodigy

Theme: Cancer, Mercury/Venus

Cancer:
Cancers thrive as children since they love their family ties. They attract support from parents and authoritative figures. The Sun is strong in Cancer in both harmonics showing an especially strong identification with parents and family background.

Mars in Libra (1 in 1,228 - 9H):
This indicates great force in mental pursuits and the need to assert oneself independently.

Mercury & Venus aspects:
An emphasis on Mercury and Venus shows a person to be highly intelligent, verbal and relatable.

Moon's North Node in the 3rd House (1 in 1,413):
The North node in this position indicates that the life path is to increase one's intelligence and ability to communicate ideas.

Signs

Size: 131	1st Harmonic	9th Harmonic	Houses	Rising Sign
Moon	Libra		2nd	
Sun	Cancer	Cancer	8th	
Mercury			4th	Lunar Phase
Venus	Aries	Aquarius		
Mars	Cancer, Pisces	Libra		☉/☽ Midpoint
Jupiter				Taurus
Saturn	Aquarius	Taurus	5th	Moon's Nodes
TOTAL	Cancer, Pisces	Cancer	9th	Cancer (S)
(weak)	Taurus	Gemini		3rd

1H	♈	♉	♊	♋	♌	♍	♎	♏	♐	♑	♒	♓
Moon	6	6	13	14	7	12	18	8	12	6	15	14
Sun	13	13	9	18	13	5	15	4	9	9	9	14
Mercury	10	14	10	10	9	14	10	11	8	11	11	13
Venus	17	10	14	8	13	13	10	15	5	7	11	8
Mars	12	4	16	19	10	10	12	11	11	5	7	14
Jupiter	15	7	6	13	13	10	15	14	12	10	8	8
Saturn	11	7	5	11	7	8	6	13	12	16	19	16
TOTAL	84	61	73	93	72	72	86	76	69	64	80	87
Rising	7	6	11	13	18	16	15	20	7	8	6	4
N. Node	9	12	6	11	7	9	12	13	14	16	14	8

9H	♈	♉	♊	♋	♌	♍	♎	♏	♐	♑	♒	♓
Moon	15	10	7	10	9	10	12	11	10	9	15	13
Sun	10	9	8	16	9	10	12	10	13	11	14	9
Mercury	14	15	11	9	8	12	6	10	13	11	10	12
Venus	12	9	11	12	5	13	14	11	9	9	16	10
Mars	8	7	7	14	12	12	21	12	12	8	10	8
Jupiter	10	13	6	15	13	13	11	10	9	11	8	12
Saturn	9	18	12	15	7	16	10	10	7	6	10	11
TOTAL	78	81	62	91	63	86	86	74	73	65	83	75

Houses	1st	2nd	3rd	4th	5th	6th	7th	8th	9th	10th	11th	12th
Moon	8	17	13	12	10	10	10	5	15	11	15	5
Sun	5	12	12	13	9	3	7	17	13	13	15	12
Mercury	8	11	7	18	9	5	9	12	13	15	11	13
Venus	13	11	13	11	8	9	9	11	15	8	7	16
Mars	13	11	12	9	7	14	10	6	13	13	15	8
Jupiter	15	11	12	7	9	8	6	11	13	9	16	14
Saturn	10	8	11	11	19	12	12	11	9	8	10	10
TOTAL	72	81	80	81	71	61	63	73	91	77	89	78
N. Node	9	12	16	15	8	11	12	11	10	10	7	10

Planets

Aspects	Parallels	4 Axis	Direction	☉/☽ Midpoint
Mercury & Venus	Moon-Mars	Pluto (Asc)	Jupiter (D)	Mars
Moon-Uranus		Mars (Dsc)		
Sun-Venus		Mercury (MC)		
Mercury-Chiron		Chiron (MC)		Moon's Nodes
Venus-Uranus				Neptune (N)

CHILDREN – PARENTS OF AT LEAST 4

Theme: Aries/Taurus/Aquarius, Mercury/Venus

Taurus (1 in 6,803):
Taureans are very magnetic, stable, and strong. Venus is the strongest planet in Taurus (1 in 370), indicating a love of long-term relationships and stability.

Aries:
Aries strong suggests that being assertive is a definite advantage for having offspring. Aries and Taurus are the first two signs of Spring, when nature is reborn.

Leo & Aquarius in the 9th Harmonic:
They are inwardly (9H) very sociable and romantic, since Leo and Aquarius represent the Axis of Charisma. Mars strong in Leo endows a person with high sexual confidence (1 in 437).

Mercury in the 9th House (1 in 1,209) & Sun in the 9th House (1 in 164):
The 9th house is a place of good fortune and optimism.

Venus aspects:
Venusian people value relationships with others and enjoy surrounding themselves with many people. They are also very skilled at creating and preserving harmonious relations. Venus with Saturn represents responsibility and this aspect causes them to be very careful about choosing a partner (1 in 398). They prefer stability and predictability in relationships, thus providing the perfect conditions for having a large family.

Signs

Size: 340	1st Harmonic	9th Harmonic	Houses	Rising Sign
Moon		Capricorn, Aquarius		Taurus
Sun	Aries	Aquarius	9th	
Mercury	Aries, Taurus	Libra	9th	Lunar Phase
Venus	Aries, Taurus		5th	Taurus, Aquar
Mars	Sagittarius	Leo	11th	☉/☽ Midpoint
Jupiter	Aries			Cancer, Capric
Saturn	Scorpio			Moon's Nodes
TOTAL	Aries, Taurus	Leo, Aquarius	5th, 9th	Virgo, Capric
(weak)	Leo, Virgo, Capr	Gemini		11th

1H	♈	♉	♊	♋	♌	♍	♎	♏	♐	♑	♒	♓
Moon	22	34	23	31	25	32	20	32	30	28	36	27
Sun	38	36	34	28	27	21	21	32	23	19	31	30
Mercury	38	34	24	31	20	20	29	27	31	27	27	32
Venus	40	44	27	33	18	29	19	28	23	16	36	27
Mars	26	30	28	23	31	20	30	34	38	30	21	29
Jupiter	35	29	23	26	24	26	38	32	34	27	20	26
Saturn	22	34	23	27	22	20	29	39	32	27	38	27
TOTAL	221	241	182	199	167	168	186	224	211	174	209	198
Rising	16	27	25	31	37	36	30	31	33	31	23	20
N. Node	29	25	27	23	25	40	34	23	27	39	26	22

9H	♈	♉	♊	♋	♌	♍	♎	♏	♐	♑	♒	♓
Moon	26	29	20	30	31	25	27	31	23	40	37	21
Sun	30	26	17	20	25	32	33	28	23	34	42	30
Mercury	20	29	29	34	33	27	42	23	26	22	26	29
Venus	26	21	25	33	25	28	33	30	24	32	30	33
Mars	25	26	28	32	42	21	19	33	35	30	24	25
Jupiter	30	32	22	23	35	32	25	30	34	17	35	25
Saturn	27	26	24	36	31	33	24	28	20	25	35	31
TOTAL	184	189	165	208	222	198	203	203	185	200	229	194

Houses	1st	2nd	3rd	4th	5th	6th	7th	8th	9th	10th	11th	12th
Moon	33	29	33	30	28	27	31	26	25	26	22	30
Sun	26	29	33	32	27	20	16	19	37	37	36	28
Mercury	28	28	31	31	28	20	20	19	41	30	38	26
Venus	35	38	27	19	36	24	20	20	28	30	32	31
Mars	37	24	26	23	26	25	31	21	16	37	39	35
Jupiter	32	33	30	28	27	24	20	35	31	26	25	29
Saturn	29	32	21	30	36	29	33	28	33	28	18	23
TOTAL	220	213	201	193	208	169	171	168	211	214	210	202
N. Node	32	29	29	32	27	22	33	22	19	26	37	32

Planets

Aspects	Parallels	4 Axis	Direction	☉/☽ Midpoint
Venus	Sun-Neptune	Mercury (MC)	Mercury (D)	
Sun-Mercury	Moon-Mercury		Venus (Rx)	
Venus-Saturn			Jupiter (D)	
			Chiron (D)	Moon's Nodes
				Sun (N)

CHILDREN – NONE

Theme: Virgo, Saturn

Virgo:
Virgo is strong in the 9th harmonic (1 in 1,940), especially for Saturn (1 in 2,518). Virgos have a tendency for introversion, being critical and serious. They also tend to occupy themselves with work and hobbies. These traits are not conducive to meeting people and having children.

Saturn:
A strong Saturn in the birth chart indicates much seriousness and caution in the person's life. These traits make a person less likely to attract partners.

Moon in the 7th House (1 in 342) & Moon on the Descendant (1 in 173):
Moon in the 7th house and on the Descendant indicates shyness and inhibition in relationships.

Mars on the Ascendant (1 in 409):
People with Mars on the Ascendant are often perceived as hostile and aggressive.

9th House is weak (1 in 8,013):
This house is a sign of abundance and good fortune. People with no children rarely have planets in the 9th house.

Signs

Size: 275	1st Harmonic	9th Harmonic	Houses	Rising Sign
Moon	Pisces	Virgo	7th	Aquarius
Sun		Libra	4th	
Mercury		Virgo		Lunar Phase
Venus	Taurus, Aquarius	Pisces		
Mars	Aries	Scorpio	12th	☉/☽ Midpoint
Jupiter		Libra	6th	
Saturn	Virgo, Sagittarius	Virgo	10th, 12th	Moon's Nodes
TOTAL	Taurus	Virgo	7th	Virgo (S)
(weak)	Scorpio	Gemini	9th	Cancer, 3rd

1H	♈	♉	♊	♋	♌	♍	♎	♏	♐	♑	♒	♓
Moon	21	25	19	26	24	25	24	20	19	21	15	36
Sun	26	30	28	27	19	21	16	18	17	20	29	24
Mercury	22	26	24	27	19	19	23	14	26	31	21	23
Venus	24	33	18	29	19	25	17	17	18	19	34	22
Mars	27	23	31	24	27	21	26	22	20	17	17	20
Jupiter	22	18	22	24	25	21	27	26	31	18	25	16
Saturn	19	24	14	15	27	30	29	24	33	19	19	22
TOTAL	161	179	156	172	160	162	162	141	164	145	160	163
Rising	16	18	23	22	34	25	23	26	22	22	26	18
N. Node	26	22	27	31	19	26	18	19	18	21	16	32

9H	♈	♉	♊	♋	♌	♍	♎	♏	♐	♑	♒	♓
Moon	22	24	21	21	28	33	19	14	18	19	30	26
Sun	19	20	16	30	19	26	31	22	14	27	22	29
Mercury	21	22	20	16	20	34	27	20	26	24	17	28
Venus	18	23	21	20	19	19	26	25	27	19	26	32
Mars	28	24	20	21	27	22	17	31	27	18	19	21
Jupiter	25	26	19	21	17	27	31	24	25	18	23	19
Saturn	28	24	14	23	19	39	23	20	24	30	18	13
TOTAL	161	163	131	152	149	200	174	156	161	155	155	168

Houses	1st	2nd	3rd	4th	5th	6th	7th	8th	9th	10th	11th	12th
Moon	29	23	24	25	21	19	36	20	16	16	20	26
Sun	28	18	24	34	21	22	21	21	10	29	29	18
Mercury	20	27	25	27	23	22	24	18	15	21	32	21
Venus	26	22	26	22	27	25	23	15	24	22	19	24
Mars	25	26	23	20	24	19	26	21	13	23	23	32
Jupiter	28	18	26	15	27	31	23	22	16	21	21	27
Saturn	23	16	25	22	21	17	26	29	16	31	18	31
TOTAL	179	150	173	165	164	155	179	146	110	163	162	179
N. Node	21	27	35	29	12	25	25	18	19	21	17	26

Planets

Aspects	Parallels	4 Axis	Direction	☉/☽ Midpoint
Saturn		Mars (Asc)	Chiron (D)	
Sun-Venus		Saturn (Asc)		
		Moon (Dsc)		
		Neptune (Dsc)		Moon's Nodes
				Mercury (S)
				Saturn (N)

Comedians

Theme: Taurus/Scorpio/Aquarius, Moon

Aquarius:
Aquarius is strong in the 1st harmonic (1 in 163,970), especially for the Sun (1 in 2,191) and Venus (1 in 987). Aquarians are adept thinkers and communicators who radiate charm. Their identity and goals revolve around being thoughtful and informative.

Taurus:
Taurus is strong in the 9th harmonic (1 in 256), especially for Mars (1 in 1,977). Taurus and Aquarius are the two signs most associated with Venus. This implies that comedians are primarily likable people with an excellent ability to connect with others.

Scorpio & the 8th House:
These positions enable people to see beneath the surface and know what is hidden within people. Comedians often use their innate likability to reveal our shadow to us in a way that we can accept.

Libra is weak:
Librans often find themselves in conflict with the beliefs of others. Unlike the Venusian signs (Taurus and Aquarius), which seek to create harmony, Librans often create tension by challenging the way people think.

Moon:
Lunar people are highly imaginative and, sometimes, eccentric. They are tuned into the emotions of people and can intuit what makes them laugh.

Signs

Size: 148	1st Harmonic	9th Harmonic	Houses	Rising Sign
Moon	Taurus		7th	Aquarius
Sun	Aquarius	Sagittarius, Pisces	8th	
Mercury	Sagittarius, Aquarius	Taurus	4th	Lunar Phase
Venus	Scorpio, Aquarius			Leo
Mars	Capricorn	Taurus, Aquarius	12th	☉/☽ Midpoint
Jupiter		Taurus, Sagittarius		Aries, Pisces
Saturn	Libra, Scorpio		5th, 8th	Capricorn
TOTAL	Scorpio, Capr, Aquar	Taurus	5th, 8th	Moon's Nodes
(weak)	Aries, Leo, Libra	Gemini, Libra		Cancer, 12th

1H	♈	♉	♊	♋	♌	♍	♎	♏	♐	♑	♒	♓
Moon	7	18	15	10	9	13	6	15	13	14	14	14
Sun	9	8	13	14	10	6	8	15	16	12	23	14
Mercury	7	9	11	10	11	6	10	15	20	16	19	14
Venus	8	7	11	15	11	10	5	18	14	11	24	14
Mars	6	15	6	9	14	19	13	13	12	17	14	10
Jupiter	10	12	12	13	7	14	11	15	12	17	14	11
Saturn	7	7	8	13	8	9	18	24	11	15	16	12
TOTAL	54	76	76	84	70	77	71	115	98	102	124	89
Rising	8	7	10	13	19	11	18	15	15	12	14	6
N. Node	11	10	15	24	15	12	13	5	15	10	9	9

9H	♈	♉	♊	♋	♌	♍	♎	♏	♐	♑	♒	♓
Moon	14	13	10	7	17	15	7	14	10	15	10	16
Sun	14	13	9	17	8	4	14	10	19	9	12	19
Mercury	13	18	9	9	13	16	11	12	14	9	12	12
Venus	16	12	12	11	12	12	14	11	14	12	17	5
Mars	16	23	13	13	9	14	6	10	9	11	18	6
Jupiter	11	20	8	14	9	16	8	5	19	12	9	17
Saturn	14	11	10	9	14	8	7	17	13	17	14	14
TOTAL	98	110	71	80	82	85	67	79	98	85	92	89

Houses	1st	2nd	3rd	4th	5th	6th	7th	8th	9th	10th	11th	12th
Moon	8	6	14	12	17	15	18	13	13	14	9	9
Sun	14	18	13	16	12	8	5	17	12	10	15	8
Mercury	13	15	14	18	14	7	9	11	8	16	12	11
Venus	16	16	11	17	8	10	11	13	9	13	10	14
Mars	9	16	10	15	13	7	17	11	10	8	12	20
Jupiter	13	17	16	9	15	10	7	13	14	10	11	13
Saturn	12	12	12	5	18	11	10	20	13	14	12	9
TOTAL	85	100	90	92	97	68	77	98	79	85	81	84
N. Node	13	18	10	10	16	8	15	9	12	10	9	18

Planets

Aspects	Parallels	4 Axis	Direction	☉/☽ Midpoint
Moon-Mercury		Pluto (Asc)	Pluto (Rx)	
Moon-Saturn		Moon (IC)		
Moon-Uranus				
				Moon's Nodes

Composers of Classical Music

Theme: Capricorn, Moon/Mercury/Uranus

Capricorn:
Capricorn is strong in the 1st and 9th harmonics (1 in 742 - 1H), especially for Venus (1 in 894 - 9H). Capricorns are willing to persevere in anonymity for great amounts of time in order to achieve long-lasting goals. Classical compositions have persevered for centuries as great artistic achievements (Venus). There is also a distinctly conservative nature to classical music compared with current popular music.

Moon-Uranus aspects and parallels:
Moon-Uranus aspects and parallels are frequent for composers. They have great powers of intuition and inventiveness.

Jupiter in the 4th House (1 in 419):
Composers are very productive working in private. The 4th house is the most hidden space in a birth chart. Jupiter here shows enthusiasm in solitude.

Signs

Size: 258	1st Harmonic	9th Harmonic	Houses	Rising Sign
Moon	Aquarius	Capricorn	5th	
Sun	Capricorn			
Mercury	Capricorn	Taurus, Leo		Lunar Phase
Venus	Taurus	Capricorn		
Mars	Aquarius			☉/☽ Midpoint
Jupiter	Capricorn	Virgo	4th, 10th	Pisces
Saturn	Sagittarius, Pisces	Libra		Moon's Nodes
TOTAL	Sagittar, Capric, Pisces	Capricorn	4th	Taurus
(weak)	Gemini, Leo			

1H	♈	♉	♊	♋	♌	♍	♎	♏	♐	♑	♒	♓
Moon	15	19	18	18	21	14	22	28	17	25	33	28
Sun	25	20	18	24	14	20	22	16	25	28	18	28
Mercury	23	23	12	20	21	19	19	19	26	32	23	21
Venus	17	33	20	19	18	25	13	21	25	24	20	23
Mars	22	20	17	19	22	25	22	27	23	21	26	14
Jupiter	21	18	13	26	11	23	28	25	27	29	17	20
Saturn	23	14	18	12	22	26	14	28	31	24	17	29
TOTAL	146	147	116	138	129	152	140	164	174	183	154	163
Rising	7	16	22	28	35	34	27	26	28	10	16	9
N. Node	26	27	22	11	22	22	14	22	24	25	22	21

9H	♈	♉	♊	♋	♌	♍	♎	♏	♐	♑	♒	♓
Moon	19	14	21	20	25	22	19	21	24	31	22	20
Sun	26	26	16	25	15	24	28	19	23	23	16	17
Mercury	23	29	20	22	31	20	20	14	21	22	21	15
Venus	23	21	20	21	18	18	23	19	23	35	20	17
Mars	18	22	16	23	22	24	18	22	23	19	23	28
Jupiter	18	23	26	14	20	30	21	19	18	21	21	27
Saturn	22	25	14	24	21	27	29	24	22	24	15	11
TOTAL	149	160	133	149	152	165	158	138	154	175	138	135

Houses	1st	2nd	3rd	4th	5th	6th	7th	8th	9th	10th	11th	12th
Moon	23	16	20	28	29	17	19	17	20	22	27	20
Sun	23	25	20	27	13	16	24	21	19	16	23	31
Mercury	31	24	19	21	17	17	22	23	18	18	20	28
Venus	21	27	27	16	22	14	24	19	24	20	13	31
Mars	18	26	27	18	16	19	19	18	22	27	25	23
Jupiter	21	23	17	34	27	21	13	19	14	31	16	22
Saturn	25	18	19	27	28	19	18	21	24	20	24	15
TOTAL	162	159	149	171	152	123	139	138	141	154	148	170
N. Node	14	23	21	26	27	23	19	21	23	18	21	22

Planets

Aspects	Parallels	4 Axis	Direction	☉/☽ Midpoint
Uranus	Moon-Uranus	Moon (IC)		Mercury
Moon-Uranus				
Mercury-Jupiter				
Mercury-Uranus				Moon's Nodes
Venus-Neptune				Mars (N)
				Chiron (S)

COMPUTER PROGRAMMERS

Theme: Scorpio/Capricorn, Mercury/Uranus

Scorpio & Capricorn:
These two signs are very driven and focused. Scorpios have penetrating minds that are capable of solving difficult problems, while Capricorns have a great capacity for work and discipline. Both signs are introverted, which suits computer programmers since most of their work is done independently.

Mercury:
Mercury represents logical reasoning ability.

Uranus:
Uranian people have inventive insights.

Sun in the 4th House (1 in 2,052):
They are very production working in solitude and do not seek attention.

Moon's North Node in the 6th House (1 in 1,766)
Technical people find purpose in the use of analytical abilities to do useful work.

Signs

Size: 82	1st Harmonic	9th Harmonic	Houses	Rising Sign
Moon	Aquarius	Scorpio, Capricorn	5th	Taurus
Sun	Scorpio		4th, 11th	
Mercury	Sagittarius	Gemini, Virgo, Scorpio		Lunar Phase
Venus	Scorpio	Virgo, Scorpio, Capr	1st	
Mars	Aries, Capricorn		3rd	☉/☽ Midpoint
Jupiter		Aries		Virgo, Sagitt
Saturn	Taurus	Capricorn	10th	Moon's Nodes
TOTAL	Scorpio, Aquarius	Scorpio, Capricorn		Aquarius
(weak)	Gemini			6th

1H	♈	♉	♊	♋	♌	♍	♎	♏	♐	♑	♒	♓
Moon	4	5	3	9	5	8	10	3	7	4	14	10
Sun	7	4	6	3	3	6	10	12	10	7	8	6
Mercury	7	5	2	4	5	6	9	10	13	7	10	4
Venus	7	11	2	3	3	4	7	12	10	9	5	9
Mars	11	7	5	6	3	7	6	11	6	10	6	4
Jupiter	4	8	6	7	10	5	9	5	8	8	8	4
Saturn	4	11	9	7	9	5	8	9	3	5	7	5
TOTAL	44	51	33	39	38	41	59	62	57	50	58	42
Rising	6	10	6	5	10	7	10	8	6	4	6	4
N. Node	4	5	7	7	6	10	3	4	10	5	12	9

9H	♈	♉	♊	♋	♌	♍	♎	♏	♐	♑	♒	♓
Moon	9	8	7	3	8	4	2	11	6	11	7	6
Sun	8	4	6	9	9	6	6	8	10	9	1	6
Mercury	3	9	12	7	3	11	7	12	2	3	6	7
Venus	6	4	5	8	5	11	2	12	5	11	9	4
Mars	6	9	4	6	7	7	9	6	5	9	7	7
Jupiter	11	8	6	9	7	8	4	4	6	9	5	5
Saturn	9	7	10	6	5	6	10	6	5	11	4	3
TOTAL	52	49	50	48	44	53	40	59	39	63	39	38

Houses	1st	2nd	3rd	4th	5th	6th	7th	8th	9th	10th	11th	12th
Moon	4	6	6	8	11	8	4	5	10	7	6	7
Sun	5	11	7	15	0	8	4	4	3	5	12	8
Mercury	7	9	9	10	5	6	6	4	3	5	11	7
Venus	16	7	6	8	4	7	4	8	3	10	3	6
Mars	8	6	12	5	3	8	6	5	6	8	11	4
Jupiter	4	8	9	10	6	6	7	6	4	4	10	8
Saturn	6	5	5	2	8	8	7	7	4	12	8	10
TOTAL	50	52	54	58	37	51	38	39	33	51	61	50
N. Node	6	6	6	2	6	15	7	4	9	8	5	8

Planets

Aspects	Parallels	4 Axis	Direction	☉/☽ Midpoint
Mercury-Venus			Venus (D)	
Mercury-Uranus			Jupiter (Rx)	
Venus-Neptune			Uranus (Rx)	
Jupiter-Neptune				Moon's Nodes
Uranus-Neptune				Mercury (S)
Uranus-Chiron				Pluto (N)

Corporate Tycoons

Theme: Cancer, Neptune

Cancer:
Cancer is strong in both the 1st and 9th harmonics (1 in 1,335 -1H). The Moon's South Node is also strong in Cancer (1 in 1,400). In spite of the sign's caring nature, it is still a Cardinal-power sign and seeks positions of influence. Cancer is the sign where the Sun's declination is highest each year. The Sun here represents the outer self seeking to gain recognition for being special.

Capricorn is weak:
Capricorn is the opposite sign from Cancer and seeks power sharing within a controlled group structure. They feel purpose by taking on responsibility and not for standing out. Capricorn is the sign where the Sun's declination is lowest each year. So, the outer self and its attempts to gain recognition for being special are repressed.

Neptune:
Sun-Neptune aspects (1 in 300) are the most frequent. Neptune represents dreams, in this case, of personal glory and success. This is often accomplished through improving financial conditions. Neptune also represents illusions and deception, which can be used to promote one's image, products, or services.

Uranus on the Moon's North Node (1 in 55,848):
This position indicates that a person's life purpose is to bring something new and inventive into the world.

Signs

Size: 149	1st Harmonic	9th Harmonic	Houses	Rising Sign
Moon	Cancer	Leo, Scorpio		Virgo
Sun	Gemini			
Mercury		Cancer		Lunar Phase
Venus		Cancer, Sagittarius	9th	
Mars	Gemini			☉/☽ Midpoint
Jupiter		Leo		Aries
Saturn	Scorpio	Cancer	2nd	Moon's Nodes
TOTAL	Gemini, Cancer	Cancer, Sagittarius		Cancer (S)
(weak)	Capricorn	Aries, Gemini	4th	

1H	♈	♉	♊	♋	♌	♍	♎	♏	♐	♑	♒	♓
Moon	11	7	6	20	10	14	13	9	11	16	15	17
Sun	11	17	20	17	15	8	7	12	7	5	16	14
Mercury	14	15	16	15	15	12	8	12	9	6	17	10
Venus	11	17	14	17	14	15	7	10	9	8	13	14
Mars	11	12	21	18	11	11	15	5	14	11	14	6
Jupiter	12	14	11	15	12	13	17	14	17	8	6	10
Saturn	11	12	12	14	10	13	14	19	13	7	7	17
TOTAL	81	94	100	116	87	86	81	81	80	61	88	88
Rising	10	12	9	15	10	23	17	15	14	11	10	3
N. Node	11	16	16	11	13	9	10	10	11	23	14	5

9H	♈	♉	♊	♋	♌	♍	♎	♏	♐	♑	♒	♓
Moon	10	5	15	8	20	7	6	20	15	13	16	14
Sun	11	14	9	13	11	12	17	10	13	8	15	16
Mercury	15	8	9	18	12	11	11	12	17	10	16	10
Venus	7	11	8	20	9	16	13	12	20	14	12	7
Mars	5	10	14	13	12	16	10	10	13	14	17	15
Jupiter	9	15	7	13	19	13	7	13	17	13	8	15
Saturn	12	12	10	18	11	16	10	8	8	14	13	17
TOTAL	69	75	72	103	94	91	74	85	103	86	97	94

Houses	1st	2nd	3rd	4th	5th	6th	7th	8th	9th	10th	11th	12th
Moon	10	14	13	8	16	14	13	13	13	14	8	13
Sun	13	9	13	4	14	14	10	8	13	18	14	19
Mercury	10	13	13	5	13	13	9	11	13	16	16	17
Venus	15	13	10	11	14	11	9	8	18	12	13	15
Mars	18	12	9	15	9	15	8	11	11	17	13	11
Jupiter	12	8	14	11	14	12	12	14	10	13	15	14
Saturn	11	20	8	14	9	12	8	15	13	8	16	15
TOTAL	89	89	80	68	89	91	69	80	91	98	95	104
N. Node	9	17	10	16	9	14	8	13	10	15	14	14

Planets

Aspects	Parallels	4 Axis	Direction	☉/☽ Midpoint
Neptune	Mercury-Neptune	Uranus (Asc)	Pluto (D)	
Sun-Neptune		Mercury (Dsc)	Chiron (D)	
Mercury-Venus		Mars (IC)		
Venus-Neptune		Pluto (IC)		Moon's Nodes
				Venus (N)
				Mars (N)
				Uranus (N)

CRIMINAL CAREERS

Theme: Scorpio, Mars

Scorpio:
They are strong, intense, and dangerous. Scorpios are renowned for their willingness to live outside the law in the underworld of society.

Virgo:
Virgo is strong in the 9th harmonic, especially for Jupiter (1 in 13,772) and Mercury (1 in 600). Virgos often attract difficult and harmful circumstances. Mercury in Virgo indicates highly developed analytical skills, which they use to figure out how to pull off and disguise their activities. Jupiter in Virgo reduces openness, honesty, and enthusiasm while increasing criticalness and cynicism.

Mars:
Mars is frequently in aspect with Uranus and Neptune. This is an indication of aggression which is both ingenious and deceptive.

Jupiter in Taurus (1 in 546):
They seek an abundance of material wealth.

Jupiter on the Imum Coeli (1 in 634) & Pluto on the Imum Coeli (1 in 125):
Jupiter and Pluto on the IC indicates an enthusiasm for shadowy activities done in private.

Saturn in the 9th House (1 in 2,699):
Saturn in this position indicates a restriction on optimism and good fortune.

Signs

Size: 81	1st Harmonic	9th Harmonic	Houses	Rising Sign
Moon	Scorpio		2nd, 8th	
Sun		Gemini	2nd	
Mercury	Scorpio	Virgo		Lunar Phase
Venus		Taurus		Gemini
Mars		Cancer	1st	☉/☽ Midpoint
Jupiter	Taurus	Virgo, Scorpio	4th	Gemini, Libra
Saturn	Sagittarius, Pisces	Pisces	9th	Moon's Nodes
TOTAL	Scorpio	Virgo		6th, 12th
(weak)	Aquarius		6th, 7th	

1H	♈	♉	♊	♋	♌	♍	♎	♏	♐	♑	♒	♓
Moon	7	9	9	7	5	5	7	12	7	7	4	2
Sun	6	6	7	7	5	7	9	8	9	8	3	6
Mercury	6	4	7	7	4	6	7	12	10	8	2	8
Venus	4	3	6	11	5	10	7	9	8	3	7	8
Mars	8	6	5	6	9	9	6	11	4	4	5	8
Jupiter	3	13	5	6	8	6	11	9	5	4	4	7
Saturn	3	5	4	6	3	6	10	3	13	5	10	13
TOTAL	37	46	43	50	39	49	57	64	56	39	35	52
Rising	4	6	6	11	9	8	8	11	7	4	4	3
N. Node	4	9	7	5	6	6	7	7	6	6	10	8

9H	♈	♉	♊	♋	♌	♍	♎	♏	♐	♑	♒	♓
Moon	5	9	9	5	5	10	6	7	6	8	7	4
Sun	9	3	11	8	5	6	7	5	6	5	9	7
Mercury	4	2	2	6	5	14	8	10	8	6	10	6
Venus	8	12	8	3	10	4	4	7	4	7	8	6
Mars	4	6	9	12	7	6	6	6	5	8	5	7
Jupiter	8	9	3	4	4	16	4	11	4	6	8	4
Saturn	6	4	10	7	10	5	5	4	8	8	3	11
TOTAL	44	45	52	45	46	61	40	50	41	48	50	45

Houses	1st	2nd	3rd	4th	5th	6th	7th	8th	9th	10th	11th	12th
Moon	7	11	6	9	1	7	5	11	3	7	7	7
Sun	8	12	6	8	4	3	3	7	3	9	6	12
Mercury	9	9	8	5	7	4	2	6	4	8	8	11
Venus	10	10	8	7	5	3	4	6	8	4	8	8
Mars	11	5	8	6	8	4	8	4	9	7	6	5
Jupiter	9	4	7	13	8	6	7	3	7	8	4	5
Saturn	6	6	9	6	9	4	4	4	15	5	6	7
TOTAL	60	57	52	54	42	31	33	41	49	48	45	55
N. Node	3	9	8	10	5	11	3	4	3	9	2	14

Planets

Aspects	Parallels	4 Axis	Direction	☉/☽ Midpoint
Mars-Uranus		Chiron (Dsc)		Saturn
Mars-Neptune		Mars (MC)		
		Jupiter (IC)		
		Pluto (IC)		**Moon's Nodes**
				Mercury (S)

CULINARY – CHEFS, RESTAURATEURS

Theme: Cancer, 5th/10th Houses, Moon/Venus

Cancer:
Cancer is strong in the 1st and 9th harmonics, especially for Mercury (1 in 1,348 - 9H). Cancers love food, nurturing people, and making us feel good. Mercury in Cancer indicates much thinking and learning about the culinary arts.

5th House:
The 5th house is associated with creative activities that bring enjoyment. Mercury (1 in 175) and the Moon's South Node (1 in 317) are especially frequent in this position. The Moon's North Node is also strong in Leo (1 in 1,254).

Moon:
Moon aspects are frequent here, showing its association with being mothering and nurturing. Chefs and cooks often have a strong connection with their cultural heritage.

Venus:
Venus is frequent on the 4 Axis points (1 in 283), especially the Midheaven (1 in 2,847). Venus being strong on almost all the axis points indicates that a great deal of their lives revolve around relationships and the enjoyment of pleasures.

Mars-Saturn aspects and parallels:
This planetary combination indicates that a person is able to sublimate their sexual-aggressive energies into productive work. A strong 10th house also indicates having a high drive for career success.

Size: 100	1st Harmonic	9th Harmonic	Houses	Rising Sign
		Signs		
Moon	Cancer, Libra	Taurus, Capricorn	3rd	
Sun	Cancer	Scorpio		
Mercury		Cancer	1st, 5th	Lunar Phase
Venus		Aries	1st, 10th	Libra
Mars	Virgo, Sagittarius	Cancer	5th	☉/☽ Midpoint
Jupiter	Aquarius	Aries, Pisces		Cancer
Saturn	Capricorn	Pisces	8th, 10th	Moon's Nodes
TOTAL	Aquarius	Cancer	5th, 10th	Leo, 5th (S)
(weak)	Scorpio	Virgo		10th

1H	♈	♉	♊	♋	♌	♍	♎	♏	♐	♑	♒	♓
Moon	8	9	4	13	9	8	15	6	6	11	5	6
Sun	10	8	8	13	12	4	10	5	6	6	9	9
Mercury	12	9	9	8	8	8	11	7	4	6	12	6
Venus	7	9	9	7	12	9	6	6	6	9	13	7
Mars	9	8	8	6	7	15	7	3	13	8	10	6
Jupiter	10	8	8	9	6	12	7	6	8	5	14	7
Saturn	8	11	10	10	4	7	5	8	7	14	10	6
TOTAL	64	62	56	66	58	63	61	41	50	59	73	47
Rising	5	5	11	9	15	14	12	7	9	8	3	2
N. Node	8	5	11	8	18	7	11	6	4	4	9	9

9H	♈	♉	♊	♋	♌	♍	♎	♏	♐	♑	♒	♓
Moon	4	14	12	5	3	6	7	10	8	13	7	11
Sun	6	8	9	4	6	6	8	15	11	7	11	9
Mercury	7	8	11	17	9	6	10	7	4	9	7	5
Venus	13	7	7	12	5	6	6	12	11	4	8	9
Mars	7	10	13	14	8	6	8	7	7	5	7	8
Jupiter	15	9	6	9	9	4	6	5	10	4	10	13
Saturn	8	6	4	11	11	11	4	8	7	8	8	14
TOTAL	60	62	62	72	51	45	49	64	58	50	58	69

Houses	1st	2nd	3rd	4th	5th	6th	7th	8th	9th	10th	11th	12th
Moon	5	10	13	5	7	8	10	7	9	5	10	11
Sun	10	8	7	7	11	8	4	4	10	10	13	8
Mercury	16	7	8	4	14	3	4	6	11	10	12	5
Venus	15	9	6	8	9	7	4	4	7	18	9	4
Mars	8	6	6	7	13	8	8	12	8	7	9	8
Jupiter	10	7	11	9	5	12	9	7	6	9	7	8
Saturn	6	4	8	8	7	8	6	14	5	14	11	9
TOTAL	70	51	59	48	66	54	45	54	56	73	71	53
N. Node	7	6	6	7	12	5	6	6	11	13	16	5

Planets

Aspects	Parallels	4 Axis	Direction	☉/☽ Midpoint
Moon, Mars, Chiron	Mars-Saturn	Mercury (Asc)	Venus (Rx)	Saturn
Moon-Venus		Venus (Asc)		
Moon-Uranus		Venus (MC)		
Moon-Chiron		Venus (IC)		Moon's Nodes
Mars-Saturn		Neptune (IC)		Mercury (S)
Mars-Chiron				

Dancers

Theme: Aries (1st), Pisces (12th)

Aries & Pisces:
This combination of signs represents dynamism and artistry, athletic ability used for creative expression. The Moon is frequent in Aries in both harmonics (1 in 244 - 9H), Mars is frequent in the 1st House (1 in 137), and the Sun is frequent in the 12th house (1 in 173).

Mercury in Cancer (1 in 261):
This position indicates that a person's reasoning is influenced by their emotions. Music, by effecting the emotions, plays a large role in the thoughts and actions of dancers.

Signs

Size: 112	1st Harmonic	9th Harmonic	Houses	Rising Sign
Moon	Aries	Aries		
Sun	Leo, Capr, Pisces		12th	
Mercury	Cancer	Cancer		Lunar Phase
Venus			12th	Pisces
Mars	Leo		1st, 7th	☉/☽ Midpoint
Jupiter	Taurus, Sagittarius			
Saturn		Taurus, Gemini		Moon's Nodes
TOTAL				Aries (S)
(weak)			5th, 8th	3rd

1H	♈	♉	♊	♋	♌	♍	♎	♏	♐	♑	♒	♓
Moon	14	9	14	6	4	8	13	8	9	10	7	10
Sun	6	9	9	11	15	8	6	7	5	14	7	15
Mercury	9	5	7	16	11	8	8	10	6	13	10	9
Venus	13	11	11	11	4	12	4	8	10	9	13	6
Mars	5	7	12	11	17	12	11	9	9	3	7	9
Jupiter	3	13	8	9	10	8	12	10	17	7	5	10
Saturn	5	10	9	6	10	12	13	5	10	11	11	10
TOTAL	55	64	70	70	71	68	67	57	66	67	60	69
Rising	6	5	9	13	17	8	14	9	11	6	10	4
N. Node	12	8	12	6	12	13	15	7	5	3	11	8

9H	♈	♉	♊	♋	♌	♍	♎	♏	♐	♑	♒	♓
Moon	17	8	13	6	6	6	12	12	7	14	6	5
Sun	11	9	6	12	12	10	5	10	9	11	9	8
Mercury	11	8	4	16	5	11	11	14	12	6	5	9
Venus	11	11	6	12	10	9	8	9	9	7	11	9
Mars	9	12	9	9	8	8	6	12	13	6	10	10
Jupiter	9	8	10	10	10	9	7	8	13	10	10	8
Saturn	7	15	15	6	12	10	10	8	8	8	5	8
TOTAL	75	71	63	71	63	63	59	73	71	62	56	57

Houses	1st	2nd	3rd	4th	5th	6th	7th	8th	9th	10th	11th	12th
Moon	9	12	7	12	6	12	10	7	8	9	13	7
Sun	9	13	7	8	8	9	7	5	9	7	11	19
Mercury	10	11	15	6	7	9	9	3	6	14	9	13
Venus	13	9	11	12	5	6	12	6	6	7	8	17
Mars	17	13	4	6	8	9	16	5	6	11	12	5
Jupiter	13	11	12	9	6	7	7	8	12	13	7	7
Saturn	11	8	12	5	7	6	10	7	10	11	12	13
TOTAL	82	77	68	58	47	58	71	41	57	72	72	81
N. Node	9	12	15	9	7	12	6	7	6	10	11	8

Planets

Aspects	Parallels	4 Axis	Direction	☉/☽ Midpoint
Sun-Saturn		Neptune (Dsc)		Saturn
Mars-Uranus				Pluto
				Moon's Nodes
				Mars (S)
				Neptune (N)

Depression

Theme: Cancer

Cancer:
Cancer is strong, especially for Mars (1 in 124). When life does not go well for deeply sensitive Cancers, they can become despondent. People with Mars in Cancer often experience much conflict and tension in their emotional and intimate relationships.

Aquarius & Leo are weak:
Aquarius is weak in the 1st harmonic and Leo is weak in the 9th harmonic. These two signs that represent the Axis of Charisma are weak for people who suffer depression. This suggests that, as a group, they do not feel attractive or well liked and do not experience being popular.

Mars on the Ascendant (1 in 301) & Mars on the Sun/Moon midpoint:
Conflict and separations are frequent for people with these positions.

Sun-Chiron aspects & Chiron on the Sun/Moon midpoint:
This indicates that a person's life path involves healing trauma.

Signs

Size: 158	1st Harmonic	9th Harmonic	Houses	Rising Sign
Moon	Sagittarius			
Sun	Cancer	Libra, Scorpio		
Mercury			12th	Lunar Phase
Venus	Aries	Virgo		
Mars	Cancer, Virgo			☉/☽ Midpoint
Jupiter		Cancer	1st	
Saturn	Sagittarius		9th	Moon's Nodes
TOTAL	Cancer		1st, 9th	
(weak)	Aquarius	Leo	7th, 10th	

1H	♈	♉	♊	♋	♌	♍	♎	♏	♐	♑	♒	♓
Moon	14	16	10	11	12	15	15	11	19	9	12	14
Sun	12	16	14	19	13	15	11	16	11	9	12	10
Mercury	10	16	15	14	12	12	14	17	14	11	10	13
Venus	21	10	11	14	17	10	13	16	9	13	12	12
Mars	12	7	11	24	9	23	14	8	13	15	12	10
Jupiter	16	13	12	12	8	18	18	11	16	15	9	10
Saturn	10	7	12	15	13	12	12	16	21	16	11	13
TOTAL	95	85	85	109	84	105	97	95	103	88	78	82
Rising	11	9	16	19	16	21	11	15	16	5	12	7
N. Node	14	15	16	13	14	13	10	11	12	14	11	15

9H	♈	♉	♊	♋	♌	♍	♎	♏	♐	♑	♒	♓
Moon	16	9	15	15	7	18	16	9	14	13	9	17
Sun	11	15	15	15	11	7	22	19	11	9	10	13
Mercury	11	11	13	16	9	14	14	12	13	13	16	16
Venus	14	13	12	11	17	20	10	10	14	13	11	13
Mars	14	17	15	15	9	10	15	10	14	17	13	9
Jupiter	11	16	14	21	11	10	7	16	15	15	11	11
Saturn	13	9	13	10	10	17	13	18	9	12	19	15
TOTAL	90	90	97	103	74	96	97	94	90	92	89	94

Houses	1st	2nd	3rd	4th	5th	6th	7th	8th	9th	10th	11th	12th
Moon	18	12	14	12	15	13	16	12	14	9	12	11
Sun	19	15	13	11	10	12	5	11	14	16	16	16
Mercury	17	13	16	15	6	11	9	7	16	12	15	21
Venus	18	11	18	4	16	10	7	13	14	10	19	18
Mars	19	14	10	18	8	12	9	18	11	9	12	18
Jupiter	21	12	10	14	13	12	12	13	15	10	15	11
Saturn	11	13	18	16	12	8	11	12	21	10	16	10
TOTAL	123	90	99	90	80	78	69	86	105	76	105	105
N. Node	16	13	14	15	9	9	13	15	8	16	13	17

Planets

Aspects	Parallels	4 Axis	Direction	☉/☽ Midpoint
Sun-Chiron	Mercury-Pluto	Mars (Asc)		Mars
Venus-Jupiter		Jupiter (Asc)		Chiron
		Uranus (Asc)		
		Pluto (Asc)		Moon's Nodes

Directors (Film & Stage)

Theme: Taurus, 7th/8th Houses

Taurus:
Taureans patiently pursue projects in order to achieve tangible goals. They have a good sense of taste and proportion. Mercury is the most frequent planet in Taurus (1 in 3,394 - 9H), indicating calm thinking and artistic ability.

7th & 8th Houses:
These houses show an aptitude for working in close relationships with people. The 8th house is most frequent (1 in 473).

Sun aspects are infrequent (1 in 247):
The Sun functions more powerfully when it is alone and not in any aspect relationship with other planetary bodies. This indicates that they strive to be special and be in charge of their own lives. Politicians, Salespeople, Sports Coaches, and the Wealthy also have a solitary Sun.

Signs

Size: 149	1st Harmonic	9th Harmonic	Houses	Rising Sign
Moon	Gemini	Leo, Virgo		Pisces
Sun	Taurus	Aries		
Mercury		Taurus		Lunar Phase
Venus		Taurus	8th	
Mars	Taurus, Capricorn		6th, 10th	☉/☽ Midpoint
Jupiter		Taurus, Pisces	7th	Leo
Saturn	Pisces			Moon's Nodes
TOTAL		Taurus	7th, 8th	Gemini, Cancer
(weak)			4th	7th, 8th

1H	♈	♉	♊	♋	♌	♍	♎	♏	♐	♑	♒	♓
Moon	12	13	18	14	11	14	10	8	14	11	17	7
Sun	16	19	13	11	15	9	15	7	12	7	14	11
Mercury	13	16	13	9	15	15	7	13	10	12	16	10
Venus	14	16	10	12	13	11	9	15	11	13	13	12
Mars	12	17	15	14	13	15	12	6	10	17	11	7
Jupiter	9	8	8	15	13	14	14	17	14	12	13	12
Saturn	12	6	11	10	16	14	13	16	12	12	9	18
TOTAL	88	95	88	85	96	92	80	82	83	84	93	77
Rising	4	6	8	15	16	20	13	16	17	14	8	12
N. Node	14	8	19	19	9	6	15	12	12	12	12	11

9H	♈	♉	♊	♋	♌	♍	♎	♏	♐	♑	♒	♓
Moon	13	15	12	3	19	18	12	9	12	12	11	13
Sun	20	5	16	12	8	17	13	17	13	9	12	7
Mercury	11	24	9	14	14	14	11	12	8	11	14	7
Venus	15	18	10	10	15	12	10	11	13	15	9	11
Mars	9	17	16	13	11	13	17	9	7	13	17	7
Jupiter	9	18	9	8	14	13	11	7	13	14	13	20
Saturn	13	9	13	14	10	11	13	17	13	16	10	10
TOTAL	90	106	85	74	91	98	87	82	79	90	86	75

Houses	1st	2nd	3rd	4th	5th	6th	7th	8th	9th	10th	11th	12th
Moon	10	9	12	10	17	15	10	17	9	17	13	10
Sun	17	10	17	5	11	10	14	15	12	11	13	14
Mercury	14	13	11	8	12	10	15	15	12	5	19	15
Venus	19	14	11	8	14	10	13	18	9	11	9	13
Mars	16	13	12	5	9	17	15	13	5	19	18	7
Jupiter	9	17	14	16	8	8	18	15	13	10	11	10
Saturn	14	14	15	14	11	10	16	14	9	12	11	9
TOTAL	99	90	92	66	82	80	101	107	69	85	94	78
N. Node	16	13	10	10	9	10	18	21	11	9	12	10

Planets

Aspects	Parallels	4 Axis	Direction	☉/☽ Midpoint
Mercury-Venus				Mars
Mars-Neptune				
Jupiter-Neptune				
				Moon's Nodes
				Venus

DRUG ABUSERS

Theme: Scorpio, Sun/Mars

Scorpio & the 8th House:
Mars in Scorpio is strong in the 1st and 9th harmonics (1 in 452 - 9H) and the 8th house (1 in 379). This suggests many struggles involving emotional intensity and prohibited activities. People with this position often get entangled in destructive situations and relationships.

Virgo:
Virgo is strongly associated with people who have serious health difficulties.

Sagittarius is weak:
Drug abusers lack the enthusiasm and philosophical outlook of Sagittarius.

Mars on the Moon's South Node (1 in 281):
This position indicates that a person has a history of conflict in their lives that needs to be overcome.

Signs

Size: 413	1st Harmonic	9th Harmonic	Houses	Rising Sign
Moon	Aries, Virgo	Scorpio, Capricorn		Taurus
Sun				
Mercury	Taurus, Leo		7th	**Lunar Phase**
Venus	Virgo		10th	
Mars	Scorpio	Scorpio	5th, 8th	☉/☽ Midpoint
Jupiter				Scorpio
Saturn		Aries	9th	**Moon's Nodes**
TOTAL	Virgo	Capricorn	8th, 10th	
(weak)	Sagittarius, Capr	Sagittarius	2nd, 3rd	

1H	♈	♉	♊	♋	♌	♍	♎	♏	♐	♑	♒	♓
Moon	46	36	31	35	28	44	32	36	26	22	40	37
Sun	30	33	43	36	42	31	35	35	30	35	32	31
Mercury	25	42	35	26	43	37	35	34	37	37	32	30
Venus	34	34	35	37	30	49	29	37	29	30	42	27
Mars	32	27	31	34	42	47	39	45	32	24	28	32
Jupiter	38	34	38	33	36	39	34	41	28	37	26	29
Saturn	22	35	35	30	34	42	36	37	40	27	41	34
TOTAL	227	241	248	231	255	289	240	265	222	212	241	220
Rising	24	30	37	36	41	46	42	48	41	19	30	19
N. Node	31	33	40	30	41	37	34	37	35	21	40	34

9H	♈	♉	♊	♋	♌	♍	♎	♏	♐	♑	♒	♓
Moon	31	42	28	32	34	35	36	43	24	49	27	32
Sun	27	37	41	29	36	37	38	27	32	40	33	36
Mercury	40	32	31	36	43	38	42	24	32	37	25	33
Venus	37	31	30	39	30	31	33	35	36	38	36	37
Mars	32	35	38	41	39	23	28	51	26	39	29	32
Jupiter	33	39	37	42	39	30	30	34	34	29	36	30
Saturn	45	38	33	32	22	40	28	38	32	42	33	30
TOTAL	245	254	238	251	243	234	235	252	216	274	219	230

Houses	1st	2nd	3rd	4th	5th	6th	7th	8th	9th	10th	11th	12th
Moon	31	38	29	34	29	35	33	35	31	38	44	36
Sun	35	30	27	33	25	29	35	34	35	43	45	42
Mercury	34	27	30	34	30	25	42	36	31	42	34	48
Venus	35	33	27	23	31	36	36	35	36	47	35	39
Mars	41	30	28	29	42	37	29	49	32	36	22	38
Jupiter	40	30	22	33	42	27	39	35	42	40	36	27
Saturn	38	33	33	32	36	29	35	32	44	42	34	25
TOTAL	254	221	196	218	235	218	249	256	251	288	250	255
N. Node	32	34	33	29	43	32	32	34	34	39	36	35

Planets

Aspects	Parallels	4 Axis	Direction	☉/☽ Midpoint
Moon-Venus		Mercury (Dsc)	Jupiter (D)	Neptune
Sun-Mars		Neptune (Dsc)		
		Sun (MC)		
		Mars (MC)		Moon's Nodes
				Sun (N)
				Mercury (S)
				Mars (S)

EDUCATION LOW (HIGH SCHOOL DROPOUTS)

Theme: Pisces, Mercury

Pisces:
Pisceans are very reclusive people who often prefer to teach themselves what interests them. People who drop out of school frequently have Mars in Pisces (1 in 361), Saturn in Pisces (1 in 271 - 9H), and are in the Pisces lunar phase (1 in 2,440), the last 30 degrees of the Lunation cycle.

Gemini & the 3rd House are weak:
This indicates a lack of intellectual confidence or curiosity.

Mercury:
Mercury describes the way we learn. People with a strong Mercury in their birth chart are highly influenced by their early learning experiences. Their reasoning abilities can be either be highly developed or undeveloped. Mercury-Jupiter is indicative of a mind that needs freedom to explore. Mercury-Saturn shows a serious mind or a late-bloomer. Mercury-Moon and Mercury-Neptune reveals an abundant imagination.

Signs

Size: 105	1st Harmonic	9th Harmonic	Houses	Rising Sign
Moon				Aquarius
Sun	Taurus			
Mercury	Aries	Gemini		Lunar Phase
Venus	Aries			Pisces
Mars	Pisces			☉/☽ Midpoint
Jupiter	Aquarius			
Saturn	Scorpio	Pisces		Moon's Nodes
TOTAL	Aquarius, Pisces		8th	Cancer
(weak)	Gemini	Taurus	3rd	Sagittarius

1H	♈	♉	♊	♋	♌	♍	♎	♏	♐	♑	♒	♓
Moon	9	12	6	11	11	7	10	6	4	5	13	11
Sun	10	14	5	10	11	3	11	7	10	4	12	8
Mercury	14	9	5	5	11	10	8	8	10	6	9	10
Venus	15	10	6	8	8	14	4	11	5	6	7	11
Mars	4	9	11	8	7	8	9	9	7	10	9	14
Jupiter	11	3	5	6	12	9	6	13	7	7	15	11
Saturn	7	7	4	9	3	12	5	17	7	12	13	9
TOTAL	70	64	42	57	63	63	53	71	50	50	78	74
Rising	3	6	9	11	15	12	9	9	7	10	11	3
N. Node	7	8	7	15	9	6	5	11	14	6	8	9

9H	♈	♉	♊	♋	♌	♍	♎	♏	♐	♑	♒	♓
Moon	12	7	11	10	9	11	9	6	8	9	10	3
Sun	7	5	8	9	10	10	11	11	10	9	8	7
Mercury	5	4	14	11	7	9	10	10	5	10	9	11
Venus	10	7	9	8	7	12	13	5	6	9	11	8
Mars	11	11	8	12	9	5	10	11	10	7	6	5
Jupiter	7	7	7	9	12	11	6	10	9	10	5	12
Saturn	13	3	9	11	11	4	6	9	5	9	9	16
TOTAL	65	44	66	70	65	62	65	62	53	63	58	62

Houses	1st	2nd	3rd	4th	5th	6th	7th	8th	9th	10th	11th	12th
Moon	11	12	4	5	5	7	11	11	9	7	11	12
Sun	6	9	8	7	8	8	7	11	10	8	13	10
Mercury	10	8	10	6	7	7	9	11	11	4	14	8
Venus	9	13	9	5	6	9	7	11	10	10	6	10
Mars	11	12	5	11	12	7	5	10	4	12	8	8
Jupiter	9	11	8	10	7	6	10	5	8	9	13	9
Saturn	9	9	6	8	9	10	4	11	9	10	7	13
TOTAL	65	74	50	52	54	54	53	70	61	60	72	70
N. Node	10	12	5	15	7	4	9	8	12	11	2	10

Planets

Aspects	Parallels	4 Axis	Direction	☉/☽ Midpoint
Mercury & Venus	Moon-Mercury	Uranus (IC)		Mercury
Mercury-Jupiter	Mercury-Jupiter			Chiron
Mercury-Saturn				
Mercury-Neptune				Moon's Nodes
Moon-Venus				Mercury (N)
Sun-Venus				Venus (N)

ENGINEERS

Theme: Mercury/Jupiter/Uranus

Mercury:
Mercury aspects are frequent. Engineers are highly mentally oriented people who love learning, solving problems, and do well in school. Mercury with Jupiter and Uranus indicates a high confidence about their intelligence and a love for exploration and invention.

Uranus:
Uranus represents originality and inventiveness. Engineers solve unique problems for which there is no precedent.

Mars in Gemini (1 in 239):
This position indicates being confident and assertive with problem solving activities.

7th House is weak (1 in 5,436):
The 7th house is the place that describes our relationships. Engineers spend a great deal of time working in isolation.

Signs

Size: 141	1st Harmonic	9th Harmonic	Houses	Rising Sign
Moon				Taurus
Sun		Virgo	12th	Leo
Mercury				Lunar Phase
Venus	Aries		1st	Cancer
Mars	Gemini		12th	☉/☽ Midpoint
Jupiter				Scorpio
Saturn		Virgo	6th	Moon's Nodes
TOTAL				Leo
(weak)	Libra		7th	6th, 12th

1H	♈	♉	♊	♋	♌	♍	♎	♏	♐	♑	♒	♓
Moon	7	13	13	7	11	15	14	11	11	14	13	12
Sun	16	14	10	13	15	10	3	13	14	12	11	10
Mercury	12	15	13	9	11	13	5	11	16	16	10	10
Venus	18	12	13	14	9	14	6	12	11	4	15	13
Mars	10	7	21	11	9	17	13	12	13	8	9	11
Jupiter	16	9	7	12	11	11	16	12	7	13	11	16
Saturn	12	14	12	11	6	13	14	12	13	12	13	9
TOTAL	91	84	89	77	72	93	71	83	85	79	82	81
Rising	8	13	12	13	23	19	13	9	8	11	4	8
N. Node	17	6	4	13	20	10	11	17	11	11	7	14

9H	♈	♉	♊	♋	♌	♍	♎	♏	♐	♑	♒	♓
Moon	14	8	12	13	16	9	8	13	11	8	12	17
Sun	6	8	10	11	9	19	13	12	15	17	12	9
Mercury	15	8	9	15	7	8	16	9	11	13	17	13
Venus	6	10	7	12	15	15	6	17	15	16	15	7
Mars	16	11	14	14	14	7	12	12	11	5	15	10
Jupiter	10	17	11	12	7	15	13	12	10	11	10	13
Saturn	11	14	11	11	12	18	12	12	13	10	11	6
TOTAL	78	76	74	88	80	91	80	87	86	80	92	75

Houses	1st	2nd	3rd	4th	5th	6th	7th	8th	9th	10th	11th	12th
Moon	8	11	16	15	11	16	13	11	13	8	17	2
Sun	17	14	10	12	11	3	6	12	9	12	14	21
Mercury	18	14	11	12	9	7	6	8	12	17	11	16
Venus	19	17	10	9	10	11	3	8	14	13	12	15
Mars	12	15	8	10	5	14	8	13	8	17	12	19
Jupiter	10	11	13	16	9	10	6	16	14	15	9	12
Saturn	12	6	13	6	9	19	5	16	12	13	15	15
TOTAL	96	88	81	80	64	80	47	84	82	95	90	100
N. Node	10	9	9	12	17	19	11	7	10	11	7	19

Planets

Aspects	Parallels	4 Axis	Direction	☉/☽ Midpoint
Moon-Sun		Jupiter (Asc)	Jupiter (D)	Neptune
Mercury-Jupiter		Uranus (MC)	Uranus (D)	
Mercury-Uranus		Chiron (MC)		
Mercury-Neptune				**Moon's Nodes**
Mercury-Chiron				Venus (N)

FASHION DESIGNERS

Theme: Cancer/Pisces

Cancer & Pisces:
These signs are creative and stylish. Fashion Designers use their sensitivity and artistry to appeal to the tastes of consumers.

Venus on the Sun/Moon midpoint (1 in 1,538):
Venus represents artistry and beauty, as well as an interest in being with attractive people. Venus on the Sun/Moon midpoint shows that a person's central energies are focused on Venusian activities.

Mars Retrograde (1 in 1,987):
This indicates a lack of aggressive, masculine energy.

Signs

Size: 113	1st Harmonic	9th Harmonic	Houses	Rising Sign
Moon	Sagittarius		7th	Cancer
Sun	Taurus, Cancer, Pisces	Cancer	6th	
Mercury	Taurus	Pisces		Lunar Phase
Venus	Cancer		12th	
Mars	Aries			☉/☽ Midpoint
Jupiter	Virgo	Cancer		
Saturn				Moon's Nodes
TOTAL	♉ ♋ ♌ ♓	Cancer, Leo	6th, 12th	Pisces
(weak)	Libra, Scorpio, Capr	Taurus		

1H	♈	♉	♊	♋	♌	♍	♎	♏	♐	♑	♒	♓
Moon	7	12	8	14	8	10	9	4	17	6	5	13
Sun	5	17	7	15	13	9	4	4	13	6	6	14
Mercury	9	15	9	10	7	14	4	6	14	6	8	11
Venus	12	9	11	16	12	10	5	9	3	10	9	7
Mars	13	11	11	10	14	10	9	7	9	3	7	9
Jupiter	11	9	4	5	13	16	10	13	8	5	9	10
Saturn	6	6	7	10	12	10	7	10	10	8	15	12
TOTAL	63	79	57	80	79	79	48	53	74	44	59	76
Rising	4	9	9	18	16	6	17	12	5	9	4	4
N. Node	7	8	11	12	11	8	12	11	5	7	7	14

9H	♈	♉	♊	♋	♌	♍	♎	♏	♐	♑	♒	♓
Moon	12	8	11	11	14	5	10	8	8	8	10	8
Sun	5	8	9	17	11	9	8	12	9	7	13	5
Mercury	8	7	12	10	11	7	5	8	10	8	11	16
Venus	9	6	12	13	14	7	7	7	12	10	11	5
Mars	11	13	8	11	9	14	9	10	8	8	8	4
Jupiter	7	9	12	15	14	8	7	9	12	8	7	5
Saturn	8	9	13	7	9	12	12	8	9	8	6	12
TOTAL	60	60	77	84	82	62	58	62	68	57	66	55

Houses	1st	2nd	3rd	4th	5th	6th	7th	8th	9th	10th	11th	12th
Moon	6	14	4	8	10	11	15	5	12	7	10	11
Sun	9	9	13	6	4	13	4	10	9	10	13	13
Mercury	11	11	9	5	6	12	6	9	8	13	11	12
Venus	4	11	8	7	10	9	8	6	10	11	11	18
Mars	14	9	12	8	5	11	5	11	11	6	9	12
Jupiter	4	13	8	12	10	11	8	9	8	9	7	14
Saturn	12	11	9	13	9	11	13	5	9	3	8	10
TOTAL	60	78	63	59	54	78	59	55	67	59	69	90
N. Node	7	7	10	8	6	12	9	6	10	14	15	9

Planets

Aspects	Parallels	4 Axis	Direction	☉/☽ Midpoint
Moon-Uranus		Mars (Asc)	Mars (Rx)	Venus
		Pluto (Asc)		Chiron
		Mercury (MC)		
		Uranus (MC)		**Moon's Nodes**
				Moon (N)
				Saturn (N)

203

FIGHTERS – BOXERS, MARTIAL ARTISTS, AND WRESTLERS

Theme: Virgo (6th), Scorpio (8th)

Scorpio & Capricorn:
Capricorn is strong in the 1st harmonic, especially for the Sun (1 in 1,267) and Mercury (1 in 10,135). Scorpio is strong in the 9th harmonic. These signs represent the life of fighters: struggle and suffering, will power and endurance, discipline and work ethic. Fighters, like Murderers, have Scorpio strong in the 9th harmonic, meaning they are both inwardly intense and inclined towards the dark side of life. Outwardly (1H), however, Fighters are strong in Capricorn instead of Taurus (Murderers are strong in Taurus and weak in Capricorn). Fighters seek an outer expression for their intensity in a form that is socially respectable (Capricorn) instead of merely personally satisfying (Taurus).

6th House:
The 6th house represents health problems and fighters are constantly being injured. The Sun (1 in 561) and Mercury (1 in 341) are most frequent.

Mars-Chiron aspects:
This aspect indicates a competitive drive to win (Mars) by wounding others (Chiron), and enduring frequent injuries as well.

Signs

Size: 103	1st Harmonic	9th Harmonic	Houses	Rising Sign
Moon	Gemini, Pisces	Libra, Scorpio	3rd	Aries
Sun	Capricorn	Virgo	5th, 6th, 8th	
Mercury	Capricorn		5th, 6th	Lunar Phase
Venus			8th	
Mars		Leo, Virgo	8th	☉/☽ Midpoint
Jupiter			9th	Aries, Pisces
Saturn		Scorpio	11th	Aquarius
TOTAL	Capricorn	Scorpio	5th, 6th	Moon's Nodes
(weak)	Libra		7th	

1H	♈	♉	♊	♋	♌	♍	♎	♏	♐	♑	♒	♓
Moon	10	6	14	5	9	10	4	6	9	8	8	14
Sun	9	10	10	10	6	6	4	7	6	17	9	9
Mercury	6	11	7	5	11	7	5	6	7	20	10	8
Venus	11	9	10	12	3	8	6	5	11	8	13	7
Mars	11	12	7	12	9	7	5	8	11	6	11	4
Jupiter	5	10	9	8	10	11	11	9	9	7	3	11
Saturn	14	8	7	8	1	8	11	10	11	7	10	8
TOTAL	66	66	64	60	49	57	46	51	64	73	64	61
Rising	9	6	5	12	13	9	13	8	13	5	9	1
N. Node	12	6	4	9	8	11	11	8	8	8	11	7

9H	♈	♉	♊	♋	♌	♍	♎	♏	♐	♑	♒	♓
Moon	8	7	5	9	6	11	15	15	4	6	11	6
Sun	5	3	11	8	8	16	6	11	8	11	7	9
Mercury	9	11	9	10	6	8	4	8	10	8	13	7
Venus	7	11	6	11	10	7	9	8	9	7	13	5
Mars	10	5	8	7	14	14	12	12	4	5	4	8
Jupiter	9	9	9	12	8	7	6	10	11	5	10	7
Saturn	7	8	10	9	9	6	9	14	4	8	9	10
TOTAL	55	54	58	66	61	69	61	78	50	50	67	52

Houses	1st	2nd	3rd	4th	5th	6th	7th	8th	9th	10th	11th	12th
Moon	7	8	14	9	8	10	7	9	10	4	11	6
Sun	9	8	7	5	14	15	2	13	6	5	8	11
Mercury	11	5	8	9	13	13	4	7	10	9	5	9
Venus	10	8	7	12	8	8	5	12	6	11	10	6
Mars	6	8	8	4	9	11	8	13	6	12	7	11
Jupiter	8	8	10	10	9	11	6	6	14	10	9	2
Saturn	7	9	10	10	8	10	5	7	7	8	14	8
TOTAL	58	54	64	59	69	78	37	67	59	59	64	53
N. Node	5	4	10	7	13	10	7	10	13	11	6	7

Planets

Aspects	Parallels	4 Axis	Direction	☉/☽ Midpoint
Moon-Neptune	Moon-Mars	Venus (Dsc)		Saturn
Mars-Chiron		Pluto (Dsc)		
		Pluto (MC)		
		Moon (IC)		**Moon's Nodes**
		Saturn (IC)		Uranus (S)
		Neptune (IC)		

Heart Attack Victims

Theme: Aries/Scorpio, 3rd House, Saturn

Aries & Scorpio:
Scorpio is strong in the 1st harmonic and Aries is strong in the 9th harmonic (1 in 571). Aries and Scorpio are the two signs most associated with Mars. They are both intense, aggressive signs and are the most likely to view life in terms of competition and conflict. Apparently, this puts great strain on the heart.

Saturn:
Saturnine people are serious, responsible, and burdened. Moon-Saturn aspects (1 in 504) are the most frequent. This is an indication of emotional grief.

Moon-Uranus aspects and parallels:
This indicates that a person may be prone to sudden emotional shocks and disruptions to their emotional life.

Mars-Chiron parallels (1 in 1,169):
Mars in relation with Chiron is an indication of sudden injuries.

Signs

Size: 130	1st Harmonic	9th Harmonic	Houses	Rising Sign
Moon	Cancer		3rd	Gemini
Sun	Scorpio	Pisces	10th	
Mercury	Scorpio			Lunar Phase
Venus		Aries, Taurus		Gemini
Mars		Aries, Capricorn		☉/☽ Midpoint
Jupiter				
Saturn	Cancer	Sagittarius	3rd, 8th	Moon's Nodes
TOTAL	Virgo, Scorpio	Aries, Taurus	3rd	Virgo
(weak)	Aries			12th

1H	♈	♉	♊	♋	♌	♍	♎	♏	♐	♑	♒	♓
Moon	9	8	7	16	12	14	10	10	15	7	10	12
Sun	9	9	10	6	13	12	10	16	10	11	15	9
Mercury	6	10	9	10	5	13	6	18	15	15	12	11
Venus	14	12	5	13	9	13	12	11	10	10	13	8
Mars	8	6	12	10	13	18	14	10	12	7	10	10
Jupiter	6	8	12	6	6	16	13	15	14	13	9	12
Saturn	5	9	7	16	11	9	13	12	10	13	12	13
TOTAL	57	62	62	77	69	95	78	92	86	76	81	75
Rising	7	6	18	8	18	13	13	12	8	10	10	7
N. Node	7	9	13	8	15	19	12	6	10	9	15	7

9H	♈	♉	♊	♋	♌	♍	♎	♏	♐	♑	♒	♓
Moon	13	10	10	7	14	12	11	11	10	10	9	13
Sun	15	13	8	13	8	9	8	9	7	9	15	16
Mercury	13	15	10	9	10	13	7	10	7	10	15	11
Venus	17	16	7	12	15	8	13	9	8	12	6	7
Mars	18	12	9	7	14	8	10	6	15	16	2	13
Jupiter	13	14	9	13	9	9	8	12	9	12	9	13
Saturn	11	13	13	10	12	13	9	8	17	8	9	7
TOTAL	100	93	66	71	82	72	66	65	73	77	65	80

Houses	1st	2nd	3rd	4th	5th	6th	7th	8th	9th	10th	11th	12th
Moon	6	6	17	12	11	15	6	12	14	3	15	13
Sun	10	11	12	13	9	9	6	12	9	17	9	13
Mercury	7	15	11	14	6	10	7	9	12	16	13	10
Venus	9	9	15	6	12	12	14	9	12	9	14	9
Mars	11	13	8	12	9	10	11	8	7	13	13	15
Jupiter	12	9	12	12	10	15	7	12	8	12	11	10
Saturn	16	8	17	8	12	12	9	17	13	4	5	9
TOTAL	71	71	92	77	69	83	60	79	75	74	80	79
N. Node	9	14	11	10	15	6	9	11	9	5	11	20

Planets

Aspects	Parallels	4 Axis	Direction	☉/☽ Midpoint
Saturn	Mercury-Chiron	Mars (Asc)	Jupiter (D)	Saturn
Moon-Saturn	Mars-Chiron		Saturn (D)	
Moon-Uranus	Moon-Uranus			Moon's Nodes

Homosexual Men

Theme: Scorpio/Capricorn, 5th House, Jupiter

Scorpio and Capricorn:
They are burdened (Capricorn) by sexual and emotional struggles (Scorpio). The same pair of signs is strong with Fighters. This is an indication of the embattled nature of being a gay man in the 20th century.

Venus in Capricorn:
Venus in Capricorn is strong in the 9th harmonic (1 in 4,321) and weak in Cancer in the 9th harmonic (1 in 859). Venus in Capricorn represents caution and restriction in relationships (as opposed to warm, caring relationships for Cancer). This is an indication of the difficulties many homosexuals have in achieving total acceptance for their sexual preferences.

5th House:
The 5th house represents romance and creativity. The Sun is especially frequent in this space (1 in 272).

Mercury-Venus aspects (1 in 10,000):
This indicates a harmonious, considerate way of relating to people. Mercury is a gender neutral planet, indicating relationships that are less sex-role identified. It also indicates a great deal of thinking about relationships (preferences).

Jupiter-Moon aspects & Jupiter on the Imum Coeli:
Jupiter represents the need for expansion, growth, and freedom. Jupiter with the Moon and Imum Coeli indicates the urge to expand beyond early life conditions and, often move away from home.

Size: 276	1st Harmonic	9th Harmonic	Houses	Rising Sign
Moon				
Sun	Aquarius	Taurus	5th	
Mercury	Scorpio, Capricorn	Virgo	5th	Lunar Phase
Venus	Capricorn	Scorpio, Capricorn	5th	
Mars	Aries, Taurus	Scorpio		☉/☽ Midpoint
Jupiter	Scorpio	Capricorn	4th	Pisces
Saturn	Leo			Moon's Nodes
TOTAL	Scorpio, Capricorn	Scorpio, Capricorn	5th	Capricorn (S)
(weak)	Virgo	Libra		5th (S), 6th

Signs

1H	♈	♉	♊	♋	♌	♍	♎	♏	♐	♑	♒	♓
Moon	29	26	27	22	18	15	23	22	21	24	19	30
Sun	20	22	19	29	19	16	27	22	21	28	31	22
Mercury	23	21	19	24	18	16	22	34	19	36	21	23
Venus	24	26	22	21	21	28	20	20	17	30	24	23
Mars	29	28	20	26	27	17	24	23	25	10	24	23
Jupiter	28	16	19	21	14	29	24	40	30	26	15	14
Saturn	17	20	20	25	29	18	23	25	33	30	21	15
TOTAL	170	159	146	168	146	139	163	186	166	184	155	150
Rising	13	14	27	25	29	31	36	29	23	13	18	18
N. Node	14	25	27	35	17	28	29	16	21	22	22	20

9H	♈	♉	♊	♋	♌	♍	♎	♏	♐	♑	♒	♓
Moon	25	17	17	26	23	16	21	29	29	25	23	25
Sun	21	33	14	29	21	13	26	27	20	24	21	27
Mercury	15	24	21	29	15	33	19	26	26	16	30	22
Venus	20	21	29	9	19	22	16	31	23	39	25	22
Mars	27	17	26	22	26	28	19	35	22	22	16	16
Jupiter	27	14	27	24	29	22	14	21	23	31	21	23
Saturn	23	27	26	18	21	27	19	24	24	26	26	15
TOTAL	158	153	160	157	154	161	134	193	167	183	162	150

Houses	1st	2nd	3rd	4th	5th	6th	7th	8th	9th	10th	11th	12th
Moon	22	25	27	24	27	22	15	20	24	18	27	25
Sun	23	23	28	24	31	17	21	20	19	19	23	28
Mercury	21	29	24	26	31	27	16	19	16	18	28	21
Venus	28	23	29	20	29	22	21	23	17	21	25	18
Mars	22	23	22	24	21	22	25	26	24	24	22	21
Jupiter	26	14	29	33	21	21	20	22	26	25	21	18
Saturn	27	13	22	23	22	28	26	23	23	28	17	24
TOTAL	169	150	181	174	182	159	144	153	149	153	163	155
N. Node	17	24	29	25	15	31	23	12	24	22	34	20

Planets

Aspects	Parallels	4 Axis	Direction	☉/☽ Midpoint
Saturn		Jupiter (IC)	Mercury (D)	Venus
Moon-Jupiter				
Mercury-Venus				
Jupiter-Saturn				Moon's Nodes

INFANT MORTALITY

Theme: Cancer, 6th House

Cancer:
Cancer is strong and Capricorn is weak in the 1st harmonic. Cancer is the sign most strongly associated with childhood. Capricorn is most associated with adulthood. These children do not become adults.

6th House (1 in 2,821):
The 6th house represents health issues. Almost every planet is frequent here, especially Mercury (1 in 1,133).

Moon-Saturn aspects and parallels:
This planetary combination is an indication of emotional despondency and early life difficulties.

Chiron on the Moon's North Node (1 in 609):
This position indicates that there is a wound in attaining a life purpose.

Signs

Size: 179	1st Harmonic	9th Harmonic	Houses	Rising Sign
Moon				Cancer
Sun			6th	Lunar Phase
Mercury	Cancer, Pisces		4th, 6th	Gemini
Venus		Sagittarius	6th	Aquarius
Mars	Sagittarius	Pisces	10th	☉/☽ Midpoint
Jupiter	Aries, Taurus, Cancer	Taurus		
Saturn	Gemini, Cancer, Sag		5th, 6th	Moon's Nodes
TOTAL	Cancer	Taurus	6th	Gemini
(weak)	Capricorn, Aquarius	Libra	1st, 9th, 12th	6th (S)

1H	♈	♉	♊	♋	♌	♍	♎	♏	♐	♑	♒	♓
Moon	11	14	16	14	15	12	17	14	19	12	18	17
Sun	20	13	17	12	10	20	16	18	9	13	12	19
Mercury	15	11	12	22	10	11	17	21	12	10	14	24
Venus	19	17	14	14	15	21	14	17	14	11	10	13
Mars	14	9	20	23	18	16	16	10	22	11	9	11
Jupiter	21	24	7	22	20	22	17	21	1	7	5	12
Saturn	13	19	27	23	17	4	2	5	33	10	18	8
TOTAL	113	107	113	130	105	106	99	106	110	74	86	104
Rising	4	7	16	27	21	25	14	16	22	11	10	6
N. Node	7	14	55	16	20	11	12	10	6	5	8	15

9H	♈	♉	♊	♋	♌	♍	♎	♏	♐	♑	♒	♓
Moon	18	12	7	18	10	13	13	20	19	15	16	18
Sun	13	18	17	14	15	18	15	14	8	15	14	18
Mercury	15	15	10	18	12	18	18	16	10	16	16	15
Venus	19	19	9	16	19	15	9	14	23	9	13	14
Mars	19	19	17	12	9	16	12	14	12	13	15	21
Jupiter	17	23	14	16	10	10	11	11	18	18	14	17
Saturn	16	19	18	10	18	16	10	18	18	15	16	5
TOTAL	117	125	92	104	93	106	88	107	108	101	104	108

Houses	1st	2nd	3rd	4th	5th	6th	7th	8th	9th	10th	11th	12th
Moon	11	16	14	13	17	12	16	19	15	11	20	15
Sun	14	17	17	20	14	21	17	13	9	18	8	11
Mercury	15	17	15	22	11	24	16	14	9	16	11	9
Venus	8	20	19	17	16	19	14	13	13	14	13	13
Mars	15	13	16	13	10	16	14	18	12	23	16	13
Jupiter	10	18	15	15	10	15	18	20	12	14	19	13
Saturn	16	18	11	11	22	23	12	11	9	14	16	16
TOTAL	89	119	107	111	100	130	107	108	79	110	103	90
N. Node	5	17	18	5	17	12	17	16	19	15	17	21

Planets

Aspects	Parallels	4 Axis	Direction	☉/☽ Midpoint
Moon-Saturn	Moon-Saturn	Neptune (MC)	Chiron (D)	
Sun-Jupiter		Venus (IC)		
Venus-Jupiter				
				Moon's Nodes
				Sun (N)
				Chiron (N)

INHERITANCE

Theme: Capricorn, 8th House, Venus

Capricorn (1 in 555):
Capricorns often have a strong relationship with their parents since they identify with the parental role. They would be likely to continue any business in which their family is involved. Mercury (1 in 1,254) and Venus (1 in 1,072) are strong in Capricorn. This suggests having a close connection with parents and traditional ideas and values.

8th House:
The 8th house represents wealth gained from other people such as inheritances.

Venus-Saturn aspects (1 in 8,196):
This indicates having long-term stable relationships and gaining benefits from elders, as well as being reliable with money.

Sun on the Imum Coeli (1 in 9,640):
The Sun and Venus are frequent on the Imum Coeli. These beneficial planets here indicate pleasure and success derived through one's home environment.

Signs

Size: 123	1st Harmonic	9th Harmonic	Houses	Rising Sign
Moon	Leo	Gemini		Taurus
Sun	Aquarius	Aquarius		
Mercury	Capricorn		4th	Lunar Phase
Venus	Capricorn		8th	
Mars	Aries			☉/☽ Midpoint
Jupiter				Aquarius
Saturn	Gemini	Gemini		Moon's Nodes
TOTAL	Capricorn	Virgo	8th	8th, 10th
(weak)		Leo	2nd	

1H	♈	♉	♊	♋	♌	♍	♎	♏	♐	♑	♒	♓
Moon	8	10	8	11	17	11	5	10	10	11	9	13
Sun	11	7	8	8	5	9	12	13	10	12	18	10
Mercury	4	11	4	8	6	9	10	11	15	21	11	13
Venus	8	9	10	8	7	15	7	7	10	18	15	9
Mars	15	5	10	14	7	12	11	12	8	8	10	11
Jupiter	10	12	8	8	10	8	14	16	11	8	11	7
Saturn	11	2	14	5	10	11	13	11	13	16	10	7
TOTAL	67	56	62	62	62	75	72	80	77	94	84	70
Rising	9	13	7	6	7	13	15	16	15	10	10	2
N. Node	14	13	9	10	9	10	8	9	10	11	10	10

9H	♈	♉	♊	♋	♌	♍	♎	♏	♐	♑	♒	♓
Moon	6	11	19	14	9	11	6	10	6	12	8	11
Sun	10	11	6	8	8	13	10	11	9	10	16	11
Mercury	10	8	10	13	6	15	10	11	11	10	10	9
Venus	7	9	13	13	6	9	7	10	8	15	13	13
Mars	6	11	8	10	12	11	13	9	12	8	13	10
Jupiter	13	15	6	6	7	12	14	6	14	13	8	9
Saturn	12	9	16	8	8	14	12	10	10	12	4	8
TOTAL	64	74	78	72	56	85	72	67	70	80	72	71

Houses	1st	2nd	3rd	4th	5th	6th	7th	8th	9th	10th	11th	12th
Moon	15	5	12	10	11	9	8	12	10	7	12	12
Sun	4	14	14	11	8	10	12	12	7	6	12	13
Mercury	9	13	12	16	6	7	13	12	9	7	12	7
Venus	15	7	11	14	12	4	10	14	11	8	8	9
Mars	7	8	11	10	12	10	11	14	14	10	8	8
Jupiter	6	5	13	12	7	10	9	12	9	15	10	15
Saturn	11	9	7	10	11	11	12	8	13	9	8	14
TOTAL	67	61	80	83	67	61	75	84	73	62	70	78
N. Node	12	9	6	9	8	8	9	15	13	17	9	8

Planets

Aspects	Parallels	4 Axis	Direction	☉/☽ Midpoint
Venus-Saturn	Mercury-Venus	Neptune (Dsc)	Pluto (Rx)	
		Pluto (Dsc)		
		Sun (IC)		
		Venus (IC)		Moon's Nodes
				Mars (N)

INSTRUMENTALISTS

Theme: Sagittarius, Neptune

Sagittarius:
Sagittarians are renowned for their enthusiasm and ability to spread joy. For this reason, they are known as the bard. Instrumentalists are frequently born during the Sagittarius lunar phase (1 in 199) - the ninth 30 degree section of the Lunation cycle.

Venus in Taurus (1 in 527):
This a beneficial position indicating talent in the arts used for the enjoyment of the senses.

Jupiter in Cancer (1 in 1,197):
This position shows an expanded capacity for sensitivity and rich feelings.

Neptune:
Neptune represents dreams and illusions. Instrumentalists have the ability to send us into a state of reverie with their music. Neptune aspects and Venus-Neptune parallels (1 in 771) are most frequent.

Signs

Size: 418	1st Harmonic	9th Harmonic	Houses	Rising Sign
Moon	Pisces			Aries
Sun	Sagittarius		7th	
Mercury		Leo, Sagittarius	1st	Lunar Phase
Venus	Taurus	Capricorn		Sagittarius
Mars	Aries	Scorpio	3rd, 10th	☉/☽ Midpoint
Jupiter	Cancer	Virgo	12th	
Saturn	Gemini	Virgo		Moon's Nodes
TOTAL		Sagittarius	1st, 10th	7th
(weak)			3rd	

1H	♈	♉	♊	♋	♌	♍	♎	♏	♐	♑	♒	♓
Moon	36	30	29	33	31	26	37	42	31	39	40	44
Sun	42	35	29	33	36	32	40	25	43	36	24	43
Mercury	37	40	19	36	27	31	42	39	36	42	36	33
Venus	31	53	29	36	29	36	31	35	33	35	30	40
Mars	40	34	41	31	45	43	39	41	33	25	24	22
Jupiter	26	31	30	51	34	35	34	37	36	27	41	36
Saturn	40	28	45	27	40	34	32	38	34	38	27	35
TOTAL	252	251	222	247	242	237	255	257	246	242	222	253
Rising	27	26	33	30	54	38	53	47	38	28	24	20
N. Node	44	36	42	33	36	43	32	32	27	25	31	37

9H	♈	♉	♊	♋	♌	♍	♎	♏	♐	♑	♒	♓
Moon	38	31	35	34	33	42	30	38	35	32	35	35
Sun	33	40	37	23	34	30	38	39	34	30	40	40
Mercury	29	42	34	43	45	24	28	32	46	37	37	21
Venus	37	28	36	35	24	28	36	36	37	46	35	40
Mars	39	36	35	31	39	28	29	44	40	29	39	29
Jupiter	31	38	28	40	27	48	34	33	40	30	38	31
Saturn	33	35	28	29	39	52	36	34	40	43	25	24
TOTAL	240	250	233	235	241	252	231	256	272	247	249	220

Houses	1st	2nd	3rd	4th	5th	6th	7th	8th	9th	10th	11th	12th
Moon	38	31	32	39	37	24	34	29	39	39	43	33
Sun	47	36	29	37	25	32	41	29	22	42	37	41
Mercury	56	32	30	39	27	35	38	23	32	31	35	40
Venus	42	42	30	27	30	30	36	33	35	34	33	46
Mars	27	23	45	39	37	30	30	36	35	50	38	28
Jupiter	36	32	27	36	33	32	31	35	35	43	33	45
Saturn	43	40	30	44	40	25	30	40	30	35	34	27
TOTAL	289	236	223	261	229	208	240	225	228	274	253	260
N. Node	32	29	35	41	30	29	45	36	39	30	38	34

Planets

Aspects	Parallels	4 Axis	Direction	☉/☽ Midpoint
Neptune	Venus-Neptune		Chiron (D)	Saturn
Sun-Neptune				
Mercury-Neptune				
Moon-Uranus				Moon's Nodes
				Moon (N)
				Mars (S)

INTELLIGENCE QUOTIENT (HIGH IQ)

Theme: Libra

This group is composed of people with Mensa level IQ's (130 and above), the top 2% of the population.

Libra:
Libra, being the Cardinal Air sign, represents active intelligence. The Moon's South Node is frequent in Libra and the 7th house, indicating a background in intellectual abilities. Aquarius, the Fixed Air sign, is also frequent and represents enjoyment of intellectual activities.

Moon in Fire signs:
Moon is Sagittarius (1 in 2,738) in the 1st harmonic and Leo (1 in 675) in the 9th harmonic. This is indicative of having a positive family background that instilled confidence in the person. They have good emotional resilience and buoyancy.

Mercury in Fixed Houses (1 in 152):
This indicates pleasure through learning and communication.

Signs

Size: 135	1st Harmonic	9th Harmonic	Houses	Rising Sign
Moon	Sagittarius	Leo		Pisces
Sun	Pisces	Cancer, Libra		
Mercury		Pisces		Lunar Phase
Venus			2nd	
Mars	Gemini			☉/☽ Midpoint
Jupiter	Leo			
Saturn	Virgo, Libra	Taurus	7th	Moon's Nodes
TOTAL	Libra, Aquarius			Libra (S)
(weak)	Taurus			7th (S)

1H	♈	♉	♊	♋	♌	♍	♎	♏	♐	♑	♒	♓
Moon	7	15	10	8	7	9	14	10	22	8	13	12
Sun	13	4	8	15	10	12	11	14	9	6	16	17
Mercury	6	8	8	10	8	13	11	17	9	15	15	15
Venus	10	10	12	13	10	12	12	12	10	7	17	10
Mars	11	3	17	17	13	11	12	14	13	9	10	5
Jupiter	12	9	6	9	17	10	17	12	14	11	9	9
Saturn	11	8	10	9	7	16	20	7	8	13	15	11
TOTAL	70	57	71	81	72	83	97	86	85	69	95	79
Rising	7	7	14	11	8	11	16	14	13	14	9	11
N. Node	18	8	11	7	15	13	8	11	8	10	16	10

9H	♈	♉	♊	♋	♌	♍	♎	♏	♐	♑	♒	♓
Moon	11	6	10	6	21	10	16	14	11	8	13	9
Sun	14	13	6	17	4	9	17	12	11	11	11	10
Mercury	6	11	10	10	15	14	5	7	12	14	13	18
Venus	9	15	12	15	4	11	10	15	11	10	9	14
Mars	11	8	12	10	15	7	12	13	9	12	13	13
Jupiter	10	16	8	13	10	12	13	6	13	12	14	8
Saturn	11	17	11	12	12	14	9	9	5	13	12	10
TOTAL	72	86	69	83	81	77	82	76	72	80	85	82

Houses	1st	2nd	3rd	4th	5th	6th	7th	8th	9th	10th	11th	12th
Moon	8	11	12	9	15	10	11	15	11	8	13	12
Sun	14	16	13	11	9	9	9	10	10	11	9	14
Mercury	14	17	12	13	12	4	9	12	9	7	17	9
Venus	14	19	12	10	5	11	8	13	10	5	13	15
Mars	9	12	8	11	8	14	12	12	7	14	14	14
Jupiter	14	11	13	9	7	13	11	7	12	15	16	7
Saturn	13	9	15	5	13	5	17	10	9	14	10	15
TOTAL	86	95	85	68	69	66	77	79	68	74	92	86
N. Node	19	10	13	8	12	14	10	9	10	5	14	11

Planets

Aspects	Parallels	4 Axis	Direction	☉/☽ Midpoint
Mercury-Mars	Mars-Chiron	Uranus (Asc)	Chiron (D)	Neptune
		Pluto (Dsc)		
				Moon's Nodes
				Venus (S)

JOURNALISTS

Theme: Virgo (6th)

Virgo & the 6th House:
Virgo and the 6th house represent analytical work. Journalists research and write in order to present clear information to the public. The Sun (1 in 438) and Mercury (1 in 317) are the most frequent planets in the 6th house.

Mercury is frequent on the Moon's South Node (1 in 865) and the Descendant:
Journalists have a strong background in writing and communication, and are adept at getting others to share information with them.

Neptune is frequent on the Ascendant (1 in 1,437) and the Descendant:
Neptune represents creative illusions. Neptune frequent on the Asc-Dsc axis reveals that many Journalists have high idealism, yet a tendency to embellish information gained from dubious sources.

Signs

Size: 153	1st Harmonic	9th Harmonic	Houses	Rising Sign
Moon	Gemini	Taurus, Sagittarius		
Sun		Cancer	6th	
Mercury		Pisces	6th	Lunar Phase
Venus	Virgo, Sagittarius			Pisces
Mars	Aquarius	Aries	7th	☉/☽ Midpoint
Jupiter		Taurus	9th	
Saturn	Scorpio	Virgo	1st, 10th	Moon's Nodes
TOTAL			6th	Leo
(weak)			1st	7th

1H	♈	♉	♊	♋	♌	♍	♎	♏	♐	♑	♒	♓
Moon	15	11	20	7	14	12	15	11	10	16	9	13
Sun	12	17	11	10	15	11	13	12	11	17	13	11
Mercury	10	12	9	10	8	15	16	12	14	16	15	16
Venus	16	16	9	15	12	19	4	13	19	10	11	9
Mars	7	10	11	14	17	12	20	10	13	11	16	12
Jupiter	12	9	11	15	18	16	14	12	7	17	10	12
Saturn	11	11	11	12	8	12	14	25	13	13	10	13
TOTAL	83	86	82	83	92	97	96	95	87	100	84	86
Rising	5	10	9	13	17	13	19	18	16	16	11	6
N. Node	10	10	15	15	20	11	8	14	10	13	11	16

9H	♈	♉	♊	♋	♌	♍	♎	♏	♐	♑	♒	♓
Moon	12	19	11	12	9	12	13	16	20	13	8	8
Sun	10	10	11	19	14	18	12	15	10	13	12	9
Mercury	8	15	10	10	12	13	10	15	10	17	13	20
Venus	10	15	15	15	12	14	12	13	10	10	14	13
Mars	19	7	15	13	10	11	12	8	15	14	17	12
Jupiter	16	19	8	12	15	12	14	13	6	11	13	14
Saturn	11	15	13	11	15	22	3	13	13	9	16	12
TOTAL	86	100	83	92	87	102	76	93	84	87	93	88

Houses	1st	2nd	3rd	4th	5th	6th	7th	8th	9th	10th	11th	12th
Moon	11	17	14	10	14	11	12	13	10	16	7	18
Sun	12	18	14	18	9	20	11	7	9	12	11	12
Mercury	6	18	17	16	12	20	10	5	14	8	16	11
Venus	8	12	17	13	14	12	13	14	8	12	17	13
Mars	12	13	16	13	9	15	20	11	8	15	10	11
Jupiter	10	13	10	15	15	10	16	12	18	13	14	7
Saturn	19	13	9	7	11	15	14	11	14	18	12	10
TOTAL	78	104	97	92	84	103	96	73	81	94	87	82
N. Node	13	13	10	13	8	13	18	14	14	12	13	12

Planets

Aspects	Parallels	4 Axis	Direction	☉/☽ Midpoint
Sun-Chiron		Neptune (Asc)	Venus (D)	
Jupiter-Saturn		Mercury (Dsc)		
		Neptune (Dsc)		
		Chiron (MC)		**Moon's Nodes**
				Mercury (S)
				Pluto (N)

LAWYERS

Theme: Libra (7th), Aquarius (11th), Neptune

Leo:
Leo is strong (1 in 772), especially for the Sun (1 in 266). Leos have great personal magnetism, charisma, and leadership ability. Most of the lawyers in this group lived in the 20th century. This was a time when the ultimate career success was in being a doctor or lawyer. Therefore, lawyers were greatly looked up to as would befit a Leo.

7th & 11th Houses:
The 7th and 11th houses are both Air houses. Lawyers are masters of communication and representation of information. Venus (1 in 187) and Jupiter (1 in 394) are the most frequent planets in the 11th house.

Moon on the Descendant (1 in 16,265):
Lawyers encounter people who are in need of their support and are the protectors of their client's interests.

Mercury-Neptune aspects and parallels:
Neptune represents illusions and deception. Mercury with Neptune shows that many lawyers may misrepresent information.

Mars-Chiron parallels (1 in 16,075):
This indicates a likelihood of experiencing sudden crisis situations.

Signs

Size: 176	1st Harmonic	9th Harmonic	Houses	Rising Sign
Moon	Scorpio	Leo	6th, 7th	Aquarius
Sun	Leo, Pisces		2nd	
Mercury	Virgo	Libra		Lunar Phase
Venus		Capricorn	11th	Gemini, Libra
Mars		Gemini	7th	☉/☽ Midpoint
Jupiter	Sagittarius		11th	Aquarius
Saturn		Capricorn	6th, 7th	Moon's Nodes
TOTAL	Leo		7th, 11th	Cancer
(weak)	Gemini		5th, 12th	

1H	♈	♉	♊	♋	♌	♍	♎	♏	♐	♑	♒	♓
Moon	14	17	10	18	16	11	9	21	12	17	18	13
Sun	15	16	9	7	25	20	10	12	11	16	12	23
Mercury	14	12	9	9	19	22	14	9	15	16	17	20
Venus	20	12	14	16	15	21	13	9	10	13	20	13
Mars	12	18	17	12	21	18	12	16	12	16	12	10
Jupiter	13	17	10	11	18	12	18	11	23	15	11	17
Saturn	9	13	11	18	18	19	18	16	17	12	10	15
TOTAL	97	105	80	91	132	123	94	94	100	105	100	111
Rising	12	12	19	16	12	20	15	14	23	12	16	5
N. Node	15	16	19	23	14	8	12	10	13	15	16	15

9H	♈	♉	♊	♋	♌	♍	♎	♏	♐	♑	♒	♓
Moon	12	8	14	20	23	15	10	16	17	11	14	16
Sun	15	19	12	17	14	19	11	18	8	14	14	15
Mercury	12	12	20	18	10	9	21	14	13	15	18	14
Venus	13	17	15	18	10	12	14	16	14	21	14	12
Mars	9	16	23	17	14	19	11	14	12	18	11	12
Jupiter	15	17	11	10	20	17	17	10	11	15	15	18
Saturn	20	18	13	10	15	14	10	10	16	21	18	11
TOTAL	96	107	108	110	106	105	94	98	91	115	104	98

Houses	1st	2nd	3rd	4th	5th	6th	7th	8th	9th	10th	11th	12th
Moon	14	12	14	8	17	21	24	12	15	16	12	11
Sun	15	25	10	17	7	12	12	17	12	21	15	13
Mercury	18	21	15	13	7	14	13	12	13	20	14	16
Venus	12	17	15	13	15	9	13	18	13	11	26	14
Mars	11	15	19	14	8	14	23	15	12	13	18	14
Jupiter	13	15	11	13	14	13	15	21	10	18	25	8
Saturn	12	10	13	8	13	22	21	13	17	16	18	13
TOTAL	95	115	97	86	81	105	121	108	92	115	128	89
N. Node	17	10	16	14	19	12	14	17	11	15	14	17

Planets

Aspects	Parallels	4 Axis	Midpoints	Direction
Neptune	Mars-Chiron	Moon (Dsc)		
Mercury-Neptune	Mercury-Neptune	Saturn (Dsc)		
Venus-Neptune		Pluto (Dsc)		
		Chiron (MC)		**Moon's Nodes**
				Moon (S)

LESBIANS

Theme: Leo, 7th House, Pluto/Chiron

Leo:
Leo is very strong (1 in 245,664) in the 1st harmonic. Lesbian women are influenced by Leo pride and confidence, avoiding taking a submissive role in relationships.

7th House:
A strong 7th house indicates that relationships are a central concern for lesbians.

Jupiter in Sagittarius (1 in 2,214):
Jupiter in Sagittarius shows great confidence in pursuing freedom and intellectual growth.

Mercury in the 11th House (1 in 1,497):
This position indicates the ability to share ideas that influence many people. Many people in the lesbian study group were involved in feminist activism.

Chiron & Pluto aspects (1 in 691 - Pluto):
Chiron is symbolic of the wound caused by gender division and the oppression of women, which has caused great suffering in our relationships. Pluto represents hidden, often manipulative, power dynamics in relationships.

Signs

Size: 213	1st Harmonic	9th Harmonic	Houses	Rising Sign
Moon		Scorpio		Gemini
Sun	Leo	Leo		
Mercury			11th	**Lunar Phase**
Venus	Aries, Cancer		12th	Capricorn
Mars			7th	**☉/☽ Midpoint**
Jupiter	Sagittarius	Taurus, Sagittarius		
Saturn	Cancer, Leo, Virgo	Virgo	7th	**Moon's Nodes**
TOTAL	Leo		7th, 11th, 12th	Taurus
(weak)	Capr, Aqu, Pisces	Libra	2nd, 9th	7th, 12th

1H	♈	♉	♊	♋	♌	♍	♎	♏	♐	♑	♒	♓
Moon	18	22	20	12	20	19	16	16	18	16	23	13
Sun	17	22	17	20	26	15	15	21	16	17	9	18
Mercury	20	18	16	18	23	15	18	19	21	16	12	17
Venus	25	13	18	28	17	24	17	16	13	10	19	13
Mars	14	17	21	19	28	28	22	19	12	15	8	10
Jupiter	14	17	17	16	17	17	19	23	33	11	15	14
Saturn	12	11	19	25	41	28	27	12	11	10	6	11
TOTAL	120	120	128	138	172	146	134	126	124	95	92	96
Rising	9	16	27	22	19	26	19	17	17	21	14	6
N. Node	20	31	22	19	17	13	10	13	12	16	22	18

9H	♈	♉	♊	♋	♌	♍	♎	♏	♐	♑	♒	♓
Moon	18	18	17	16	22	10	16	26	22	17	15	16
Sun	22	19	15	20	24	18	12	22	15	9	16	21
Mercury	19	18	19	15	20	22	16	14	20	11	17	22
Venus	20	16	21	21	22	14	8	17	16	22	13	23
Mars	15	15	21	18	14	14	18	23	14	18	23	20
Jupiter	14	29	12	16	20	16	20	16	26	15	17	12
Saturn	23	18	19	14	16	26	15	17	20	22	15	8
TOTAL	131	133	124	120	138	120	105	135	133	114	116	122

Houses	1st	2nd	3rd	4th	5th	6th	7th	8th	9th	10th	11th	12th
Moon	17	16	13	13	21	17	19	14	20	21	19	23
Sun	22	8	20	16	15	17	19	11	16	16	26	27
Mercury	21	12	19	14	15	21	16	11	14	11	33	26
Venus	17	14	16	17	19	17	17	12	12	17	24	31
Mars	21	16	22	23	13	14	24	18	11	22	11	18
Jupiter	16	19	15	20	16	21	19	18	10	17	21	21
Saturn	14	12	15	20	20	13	26	17	14	14	26	22
TOTAL	128	97	120	123	119	120	140	101	97	118	160	168
N. Node	17	17	13	16	14	20	26	17	18	11	18	26

Planets

Aspects	Parallels	4 Axis	Direction	☉/☽ Midpoint
Pluto & Chiron		Venus (Asc)	Mercury (Rx)	
Moon-Pluto		Jupiter (Dsc)	Uranus (D)	
Sun-Pluto			Pluto (D)	
Venus-Chiron				Moon's Nodes
				Moon (S)
				Chiron (N)
				Chiron (S)

LIFESPAN: LONG (80 YEARS OR OLDER)

Theme: Taurus/Pisces, Moon

Taurus:
Taureans are calm and physically strong, causing reduced tension and physical health. Mars is frequent in the 2nd House (1 in 335), a position showing good physical ability.

Pisces:
Pisces is frequent in the 1st harmonic (1 in 16,536), especially for Mercury (1 in 1,090) and the Sun (1 in 300). Pisceans, representing the last sign, enjoy the retirement phase of life. Isolation and time for reflection are welcomed, unlike for most other sign types. The Sun/Moon midpoint is also frequent in Pisces (1 in 2,856).

Moon:
The Moon is the most influential planet for long-lived people. Lunar people are often highly connected with their home and family.

Mars aspects are infrequent & Mars is infrequent on the 4 Axis Points:
Mars is a common indicator of physical harm. Therefore, these people avoid harm's way. The Short Lifespan group has frequent Mars aspects.

Sun on the Imum Coeli (1 in 416):
This is an indication that much of a person's life is enjoyed in private.

Signs

Size: 393	1st Harmonic	9th Harmonic	Houses	Rising Sign
Moon			2nd, 7th	Scorpio
Sun	Aries, Taurus, Pisces	Leo		Lunar Phase
Mercury	Capricorn, Pisces		4th	
Venus	Taurus, Pisces		6th	☉/☽ Midpoint
Mars	Taurus, Aquarius	Cancer	2nd, 5th	Pisces
Jupiter	Aries	Scorpio		Moon's Nodes
Saturn				Pisces
TOTAL	Taurus, Pisces		7th	1st, 11th
(weak)	Gemini, Libra		8th, 11th	2nd (S)

1H	♈	♉	♊	♋	♌	♍	♎	♏	♐	♑	♒	♓
Moon	30	40	23	34	41	35	27	33	27	29	33	41
Sun	47	44	25	26	27	25	26	30	26	38	32	47
Mercury	39	29	28	27	24	28	26	28	28	45	42	49
Venus	32	45	32	40	22	31	21	29	37	26	34	44
Mars	30	43	24	31	41	35	30	33	35	30	37	24
Jupiter	41	19	27	36	33	38	33	28	43	36	28	31
Saturn	28	26	30	29	29	31	30	41	41	39	33	36
TOTAL	247	246	189	223	217	223	193	222	237	243	239	272
Rising	13	20	27	43	39	38	46	55	35	32	22	23
N. Node	40	32	29	29	38	28	31	31	36	31	26	42

9H	♈	♉	♊	♋	♌	♍	♎	♏	♐	♑	♒	♓
Moon	38	31	37	33	26	24	33	39	34	38	33	27
Sun	40	39	23	29	43	38	33	40	26	24	35	23
Mercury	41	28	26	32	30	28	33	34	38	32	30	41
Venus	26	33	36	34	29	33	23	29	40	33	36	41
Mars	29	29	35	42	32	35	33	34	31	27	29	37
Jupiter	35	36	29	32	38	28	29	44	32	31	34	25
Saturn	38	39	27	37	35	30	39	29	36	29	23	31
TOTAL	247	235	213	239	233	216	223	249	237	214	220	225

Houses	1st	2nd	3rd	4th	5th	6th	7th	8th	9th	10th	11th	12th
Moon	34	42	36	27	32	36	48	18	32	31	26	31
Sun	37	34	37	40	28	24	30	31	28	33	38	33
Mercury	36	36	34	44	27	26	30	32	29	33	36	30
Venus	35	32	36	35	29	38	30	28	32	38	22	38
Mars	41	47	32	29	40	25	39	32	24	25	33	26
Jupiter	35	32	40	37	25	38	33	26	27	38	28	34
Saturn	42	32	40	29	41	27	28	23	38	25	29	39
TOTAL	260	255	255	241	222	214	238	190	210	223	212	231
N. Node	45	29	30	29	27	29	30	42	30	27	42	33

Planets

Aspects	Parallels	4 Axis	Direction	☉/☽ Midpoint
Moon-Saturn	Moon-Pluto	Moon (Dsc)	Saturn (D)	
Mercury-Saturn		Venus (Dsc)	Chiron (D)	
		Sun (IC)		
				Moon's Nodes
				Moon (N)
				Mercury (N)
				Venus (N)
				Pluto (S)

Lifespan: Short (29 years or younger)

Theme: Scorpio (8th)

Scorpio:
Scorpios are renowned for living on the dangerous side of life.

Jupiter in Virgo:
Jupiter in Virgo is strong in the 1st and 9th harmonics (1 in 992 – 9H). Enthusiasm and growth (Jupiter) are muted in highly critical Virgo. This also suggests increased difficulties with health issues.

Mars aspects (1 in 106):
Mars is a common indicator for physical harm. The person's level of aggressiveness increases, causing many more dangerous situations to occur. Interestingly, the Life Long group has very infrequent Mars aspects.

Pluto in the 8th House (1 in 1,589):
Pluto is an indication of danger in the house of death.

Signs

Size: 197	1st Harmonic	9th Harmonic	Houses	Rising Sign
Moon	Gemini	Scorpio	3rd	
Sun		Taurus	3rd, 8th, 10th	
Mercury		Cancer	10th	Lunar Phase
Venus	Scorpio	Leo	9th	Libra
Mars	Pisces		8th, 9th	☉/☽ Midpoint
Jupiter	Virgo	Virgo		Libra
Saturn	Taurus			Moon's Nodes
TOTAL	Scorpio, Aquarius		3rd	Aries
(weak)	Aries, Cancer			Pisces

1H	♈	♉	♊	♋	♌	♍	♎	♏	♐	♑	♒	♓
Moon	19	15	25	12	12	15	15	21	15	5	22	21
Sun	14	15	14	13	16	14	21	17	16	18	22	17
Mercury	11	14	15	11	12	19	19	21	17	23	19	16
Venus	15	10	17	11	17	14	13	24	17	14	24	21
Mars	10	14	16	12	19	25	16	13	19	14	18	21
Jupiter	15	15	9	14	17	25	22	22	19	14	11	14
Saturn	12	23	13	16	19	16	16	18	16	16	20	12
TOTAL	96	106	109	89	112	128	122	136	119	104	136	122
Rising	10	11	15	18	18	21	21	21	22	14	13	13
N. Node	24	13	14	22	15	19	15	18	11	15	8	23

9H	♈	♉	♊	♋	♌	♍	♎	♏	♐	♑	♒	♓
Moon	13	15	21	15	14	12	22	26	13	14	15	17
Sun	10	26	16	20	16	21	14	15	14	14	15	16
Mercury	20	21	17	25	14	16	14	17	13	16	16	8
Venus	14	14	17	19	23	15	17	11	15	16	16	20
Mars	22	14	17	15	13	14	17	16	19	15	17	18
Jupiter	15	13	14	15	14	28	20	10	17	15	16	20
Saturn	13	20	16	16	17	16	16	10	18	13	20	22
TOTAL	107	123	118	125	111	122	120	105	109	103	115	121

Houses	1st	2nd	3rd	4th	5th	6th	7th	8th	9th	10th	11th	12th
Moon	16	15	23	17	21	14	16	8	17	14	18	18
Sun	14	15	25	18	9	16	7	21	12	29	13	18
Mercury	15	19	19	18	12	19	10	12	12	26	18	17
Venus	16	14	21	17	16	11	13	11	25	17	17	19
Mars	12	16	17	20	16	10	16	22	23	12	20	13
Jupiter	13	21	19	11	9	21	19	19	17	14	14	20
Saturn	20	16	20	16	11	11	21	15	19	14	19	15
TOTAL	106	116	144	117	94	102	102	108	125	126	119	120
N. Node	17	16	16	14	14	21	16	15	18	11	17	22

Planets

Aspects	Parallels	4 Axis	Direction	☉/☽ Midpoint
Mars & Jupiter	Mercury-Saturn	Venus (MC)	Saturn (D)	
Moon-Mars				
Jupiter-Uranus				
				Moon's Nodes
				Mercury (S)
				Pluto (S)

LITERATURE WRITERS

Theme: Aquarius, Mercury

Aquarius:
Aquarians excel in language and communication. Their keen intellect enables them to describe thoughts and experiences in new and interesting ways.

Virgo:
Writers are very analytical people, mastering the rules of language.

Mercury:
Mercury is the planet symbolizing reasoning ability, knowledge, and communication.

4th House (1 in 490):
The 4th house is one's home, the place of privacy. Mercury is the most frequent planet in this house (1 in 474), indicating that writing is done when in solitude.

Sun on the Imum Coeli (1 in 128)
This is an indication that a person prefers to work in private.

Signs

Size: 331	1st Harmonic	9th Harmonic	Houses	Rising Sign
Moon		Aquarius, Pisces		Virgo
Sun	Aquarius	Cancer, Virgo	4th	
Mercury	Aquarius	Capricorn	2nd, 4th	Lunar Phase
Venus	Pisces	Cancer		Virgo
Mars	Sagittarius, Pisces			☉/☽ Midpoint
Jupiter		Aquarius		Scorpio
Saturn		Aquarius	6th	Moon's Nodes
TOTAL	Aquarius	Aquarius	2nd, 3rd, 4th	Virgo
(weak)	Taurus, Gemini	Leo	9th	8th

1H	♈	♉	♊	♋	♌	♍	♎	♏	♐	♑	♒	♓
Moon	24	22	24	29	33	27	33	22	27	27	35	29
Sun	31	22	27	17	30	34	30	21	27	29	37	27
Mercury	24	20	23	21	23	30	34	26	30	36	37	28
Venus	31	31	10	28	25	31	27	25	30	29	29	36
Mars	24	25	21	39	33	26	29	33	34	21	17	30
Jupiter	22	24	27	27	30	25	32	38	24	31	30	22
Saturn	18	18	20	18	25	33	35	37	34	32	33	29
TOTAL	174	162	152	179	199	206	220	202	206	205	218	201
Rising	16	17	13	31	36	48	44	39	31	18	22	17
N. Node	23	31	20	29	21	36	29	21	34	31	29	28

9H	♈	♉	♊	♋	♌	♍	♎	♏	♐	♑	♒	♓
Moon	33	23	21	19	28	26	28	30	18	33	36	37
Sun	19	25	29	37	21	40	23	27	24	30	29	28
Mercury	30	29	19	32	19	20	27	29	32	36	25	34
Venus	25	33	23	37	25	29	20	34	21	28	33	24
Mars	25	20	28	34	29	35	28	25	34	31	22	21
Jupiter	20	28	31	32	22	20	32	29	25	26	38	29
Saturn	25	34	24	22	28	33	26	30	20	27	39	24
TOTAL	177	192	175	213	172	203	184	204	174	211	222	197

Houses	1st	2nd	3rd	4th	5th	6th	7th	8th	9th	10th	11th	12th
Moon	27	31	29	30	23	29	29	35	20	23	28	28
Sun	35	36	32	39	22	15	27	31	14	24	32	25
Mercury	27	39	36	42	19	25	23	25	16	20	28	32
Venus	33	29	33	33	30	24	24	21	23	23	27	32
Mars	28	30	35	33	26	30	23	25	20	28	25	29
Jupiter	24	34	30	28	24	22	27	33	25	29	24	32
Saturn	30	34	32	25	30	37	15	25	27	30	17	30
TOTAL	204	233	227	230	174	182	168	195	145	177	181	208
N. Node	27	30	30	27	29	27	20	36	25	22	33	26

Planets

Aspects	Parallels	4 Axis	Direction	☉/☽ Midpoint
Mercury-Mars		Sun (IC)	Mercury (D)	
		Mars (IC)	Chiron (D)	
				Moon's Nodes
				Mercury (N)

Marriage – 15 years or more

Theme: Sagittarius/Aquarius, 1st House, Jupiter

Aquarius:
Aquarians are attractive people who like being involved in relationships of all kinds, which is of great benefit to finding romantic partners. This strongly confirms the idea that Aquarius is a Venusian sign and not Libra, which is weak for this group.

Sagittarius:
Sagittarians are full of optimism and enthusiasm. People feel confident and joyful in their presence.

Mars in Gemini (1 in 2,054):
This position indicates a great deal of discussion and argumentation, showing how important frequent communication is for lasting relationships.

1st House:
A strong 1st house indicates that a person has good confidence and presence. Jupiter is especially strong in this house (1 in 395).

Sun-Jupiter aspects, Jupiter on the Ascendant (1 in 1,217), and Jupiter Direct:
Enthusiasm and positive experiences result from having Jupiter influencing the prime indicators of the self.

Mars-Jupiter aspects and parallels:
This planetary combination gives great confidence in pursuing goals and desires.

Signs

Size: 463	1st Harmonic	9th Harmonic	Houses	Rising Sign
Moon	Sagittarius	Sagittarius	2nd	Aries
Sun	Aquarius	Aquarius		Lunar Phase
Mercury			1st, 9th	Leo
Venus	Aries, Aquarius	Sagittarius, Aquarius		☉/☽ Midpoint
Mars	Gemini		1st	Gemini
Jupiter			1st, 8th	Capricorn
Saturn	Taurus	Virgo	12th	Moon's Nodes
TOTAL	Aquarius	Sagittarius, Aquarius	1st, 9th	Virgo, Aries (S)
(weak)	Libra		3rd, 11th	10th

1H	♈	♉	♊	♋	♌	♍	♎	♏	♐	♑	♒	♓
Moon	32	42	38	37	45	42	36	39	52	36	34	30
Sun	39	45	33	37	44	24	39	38	28	41	48	47
Mercury	43	33	34	34	38	32	25	45	38	50	47	44
Venus	53	40	36	47	29	52	20	36	32	36	51	31
Mars	35	36	60	42	42	44	39	36	33	27	36	33
Jupiter	35	39	43	33	44	46	38	47	37	43	28	30
Saturn	37	51	38	36	31	30	33	43	48	34	49	33
TOTAL	274	286	282	266	273	270	230	284	268	267	293	248
Rising	31	28	28	42	41	44	41	57	52	41	33	25
N. Node	33	25	37	37	30	54	54	35	43	38	38	39

9H	♈	♉	♊	♋	♌	♍	♎	♏	♐	♑	♒	♓
Moon	28	44	31	40	45	37	29	44	48	47	43	27
Sun	40	29	31	37	22	45	41	37	46	39	53	43
Mercury	33	37	41	46	36	45	43	40	29	32	45	36
Venus	35	34	35	35	33	34	38	33	52	47	52	35
Mars	41	43	46	38	37	41	32	34	41	41	32	37
Jupiter	33	42	42	38	38	37	34	32	46	35	39	47
Saturn	38	37	32	41	38	48	35	39	36	34	39	46
TOTAL	248	266	258	275	249	287	252	259	298	275	303	271

Houses	1st	2nd	3rd	4th	5th	6th	7th	8th	9th	10th	11th	12th
Moon	39	53	38	30	33	38	46	29	44	36	33	44
Sun	48	36	41	37	31	29	30	41	40	43	35	52
Mercury	58	36	35	41	33	29	37	34	49	34	38	39
Venus	44	40	43	30	34	35	36	40	42	37	37	45
Mars	52	43	30	36	32	35	44	40	35	36	42	38
Jupiter	56	31	31	34	34	28	29	48	45	48	39	40
Saturn	35	44	30	37	34	41	36	40	41	45	29	51
TOTAL	332	283	248	245	231	235	258	272	296	279	253	309
N. Node	47	28	38	36	38	29	42	45	29	49	37	45

Planets

Aspects	Parallels	4 Axis	Direction	☉/☽ Midpoint
Uranus	Moon-Sun	Jupiter (Asc)	Jupiter (D)	Saturn
Sun-Jupiter	Mars-Jupiter	Saturn (Asc)		Neptune
Mercury-Neptune		Pluto (Dsc)		
Venus-Saturn				**Moon's Nodes**
Venus-Uranus				Mercury (N)
Mars-Jupiter				Mars (S)

Marriage — Never

Theme: Virgo, Moon

Virgo & Pisces:
Pisces is strong in the 1st harmonic, and Virgo is strong in the 9th harmonic. Many unmarried people are reclusive and skeptical about forming long-term relationships. They may also be critical of their own or other's attractiveness and likability. Saturn is strong in Virgo in the 1st and 9th harmonics, indicating that their focus may be on work. The Virgo Lunar Phase (1 in 301) is also frequent for unmarried people, which is the sixth 30 degree section of the Lunation cycle.

Moon:
The Moon is a strong influence for unmarried people. Lunar people are private, shy, and home-loving. The Moon is most frequent on the Descendant (1 in 382) symbolizing shyness and moodiness with people.

Mars on the Sun/Moon midpoint (1 in 1,131):
This is an indication of frustrations and tension in a person's life.

Signs

Size: 163	1st Harmonic	9th Harmonic	Houses	Rising Sign
Moon	Gemini		6th	Gemini
Sun		Cancer		
Mercury	Pisces			Lunar Phase
Venus	Aquarius	Sagittarius	3rd, 5th	Virgo
Mars	Aries		4th	☉/☽ Midpoint
Jupiter	Aquarius	Gemini	8th	
Saturn	Virgo	Virgo, Libra		Moon's Nodes
TOTAL	Pisces	Virgo		Cancer
(weak)	Leo		9th	

1H	♈	♉	♊	♋	♌	♍	♎	♏	♐	♑	♒	♓
Moon	9	17	20	11	18	14	11	11	14	8	14	16
Sun	16	14	18	13	5	16	11	10	13	13	17	17
Mercury	11	12	17	11	6	13	15	12	16	16	14	20
Venus	18	14	13	13	11	15	12	13	9	10	22	13
Mars	20	15	11	12	12	19	12	14	13	11	11	13
Jupiter	14	12	10	11	16	15	16	11	14	10	20	14
Saturn	17	8	6	16	10	20	18	16	17	11	11	13
TOTAL	105	92	95	87	78	112	95	87	96	79	109	106
Rising	9	9	20	13	19	19	15	15	19	7	11	7
N. Node	10	13	18	25	15	12	15	9	9	16	7	14

9H	♈	♉	♊	♋	♌	♍	♎	♏	♐	♑	♒	♓
Moon	12	16	9	18	14	14	6	14	14	12	17	17
Sun	15	7	9	20	14	14	17	16	9	12	16	14
Mercury	8	12	15	14	8	16	19	10	16	13	15	17
Venus	12	11	10	7	13	19	11	17	21	13	16	13
Mars	14	14	9	13	17	17	11	16	15	13	11	13
Jupiter	11	15	20	10	5	12	15	14	15	19	11	16
Saturn	9	11	18	10	13	20	20	9	19	16	11	7
TOTAL	81	86	90	92	84	112	99	96	109	98	97	97

Houses	1st	2nd	3rd	4th	5th	6th	7th	8th	9th	10th	11th	12th
Moon	11	16	16	10	10	20	14	15	11	11	14	15
Sun	12	16	18	14	13	15	13	13	6	16	16	11
Mercury	10	17	20	12	14	10	17	10	13	10	17	13
Venus	16	9	22	12	18	11	12	14	12	12	14	11
Mars	12	16	9	20	18	13	14	11	12	14	11	13
Jupiter	14	11	12	15	14	8	16	19	8	17	16	13
Saturn	17	15	13	9	16	15	11	10	14	17	14	12
TOTAL	92	100	110	92	103	92	97	92	76	97	102	88
N. Node	18	16	16	15	11	17	9	9	14	9	18	11

Planets

Aspects	Parallels	4 Axis	Direction	☉/☽ Midpoint
Saturn	Moon-Pluto	Moon (Dsc)	Neptune (Rx)	Mars
Moon-Jupiter		Chiron (MC)		
				Moon's Nodes
				Moon (S)
				Saturn (N)

MATHEMATICIANS & PHYSICISTS

Theme: Aries/Capricorn, Mars/Saturn

Aries & Capricorn:
Aries and Capricorn are strong in the 1st harmonic (Aries: 1 in 343; Capricorn: 1 in 428) and Scorpio is strong in the 9th harmonic. These people are outwardly (1H) both independent (Aries) and disciplined (Capricorn). Inwardly (9H), they are deep seekers of knowledge (Scorpio) This group demonstrates the technical, productive side to these signs and their corresponding planets, which are also strong: Mars, Saturn, and Pluto.

Saturn in Taurus (1 in 1,704 - 9H):
They are very hard working over long periods of time. This also indicates a conservative approach to material things.

Mars-Saturn-Pluto aspects:
Mathematicians and physicists have excellent technical ability (Mars) and persistence (Saturn) used to research and understand the hidden structures of existence (Pluto).

Mars on the Ascendant:
They are self-willed individuals who often work on their own.

Neptune Direct (1 in 2,391):
Mathematicians and physicists have the highest frequency of Neptune direct of all the study groups. This group of people is, in fact, the antithesis of Neptune: direct, honest, fact-based and truth-seeking. The suggestion here is that Neptune is more powerful when retrograde.

5th House is weak (1 in 303):
The 5th house represents the fun, romance, and the enjoyment of creative expression.

Signs

Size: 127	1st Harmonic	9th Harmonic	Houses	Rising Sign
Moon	Sagittarius	Libra		Gemini
Sun	Aries, Pisces	Virgo		Lunar Phase
Mercury	Aries, Aquarius, Pisces	Scorpio		Cancer
Venus	Aries, Taurus, Capricorn			☉/☽ Midpoint
Mars	Capricorn	Leo	10th	Aries
Jupiter		Aquarius	3rd	Aquarius
Saturn		Taurus, Pisces	12th	Moon's Nodes
TOTAL	Aries, Capricorn, Pisces	Scorpio	10th	Aries
(weak)	Gemini, Cancer, Virgo	Capricorn	5th	

1H	♈	♉	♊	♋	♌	♍	♎	♏	♐	♑	♒	♓
Moon	11	11	10	5	9	8	13	10	17	12	9	12
Sun	19	10	9	8	8	7	10	3	7	12	15	19
Mercury	16	9	5	9	7	8	6	10	8	13	19	17
Venus	16	18	8	7	10	12	3	8	7	15	9	14
Mars	11	10	5	11	16	6	15	16	9	14	8	6
Jupiter	10	9	11	9	12	11	7	15	9	14	10	10
Saturn	12	14	6	6	11	10	12	11	12	16	11	6
TOTAL	95	81	54	55	73	62	66	73	69	96	81	84
Rising	6	7	17	15	14	16	9	15	7	11	3	7
N. Node	17	10	14	11	12	12	9	12	3	5	11	11

9H	♈	♉	♊	♋	♌	♍	♎	♏	♐	♑	♒	♓
Moon	9	7	8	9	12	12	16	12	10	13	8	11
Sun	10	15	8	9	5	19	6	15	14	7	13	6
Mercury	12	7	9	11	10	10	15	17	5	6	13	12
Venus	8	11	12	13	15	9	10	14	12	6	8	9
Mars	9	11	8	13	18	11	12	12	6	8	12	7
Jupiter	9	8	8	9	11	12	9	10	9	9	19	14
Saturn	10	21	9	8	6	10	10	11	8	7	10	17
TOTAL	67	80	62	72	77	83	78	91	64	56	83	76

Houses	1st	2nd	3rd	4th	5th	6th	7th	8th	9th	10th	11th	12th
Moon	9	10	13	15	5	11	9	8	8	13	14	12
Sun	10	11	10	12	7	10	6	11	12	13	13	12
Mercury	12	13	7	14	7	10	8	6	12	15	12	11
Venus	10	12	11	11	6	8	12	9	7	14	17	10
Mars	13	6	6	11	7	11	12	10	10	16	15	10
Jupiter	9	13	16	9	7	10	12	10	8	14	10	9
Saturn	8	13	13	10	9	8	12	10	10	8	10	16
TOTAL	71	78	76	82	48	68	71	64	67	93	91	80
N. Node	13	4	10	15	9	14	12	14	7	8	12	9

Planets

Aspects	Parallels	4 Axis	Direction	☉/☽ Midpoint
Mars-Saturn		Mars (Asc)	Neptune (D)	
Mars-Pluto		Sun (MC)	Chiron (D)	
Saturn-Pluto				
				Moon's Nodes
				Moon (N)
				Venus (S)
				Saturn (N)

MEDICAL PROFESSIONALS

Theme: Virgo, Pluto

Virgo:
Virgo is the sign most associated with sickness and health. They have great analytical abilities, which can be put to use in remedying people's health problems.

Mars in Pisces (1 in 2,315 - 9H):
This placement indicates a desire to be in service to others. It also indicates serving people who have been injured or suffered violence.

Pluto:
A strong Pluto influences people towards deep, transformational experiences. Doctors and nurses get embroiled in some of the most difficult experiences of other people's lives. Pluto frequent on the Moon's South Node (1 in 1,362) indicates that these people come from a background with intense struggles.

Signs

Size: 445	1st Harmonic	9th Harmonic	Houses	Rising Sign
Moon		Libra	9th	
Sun	Gemini, Capricorn	Aquarius		
Mercury		Capricorn	11th	**Lunar Phase**
Venus	Virgo, Aquarius	Aries		Virgo
Mars	Scorpio	Pisces		☉/☽ Midpoint
Jupiter	Virgo	Gemini	4th	
Saturn	Leo, Virgo, Libra			**Moon's Nodes**
TOTAL	Virgo			Aquarius
(weak)			4th	

1H	♈	♉	♊	♋	♌	♍	♎	♏	♐	♑	♒	♓
Moon	46	34	42	34	40	35	32	27	42	38	33	42
Sun	24	38	50	35	37	41	37	40	21	48	33	41
Mercury	28	35	40	29	37	42	35	39	38	46	37	39
Venus	39	37	35	44	32	52	20	35	36	32	51	32
Mars	35	26	36	35	45	39	50	52	34	37	26	30
Jupiter	34	41	31	29	39	47	30	45	42	32	34	41
Saturn	24	20	35	37	47	56	64	42	29	33	35	23
TOTAL	230	231	269	243	277	312	268	280	242	266	249	248
Rising	28	22	38	49	57	44	56	45	42	19	21	24
N. Node	45	32	43	35	33	30	30	35	26	39	52	45

9H	♈	♉	♊	♋	♌	♍	♎	♏	♐	♑	♒	♓
Moon	29	44	38	40	35	33	47	43	30	35	38	33
Sun	40	40	32	28	34	43	36	30	32	37	49	44
Mercury	28	26	37	41	44	41	37	37	35	48	32	39
Venus	46	27	33	37	34	38	33	40	41	41	33	42
Mars	34	37	29	38	35	31	36	33	32	45	38	57
Jupiter	44	34	49	40	39	28	35	32	37	29	42	36
Saturn	37	42	43	38	36	43	35	30	37	35	41	28
TOTAL	258	250	261	262	257	257	259	245	244	270	273	279

Houses	1st	2nd	3rd	4th	5th	6th	7th	8th	9th	10th	11th	12th
Moon	39	32	33	35	39	30	39	41	50	39	35	33
Sun	47	45	33	34	26	32	40	31	26	39	48	44
Mercury	39	47	40	28	36	34	35	28	30	38	52	38
Venus	32	45	30	31	38	28	38	30	40	43	40	50
Mars	40	42	38	24	31	42	38	37	33	46	28	46
Jupiter	38	37	43	47	29	45	18	32	38	37	36	45
Saturn	40	34	43	27	29	39	37	39	35	43	40	39
TOTAL	275	282	260	226	228	250	245	238	252	285	279	295
N. Node	42	30	46	40	45	31	26	43	45	30	33	34

Planets

Aspects	Parallels	4 Axis	Direction	☉/☽ Midpoint
Moon-Jupiter	Venus-Pluto	Uranus (Asc)	Pluto (D)	
Saturn-Pluto		Sun (Dsc)		
Neptune-Pluto		Venus (MC)		
				Moon's Nodes
				Mercury (S)
				Neptune (S)
				Pluto (S)

MENTAL HANDICAP

Theme: Pisces, 8th House, Mercury

Pisces:
Pisceans are known for their lack of analytical or organizational ability. Their minds prefer abstract lines of thought and can get absorbed in imaginary realms.

8th House:
The 8th house represents shadow aspects of the psyche. Mars is especially frequent in the 8th house (1 in 391), showing conflicts in a person's psychology.

Neptune in the 3rd House (1 in 3,087):
This position indicates confusion in the person's thinking and communications.

Mercury:
Mentally handicapped people have many difficult indicators with Mercury, which represents reasoning abilities. Mercury is frequently retrograde and parallel with Chiron, showing mental troubles. Mercury conjunct the Moon's South Node indicates that a person has early life issues with their minds.

Signs

Size: 102	1st Harmonic	9th Harmonic	Houses	Rising Sign
Moon	Gemini	Gemini	10th	
Sun	Pisces		8th	Lunar Phase
Mercury	Aries	Virgo	2nd	
Venus		Pisces		☉/☽ Midpoint
Mars	Sagittarius	Aries, Cancer	5th, 8th	Capricorn
Jupiter	Libra	Taurus, Pisces		Moon's Nodes
Saturn	Taurus, Aqu, Pisces	Scorpio	8th	Leo, 5th
TOTAL	Taurus, Aqu, Pisces		8th	Aquarius
(weak)	Leo	Sagittarius		Pisces

1H	♈	♉	♊	♋	♌	♍	♎	♏	♐	♑	♒	♓
Moon	7	7	14	9	9	8	7	8	4	7	11	11
Sun	11	8	4	8	3	7	9	10	8	8	10	16
Mercury	17	9	4	2	7	7	8	13	7	11	11	6
Venus	10	13	5	2	8	7	7	12	8	10	11	9
Mars	3	9	6	9	8	9	8	11	13	11	9	6
Jupiter	3	6	11	12	5	8	16	12	11	7	5	6
Saturn	10	29	6	3	1	3	6	2	4	5	15	18
TOTAL	61	81	50	45	41	49	61	68	55	59	72	72
Rising	2	8	8	12	6	14	10	12	12	11	6	1
N. Node	10	7	4	12	14	9	3	3	1	3	17	19

9H	♈	♉	♊	♋	♌	♍	♎	♏	♐	♑	♒	♓
Moon	6	10	13	10	11	9	9	9	3	10	10	2
Sun	5	4	10	9	8	9	8	9	8	10	10	12
Mercury	7	9	8	10	8	15	6	6	9	5	8	11
Venus	7	7	8	6	11	5	5	11	7	10	10	15
Mars	13	8	6	14	10	7	10	3	8	10	7	6
Jupiter	8	15	5	9	9	5	10	9	4	9	5	14
Saturn	7	9	9	9	2	9	9	13	6	12	8	9
TOTAL	53	62	59	67	59	59	57	60	45	66	58	69

Houses	1st	2nd	3rd	4th	5th	6th	7th	8th	9th	10th	11th	12th
Moon	11	12	5	7	3	5	12	9	10	14	8	6
Sun	10	6	9	12	4	7	9	14	6	7	11	7
Mercury	4	14	7	9	6	11	8	9	8	6	8	12
Venus	10	7	9	11	5	10	6	10	6	7	12	9
Mars	9	10	9	6	13	6	7	16	4	6	10	6
Jupiter	12	10	6	5	9	10	9	4	10	12	9	6
Saturn	10	3	7	10	4	5	12	14	11	6	12	8
TOTAL	66	62	52	60	44	54	63	76	55	58	70	54
N. Node	6	7	10	11	13	10	6	8	10	10	5	6

Planets

Aspects	Parallels	4 Axis	Direction	☉/☽ Midpoint
Neptune	Mercury-Chiron		Mercury (Rx)	Mercury
Moon-Venus			Saturn (D)	
Moon-Neptune			Pluto (Rx)	**Moon's Nodes**
Venus-Saturn			Chiron (D)	Mercury (S)
Saturn-Chiron				Venus (S)
				Mars (S)
				Jupiter (S)
				Uranus (S)

MENTAL ILLNESS

Theme: Scorpio, 11th House, Moon/Mercury/Neptune

Mental Illnesses include Schizophrenia and Psychosis, not Alzheimer's or Depression.

Scorpio:
Scorpios tend to see the dark side of life and often get themselves involved in stressful, dangerous situations, which could cause mental breakdowns.

11th House (1 in 19,577):
The 11th house represents higher life goals. Mercury is especially frequent in this space (1 in 1,099), showing that their goals are to achieve mental health and aptitude.

Mercury:
Mercury is frequently Retrograde and Mercury-Neptune aspects are frequent (1 in 639). This shows how the mentally ill often have a difficult time communicating and connecting with people.

Moon & Neptune:
The Moon is associated with the subconscious mind and emotions. Behavior arising from here is often seen as irrational. Thus, the Moon's historical association with lunacy shows some truth. Neptune aspects can incline people towards fantasy, illusion, and disorientation. Moon-Neptune aspects (1 in 1,417) and Moon-Neptune parallels are the most frequent (1 in 342). The Moon is also frequent on the Moon's South Node (1 in 2,292).

Signs

Size: 152	1st Harmonic	9th Harmonic	Houses	Rising Sign
Moon		Aries, Scorpio	9th	Scorpio
Sun		Gemini	4th, 11th	
Mercury	Scorpio	Scorpio, Capricorn	5th, 11th	**Lunar Phase**
Venus	Aries		11th	
Mars	Virgo, Sagittarius			**☉/☽ Midpoint**
Jupiter	Cancer			Sagittarius
Saturn	Scorpio, Sagittarius			**Moon's Nodes**
TOTAL	Scorpio	Gemini, Scorpio	10th, 11th	3rd
(weak)	Capricorn, Pisces		3rd	

1H	♈	♉	♊	♋	♌	♍	♎	♏	♐	♑	♒	♓
Moon	11	14	11	13	10	12	12	16	14	13	17	9
Sun	16	17	11	12	10	13	17	15	14	10	10	7
Mercury	11	13	12	13	9	13	15	21	12	9	12	12
Venus	19	16	12	11	11	12	13	17	9	7	15	10
Mars	10	13	12	12	9	23	14	10	18	8	14	9
Jupiter	11	13	14	19	11	15	14	13	9	11	8	14
Saturn	7	12	9	7	15	14	9	22	20	13	15	9
TOTAL	85	98	81	87	75	102	94	114	96	71	91	70
Rising	4	12	12	13	19	15	15	25	16	10	8	3
N. Node	12	9	13	9	14	11	14	13	14	18	10	15

9H	♈	♉	♊	♋	♌	♍	♎	♏	♐	♑	♒	♓
Moon	20	8	16	12	11	12	8	20	10	13	13	9
Sun	14	12	24	12	11	13	13	12	7	10	13	11
Mercury	7	11	9	8	13	15	10	18	16	19	11	15
Venus	8	12	15	12	12	14	10	13	18	14	10	14
Mars	14	10	15	11	12	12	11	10	17	13	16	11
Jupiter	9	12	16	12	11	11	16	17	12	15	10	11
Saturn	13	10	12	11	13	16	15	14	15	11	11	11
TOTAL	85	75	107	78	83	93	83	104	95	95	84	82

Houses	1st	2nd	3rd	4th	5th	6th	7th	8th	9th	10th	11th	12th
Moon	10	14	13	12	17	16	6	8	20	14	13	9
Sun	8	13	8	22	7	8	10	11	10	15	24	16
Mercury	14	12	11	10	17	7	10	9	11	14	25	12
Venus	16	13	11	11	9	13	7	11	10	15	21	15
Mars	10	11	12	13	13	11	10	13	11	17	17	14
Jupiter	11	11	14	9	7	10	13	16	11	16	16	18
Saturn	15	18	6	10	11	13	13	8	13	16	13	16
TOTAL	84	92	75	87	81	78	69	76	86	107	129	100
N. Node	6	13	19	17	11	10	16	10	13	9	18	10

Planets

Aspects	Parallels	4 Axis	Direction	☉/☽ Midpoint
Moon & Neptune	Moon-Neptune	Jupiter (Asc)	Mercury (Rx)	Uranus
Moon-Venus	Mercury-Pluto	Saturn (Asc)		
Moon-Neptune		Pluto (MC)		
Mercury-Neptune				Moon's Nodes
				Moon (S)
				Mercury (S)

Military Personnel

Theme: Virgo/Capricorn, Neptune/Chiron

Virgo & Capricorn:
Virgo and Capricorn are frequent in the 1st and 9th harmonics. They are the hardest working and most practical signs. Both seek order and discipline. Capricorns are the most authoritative and domineering people. They desire respect that can be seen in forms such as large hierarchical organizations. Mercury in Capricorn (1 in 4,473 - 9H) shows that they think in very serious, realistic terms and respect authoritative communication.

Chiron:
Mercury-Chiron aspects are frequent, Mars-Chiron parallels are frequent (1 in 2,788), and Chiron Direct is frequent (1 in 110). The military inflict and suffer a great deal of injury.

Neptune aspects:
They are frequently subjected to illusory beliefs and expectations in the form of propaganda about "the enemy." Loss of co-workers and separation from their spouses and children is a constant theme in their lives.

Moon's North Node in the 11th House (1 in 211,732):
This position indicates that the military often see their life goals as contributing to the welfare of a larger community, their country.

	Signs			
Size: 214	1st Harmonic	9th Harmonic	Houses	Rising Sign
Moon	Capricorn	Aquarius		
Sun	Aquarius		8th	
Mercury	Capricorn, Aquarius	Capricorn, Pisces	9th	Lunar Phase
Venus			6th	Gemini
Mars		Taurus	7th	☉/☽ Midpoint
Jupiter		Capricorn	10th, 12th	
Saturn	Virgo	Aries	12th	Moon's Nodes
TOTAL	Cancer, Virgo, Capricorn	Virgo, Capricorn	9th	Virgo
(weak)		Gemini	2nd, 5th, 11th	11th

1H	♈	♉	♊	♋	♌	♍	♎	♏	♐	♑	♒	♓
Moon	18	23	18	20	12	18	20	10	16	29	13	17
Sun	11	16	14	20	21	22	14	17	15	16	28	20
Mercury	18	12	11	22	12	24	17	20	12	27	26	13
Venus	14	16	16	26	11	18	18	21	20	20	19	15
Mars	8	16	17	20	23	27	24	14	16	17	18	14
Jupiter	19	10	13	19	21	19	23	23	18	14	15	20
Saturn	19	13	16	20	14	25	15	21	18	21	14	18
TOTAL	107	106	105	147	114	153	131	126	115	144	133	117

	♈	♉	♊	♋	♌	♍	♎	♏	♐	♑	♒	♓
Rising	9	13	18	23	26	17	29	19	19	20	13	8
N. Node	18	22	15	15	19	25	18	16	14	20	14	18

9H	♈	♉	♊	♋	♌	♍	♎	♏	♐	♑	♒	♓
Moon	20	14	15	14	17	23	12	18	18	19	26	18
Sun	21	15	11	19	19	22	20	17	17	17	15	21
Mercury	6	22	14	13	13	20	19	12	17	32	17	29
Venus	12	23	14	17	17	20	15	22	20	13	23	18
Mars	15	26	13	14	8	19	22	20	20	19	18	20
Jupiter	16	11	18	23	19	20	22	15	20	26	8	16
Saturn	25	15	18	23	18	23	11	14	15	21	19	12
TOTAL	115	126	103	123	111	147	121	118	127	147	126	134

Houses	1st	2nd	3rd	4th	5th	6th	7th	8th	9th	10th	11th	12th
Moon	14	14	24	22	18	21	16	21	17	18	17	12
Sun	21	15	22	18	11	13	13	25	17	20	18	21
Mercury	22	23	20	17	11	13	15	19	25	15	17	17
Venus	16	16	23	15	13	22	10	18	20	18	17	26
Mars	18	13	15	16	11	19	24	17	20	24	19	18
Jupiter	20	14	16	15	16	19	18	13	21	24	13	25
Saturn	21	14	22	14	14	16	21	20	19	17	11	25
TOTAL	132	109	142	117	94	123	117	133	139	136	112	144
N. Node	11	18	19	22	20	15	17	14	16	13	36	13

Planets

Aspects	Parallels	4 Axis	Direction	☉/☽ Midpoint
Neptune		Sun (MC)	Neptune (Rx)	
Moon-Mercury		Venus (MC)	Chiron (D)	
Mercury-Chiron	Mars-Chiron			
Venus-Neptune				Moon's Nodes
				Mercury (N)

MODELS

Theme: Scorpio (8th)

Scorpio & the 8th House:
Scorpios have a strong sexual magnetism and the 8th house reveals a person's sexual experiences. Mars in the 8th house (1 in 831) shows that models aggressively assert their sexual energies.

Leo:
Leo is strong in the 1st harmonic (1 in 248). Leos have great personal charisma and love admiration. They are often viewed by others as attractive, successful people.

Moon on the Ascendant (1 in 200) & Venus on the Ascendant (1 in 190):
These are the two most feminine planets. Planets on the Eastern horizon show how we are seen by others.

Chiron on the Midheaven (1 in 714) & the Imum Coeli (1 in 1,093):
Chiron on Midheaven-Imum Coeli axis shows that models experience a lot of criticism within their homes and careers. They are held to high standards of physical perfection.

Sun-Neptune parallels (1 in 2,464) & Moon-Neptune aspects (1 in 1,022):
Neptune with the Sun and Moon is an indication of the glamour that models have.

Signs

Size: 130	1st Harmonic	9th Harmonic	Houses	Rising Sign
Moon			7th	
Sun		Capricorn	8th	Lunar Phase
Mercury	Leo, Virgo			Capricorn
Venus		Scorpio	8th, 12th	☉/☽ Midpoint
Mars			8th	Cancer
Jupiter	Scorpio			Virgo
Saturn	Aquarius	Taurus, Virgo	2nd	Moon's Nodes
TOTAL	Leo	Capricorn	1st, 8th, 12th	Gemini
(weak)	Sagittarius, Capricorn	Libra		Libra

1H	♈	♉	♊	♋	♌	♍	♎	♏	♐	♑	♒	♓
Moon	11	13	13	14	13	13	10	10	7	8	11	7
Sun	8	11	13	12	13	15	14	10	4	6	13	11
Mercury	8	13	6	9	18	17	11	10	8	8	9	13
Venus	9	6	13	12	13	16	14	11	4	7	15	10
Mars	13	12	13	11	17	13	13	7	7	7	8	9
Jupiter	8	13	11	8	10	8	15	21	11	10	7	8
Saturn	7	14	7	5	14	11	10	13	13	11	17	8
TOTAL	64	82	76	71	98	93	87	82	54	57	80	66
Rising	8	6	14	18	9	14	12	19	11	8	8	3
N. Node	8	10	18	12	13	9	16	10	4	10	11	9

9H	♈	♉	♊	♋	♌	♍	♎	♏	♐	♑	♒	♓
Moon	11	8	12	12	13	14	7	10	15	8	8	12
Sun	4	12	7	10	6	11	15	13	14	19	11	8
Mercury	11	12	11	15	14	8	8	8	10	10	15	8
Venus	10	7	13	9	7	12	5	17	14	11	10	15
Mars	12	13	13	16	8	6	9	11	8	13	12	9
Jupiter	11	10	11	16	10	5	13	5	12	16	13	8
Saturn	13	17	14	9	5	18	4	9	8	14	10	9
TOTAL	72	79	81	87	63	74	61	73	81	91	79	69

Houses	1st	2nd	3rd	4th	5th	6th	7th	8th	9th	10th	11th	12th
Moon	15	10	7	10	10	7	20	12	9	8	8	14
Sun	15	14	11	11	7	5	10	15	6	12	9	15
Mercury	21	8	10	11	12	8	7	11	10	10	6	16
Venus	12	9	10	12	7	7	10	16	6	11	11	19
Mars	13	12	13	7	5	5	13	20	5	13	12	12
Jupiter	8	10	13	10	12	13	11	9	14	9	10	11
Saturn	14	17	9	10	7	13	10	6	10	9	10	15
TOTAL	98	80	73	71	60	58	81	89	60	72	66	102
N. Node	15	8	13	11	9	9	12	11	11	12	8	11

Planets

Aspects	Parallels	4 Axis	Direction	☉/☽ Midpoint
Moon-Neptune	Sun-Neptune	Moon (Asc)	Venus (Rx)	
Mercury-Mars		Venus (Asc)	Saturn (D)	
		Saturn (Asc)		
		Chiron (MC)		**Moon's Nodes**
		Chiron (IC)		Jupiter (N)

Murderers

Theme: Taurus/Scorpio, 4th House

Scorpio:
Scorpios are secretive, powerful, and full of desire. The Moon is especially frequent in Scorpio, showing the tendency to form intense emotional attachments to others.

Taurus:
Taurus is strong in the 1st harmonic, especially for Mercury (1 in 341). Taureans are strong and physical. The Taurus-Scorpio Axis of Magnetism is clearly how the Fixed signs earned the reputation of being stubborn, strong, and unmovable. Murderers, as a whole, are very powerful people with strong desires.

Venus & Jupiter in Libra:
Venus and Jupiter in Libra indicate that a person easily enters relationships, but that are oppositional, often involving power struggles over beliefs.

Capricorn is weak in both the 1st & 9th Harmonics, and the 10th House is weak:
This shows a lack of responsible or cautious behavior and an inability to control their impulses.

4th House:
The 4th house is the place of private, hidden activities.

Chiron on the Midheaven (1 in 474) & the Imum Coeli (1 in 37,291):
Wounding and being wounded are strong themes in their public and private lives.

Signs

Size: 345	1st Harmonic	9th Harmonic	Houses	Rising Sign
Moon	Scorpio	Scorpio	4th	
Sun	Taurus		4th	Lunar Phase
Mercury	Taurus			Capricorn
Venus		Libra, Aquarius		
Mars	Taurus		4th	☉/☽ Midpoint
Jupiter	Libra	Libra, Scorpio		Aries
Saturn	Sagittarius	Leo	9th	Moon's Nodes
TOTAL	Taurus	Scorpio	4th	Scorpio
(weak)	Capricorn	Capricorn	10th	

1H	♈	♉	♊	♋	♌	♍	♎	♏	♐	♑	♒	♓
Moon	30	29	27	25	28	24	27	37	31	18	35	34
Sun	32	39	26	27	35	23	25	32	24	22	35	25
Mercury	23	40	19	31	28	27	28	26	33	27	32	31
Venus	34	26	25	36	24	36	21	28	18	26	39	32
Mars	23	34	33	36	38	42	21	23	30	22	20	23
Jupiter	27	26	27	33	33	38	42	26	29	24	17	23
Saturn	22	32	31	29	29	25	21	37	43	30	26	20
TOTAL	191	226	188	217	215	215	185	209	208	169	204	188
Rising	19	22	21	40	34	34	39	43	34	23	18	18
N. Node	28	20	27	27	30	37	30	38	30	25	25	28

9H	♈	♉	♊	♋	♌	♍	♎	♏	♐	♑	♒	♓
Moon	30	24	33	28	27	29	25	39	32	26	27	25
Sun	32	28	27	30	30	25	33	32	25	23	35	25
Mercury	22	33	31	33	26	31	20	29	21	35	32	32
Venus	32	23	25	18	29	27	39	29	32	27	42	22
Mars	28	33	25	37	28	31	25	31	28	22	32	25
Jupiter	21	29	35	29	24	35	37	38	28	23	20	26
Saturn	30	28	35	35	40	27	19	27	32	19	28	25
TOTAL	195	198	211	210	204	205	198	225	198	175	216	180

Houses	1st	2nd	3rd	4th	5th	6th	7th	8th	9th	10th	11th	12th
Moon	23	22	31	42	26	34	31	30	29	23	28	26
Sun	30	32	35	37	23	25	30	27	24	26	27	29
Mercury	35	37	27	33	33	23	29	26	28	23	26	25
Venus	31	34	33	29	30	30	26	32	22	18	33	27
Mars	32	28	22	36	28	31	25	34	17	30	38	24
Jupiter	27	26	28	31	31	36	24	20	27	34	30	31
Saturn	35	26	21	23	26	25	26	28	38	28	35	34
TOTAL	213	205	197	231	197	204	191	197	185	182	217	196
N. Node	25	31	37	32	34	35	25	13	31	27	39	16

Planets

Aspects	Parallels	4 Axis	Direction	☉/☽ Midpoint
Moon-Mars		Chiron (MC)	Jupiter (Rx)	
Sun-Jupiter		Uranus (IC)	Neptune (Rx)	
Mercury-Mars		Pluto (IC)		
Mercury-Neptune		Chiron (IC)		**Moon's Nodes**
				Uranus (N)

MURDER VICTIMS

Theme: 10th House

10th House:
The 10th house is the most visible space, making murder victims targets. Murderers are strongest in the opposite 4th house, where they hide, and weak in the 10th house. (The 8th house of death is also strong.)

Neptune aspects:
Neptune is the planet most associated with victimization and loss. Neptune is also associated with illusion and delusion, which is probably a strong factor for homicide victims being involved in vulnerable situations.

Venus (1 in 394) & Saturn (1 in 618) on the Midheaven:
Venus shows a person to have an attractive public image, while Saturn causes people to have a serious, burdened, or restricted public image.

Signs

Size: 435	1st Harmonic	9th Harmonic	Houses	Rising Sign
Moon	Pisces			Capricorn
Sun	Aquarius			
Mercury			10th	Lunar Phase
Venus			9th	
Mars	Taurus, Libra, Scorpio		8th	☉/☽ Midpoint
Jupiter	Taurus			Pisces
Saturn	Virgo	Aquarius	7th, 10th	Moon's Nodes
TOTAL	Aquarius		8th, 10th	Aries
(weak)			6th	10th

1H	♈	♉	♊	♋	♌	♍	♎	♏	♐	♑	♒	♓
Moon	39	45	40	29	33	34	33	36	36	27	37	46
Sun	41	30	33	41	34	33	45	27	34	34	47	36
Mercury	25	35	27	32	29	43	40	41	28	47	45	43
Venus	37	30	37	37	32	45	25	39	32	33	48	40
Mars	25	44	37	33	37	35	55	49	32	32	31	25
Jupiter	33	43	27	35	44	41	28	33	39	34	43	35
Saturn	28	40	35	31	21	47	39	42	47	45	32	28
TOTAL	228	267	236	238	230	278	265	267	248	252	283	253
Rising	16	27	31	44	49	40	54	45	39	46	28	16
N. Node	47	34	26	40	42	39	38	44	35	25	29	36

9H	♈	♉	♊	♋	♌	♍	♎	♏	♐	♑	♒	♓
Moon	36	35	44	29	37	34	32	40	34	37	35	42
Sun	31	42	37	36	41	38	44	39	31	29	33	34
Mercury	34	26	42	40	43	34	32	36	31	43	36	38
Venus	36	36	42	37	39	39	28	42	30	40	30	36
Mars	38	27	39	35	40	34	45	33	41	41	38	24
Jupiter	43	40	35	45	26	40	30	37	34	40	29	36
Saturn	43	32	37	33	37	34	28	33	41	33	48	36
TOTAL	261	238	276	255	263	253	239	260	242	263	249	246

Houses	1st	2nd	3rd	4th	5th	6th	7th	8th	9th	10th	11th	12th
Moon	36	35	44	44	33	21	35	33	39	39	40	36
Sun	44	43	40	28	31	26	33	38	29	45	47	31
Mercury	39	42	42	35	27	30	34	35	23	46	40	42
Venus	45	44	38	29	32	35	25	38	44	36	32	37
Mars	36	34	36	34	32	28	32	47	44	35	42	35
Jupiter	30	40	37	31	35	42	41	41	39	40	27	32
Saturn	34	34	43	35	38	25	50	37	33	46	29	31
TOTAL	264	272	280	236	228	207	250	269	251	287	257	244
N. Node	30	43	31	34	31	40	35	30	28	46	48	39

Planets

Aspects	Parallels	4 Axis	Direction	☉/☽ Midpoint
Neptune		Venus (MC)		
Venus-Pluto		Saturn (MC)		
Mars-Uranus				
				Moon's Nodes
				Jupiter (D)

MUSICIANS

Theme: Leo (5th), Capricorn, Pisces

The Musicians group includes all Composers, Instrumentalists, Singers, and Songwriters.

Leo & the 5th House:
Leo represents confidence in individual creative expression. Mercury in Leo (1 in 329 - 9H) shows their confidence using their hands (Instrumentalists) and voices (Singers). The 5th house is the place of joyful creative expression. Musicians express their creativity in a manner that gives them a great deal of attention and appeal.

Sun in Pisces:
The Sun in Pisces is frequent in the 1st and 9th harmonics. Pisceans are sensitive people who enjoy being artistic and creative.

Capricorn:
Capricorn is especially frequent for the Sun (1 in 193) and Venus (1 in 307 - 9H). Capricorns have great motivation to work diligently towards creative accomplishments. Taurus and Pisces are also strong in the 1st harmonic. These signs both enjoy artistry.

Size: 1,160	1st Harmonic (Signs)	9th Harmonic (Signs)	Houses	Rising Sign
Moon	Aquarius		5th	Gemini
Sun	Capricorn, Pisces	Libra, Pisces		Lunar Phase
Mercury	Taurus, Capricorn	Leo, Sagittarius		
Venus	Taurus, Pisces	Capricorn	12th	☉/☽ Midpoint
Mars			10th	Capricorn
Jupiter		Virgo	4th	Pisces
Saturn	Gemini, Leo	Virgo	4th	Moon's Nodes
TOTAL	Taurus, Capr, Pisces		5th	Aries
(weak)	Libra	Scorpio	6th	Leo

1H	♈	♉	♊	♋	♌	♍	♎	♏	♐	♑	♒	♓
Moon	91	83	82	88	99	91	91	110	101	104	112	108
Sun	105	109	91	83	94	84	103	71	103	117	87	113
Mercury	92	109	67	92	86	83	92	101	102	124	104	108
Venus	96	124	85	95	79	115	69	92	95	97	108	105
Mars	97	96	115	101	111	122	100	109	94	65	75	75
Jupiter	70	86	80	107	88	100	104	122	115	99	96	93
Saturn	85	93	118	86	114	101	89	89	93	108	87	97
TOTAL	636	700	638	652	671	696	648	694	703	714	669	699
Rising	61	68	101	118	131	130	124	110	108	86	77	46
N. Node	113	96	107	101	123	101	94	85	81	74	95	90

9H	♈	♉	♊	♋	♌	♍	♎	♏	♐	♑	♒	♓
Moon	91	89	94	109	97	102	91	95	99	97	99	97
Sun	104	96	86	98	88	99	112	91	94	89	92	111
Mercury	104	98	90	106	123	90	93	84	112	87	89	84
Venus	101	88	98	107	78	81	102	75	100	122	102	106
Mars	102	98	96	104	98	88	78	108	93	99	100	96
Jupiter	82	99	85	104	93	120	94	99	94	84	98	108
Saturn	106	104	101	88	95	118	89	81	102	108	83	85
TOTAL	690	672	650	716	672	698	659	633	694	686	663	687

Houses	1st	2nd	3rd	4th	5th	6th	7th	8th	9th	10th	11th	12th
Moon	101	97	104	96	113	84	87	75	99	95	99	110
Sun	119	108	101	98	81	73	85	95	86	92	111	111
Mercury	125	99	103	100	82	85	85	80	99	88	98	116
Venus	105	115	99	84	85	86	85	96	99	91	89	126
Mars	76	102	112	96	99	77	81	94	93	128	108	93
Jupiter	94	82	80	114	111	100	86	95	92	106	91	109
Saturn	110	87	89	114	105	85	86	98	89	95	104	98
TOTAL	730	690	688	702	676	590	595	633	657	695	700	763
N. Node	90	84	99	97	104	79	101	98	104	94	106	104

Planets

Aspects	Parallels	4 Axis	Direction	☉/☽ Midpoint
Venus-Neptune			Mercury (D)	
				Moon's Nodes
				Mercury (S)
				Pluto (N)
				Chiron (S)

Mystics

Theme: Scorpio, Mercury

Mystics include all Astrologers, Priests, Psychics, and Spiritual Teachers.

Scorpio:
Scorpio emphasizes a strong need to learn about things which are not understood and hidden from view. They see the inner side of life, have keen insight and trust the guidance of their intuition.

Mercury in Pisces (1 in 231 - 9H):
This position indicates much rumination on inner, spiritual experience. They are productive when thinking and meditating in solitude.

Venus in Cancer (1 in 274 - 9H):
They enjoy spending time at home reflecting on personal feelings and ideas.

Mercury-Uranus parallels (1 in 1,895):
Mystics have a strong intuition and original insights.

Mercury & Jupiter on the Ascendant:
They express themselves through their keen intellects and clear communication.

Signs

Size: 862	1st Harmonic	9th Harmonic	Houses	Rising Sign
Moon			7th	
Sun	Aries	Libra	8th	Lunar Phase
Mercury	Gemini	Pisces		Scorpio
Venus	Scorpio	Cancer		☉/☽ Midpoint
Mars		Aries, Scorpio		Virgo
Jupiter		Sagittarius		Moon's Nodes
Saturn	Virgo, Libra	Sagittarius		Gemini
TOTAL	Scorpio		1st, 7th, 12th	Scorpio (S)
(weak)	Capricorn		2nd	Pisces

1H	♈	♉	♊	♋	♌	♍	♎	♏	♐	♑	♒	♓
Moon	74	85	75	69	70	69	68	65	79	68	71	69
Sun	86	69	65	73	71	71	67	80	68	64	73	75
Mercury	62	60	79	57	58	66	81	89	84	68	76	82
Venus	72	80	60	86	56	84	66	87	69	51	85	66
Mars	59	72	79	78	99	84	77	75	77	58	50	54
Jupiter	50	76	56	73	82	72	80	87	84	70	68	64
Saturn	56	46	64	72	76	87	88	86	69	78	64	76
TOTAL	459	488	478	508	512	533	527	569	530	457	487	486
Rising	45	58	64	69	99	91	94	92	85	71	54	40
N. Node	81	84	91	74	57	70	57	54	71	69	69	85

9H	♈	♉	♊	♋	♌	♍	♎	♏	♐	♑	♒	♓
Moon	73	85	70	70	68	78	64	69	77	64	69	75
Sun	65	84	74	80	83	69	87	66	53	67	60	74
Mercury	64	69	70	64	63	75	73	68	71	79	75	91
Venus	71	69	75	91	72	71	56	77	74	73	70	63
Mars	86	68	60	69	72	70	61	86	72	75	65	78
Jupiter	63	70	71	78	84	68	74	62	84	69	58	81
Saturn	67	72	82	60	67	72	65	62	90	79	86	60
TOTAL	489	517	502	512	509	503	480	490	521	506	483	522

Houses	1st	2nd	3rd	4th	5th	6th	7th	8th	9th	10th	11th	12th
Moon	66	62	80	81	62	72	88	72	71	69	69	70
Sun	94	64	69	69	57	61	63	78	59	68	87	93
Mercury	87	75	77	67	68	58	67	62	64	61	84	92
Venus	84	74	78	54	70	66	67	76	67	69	81	76
Mars	80	70	61	66	76	74	76	74	60	67	69	88
Jupiter	84	63	51	76	78	60	79	71	66	74	76	84
Saturn	83	72	72	85	64	58	69	65	82	73	66	73
TOTAL	578	480	488	498	475	449	509	498	469	481	532	576
N. Node	74	68	78	69	72	80	71	60	70	69	70	81

Planets

Aspects	Parallels	4 Axis	Direction	☉/☽ Midpoint
Mercury-Venus	Moon-Jupiter	Mercury (Asc)	Venus (Rx)	
	Mercury-Uranus	Jupiter (Asc)		
		Saturn (MC)		
		Neptune (MC)		Moon's Nodes
		Moon (IC)		Venus (N)
				Mars (S)

National Socialists (Nazis)

Theme: Aries, Mars/Chiron

Aries:
Aries is the sign of leadership and aggression. At their best, Arians are pioneering and independent. At worst, they can be hurtful and insensitive.

Moon in Libra (1 in 1,862):
This shows a need to express opinions and assert beliefs. They can be emotionally detached from others.

Mars in the 7th House (1 in 10,833):
The 7th house shows how we relate to each other. With Mars, we can be aggressive to others and experience much conflict.

Chiron:
A strong influence from Chiron indicates having many experiences that require healing. Nazis have frequent Mars-Chiron parallels (1 in 413), which is a major indicator of causing and enduring injuries.

Mercury-Pluto parallels (1 in 1,970):
This is an indication of devious manipulative thinking.

Signs

Size: 129	1st Harmonic	9th Harmonic	Houses	Rising Sign
Moon	Aries, Libra, Capricorn			
Sun	Aquarius			
Mercury		Capricorn		Lunar Phase
Venus				Aries
Mars			7th	☉/☽ Midpoint
Jupiter	Aries	Virgo, Libra	3rd	Aquarius
Saturn	Virgo	Sagittarius	12th	Moon's Nodes
TOTAL	Aries, Aquarius			Taurus
(weak)	Gemini, Leo		1st	

1H	♈	♉	♊	♋	♌	♍	♎	♏	♐	♑	♒	♓
Moon	16	14	2	10	8	12	21	10	7	16	6	7
Sun	11	8	10	15	7	10	11	10	9	7	16	15
Mercury	13	10	7	9	12	13	6	12	9	11	14	13
Venus	9	13	12	16	7	11	6	13	12	7	14	9
Mars	13	10	11	8	7	14	9	13	11	11	12	10
Jupiter	20	5	9	9	8	8	15	14	11	4	15	11
Saturn	7	9	6	11	9	20	10	12	10	9	11	15
TOTAL	89	69	57	78	58	88	78	84	69	65	88	80
Rising	4	5	11	10	14	17	18	15	16	10	5	4
N. Node	9	20	10	7	15	15	7	7	12	10	11	6

9H	♈	♉	♊	♋	♌	♍	♎	♏	♐	♑	♒	♓
Moon	13	7	9	12	14	10	6	5	15	11	13	14
Sun	14	10	8	11	12	12	9	9	7	11	15	11
Mercury	7	12	7	12	14	12	10	10	10	17	11	7
Venus	11	12	10	8	9	6	5	11	13	14	15	15
Mars	11	12	9	13	5	7	9	14	11	11	12	15
Jupiter	10	10	14	5	12	18	16	10	8	13	4	9
Saturn	6	11	10	12	12	16	9	8	16	7	14	8
TOTAL	72	74	67	73	78	81	64	67	80	84	84	79

Houses	1st	2nd	3rd	4th	5th	6th	7th	8th	9th	10th	11th	12th
Moon	10	7	15	12	9	7	10	7	15	14	13	10
Sun	11	9	10	9	9	13	6	13	8	13	15	13
Mercury	11	14	9	9	6	12	11	8	8	13	16	12
Venus	5	11	10	13	7	11	10	12	12	8	14	16
Mars	10	10	11	13	8	11	22	5	13	8	10	8
Jupiter	8	7	18	11	13	11	9	8	14	12	11	7
Saturn	7	12	14	12	12	7	4	10	8	13	13	17
TOTAL	62	70	87	79	64	72	72	63	78	81	92	83
N. Node	14	6	8	15	16	10	11	16	14	5	10	4

Planets

Aspects	Parallels	4 Axis	Direction	☉/☽ Midpoint
Venus-Saturn	Mercury-Pluto	Neptune (Asc)	Chiron (D)	
Saturn-Chiron	Mars-Chiron	Mars (Dsc)		
				Moon's Nodes
				Mercury (N)
				Mars (S)
				Chiron (S)

Nobel Prize Winners

Theme: Taurus, Jupiter

Taurus:
Taureans patiently endure long periods of effort in order to achieve real world results. Venus is the most frequent planet in Taurus in both harmonics (1 in 901 - 1H; 1 in 742 - 9H). This is an indication of receiving tangible rewards from their work.

Aries:
Arians are achievers who set high goals for themselves. They are very independent people who ignore the preconceptions of others.

Jupiter:
Jupiter is a planet symbolizing the intellect. It influences us to enthusiastically explore fields of knowledge. Jupiter is most frequent on the Ascendant (1 in 841), where it is easily expressed by the personality.

Moon-Saturn parallels (1 in 857):
This indicates the ability to endure long periods of work without any recognition.

Signs

Size: 129	1st Harmonic	9th Harmonic	Houses	Rising Sign
Moon			2nd, 10th	
Sun	Aries	Taurus, Scorpio	4th	
Mercury	Pisces	Sagittarius		Lunar Phase
Venus	Taurus	Aries, Taurus		Sagittarius
Mars		Aries, Gemini		☉/☽ Midpoint
Jupiter		Taurus	1st	Aquarius
Saturn		Taurus, Virgo		Moon's Nodes
TOTAL	Taurus	Taurus		1st, 11th
(weak)		Capricorn		2nd (S)

1H	♈	♉	♊	♋	♌	♍	♎	♏	♐	♑	♒	♓
Moon	14	13	9	10	12	6	9	7	11	9	14	15
Sun	16	12	15	10	9	11	14	9	8	9	7	9
Mercury	9	12	11	10	9	12	13	8	14	9	6	16
Venus	10	21	11	13	12	10	7	8	11	5	12	9
Mars	6	12	9	15	10	13	17	9	10	14	9	5
Jupiter	13	10	12	6	9	12	10	11	11	14	12	9
Saturn	8	10	7	7	10	12	8	16	15	13	14	9
TOTAL	76	90	74	71	71	76	78	68	80	73	74	72
Rising	7	8	11	15	16	14	11	13	16	10	1	7
N. Node	8	9	11	11	8	12	13	12	12	9	12	12

9H	♈	♉	♊	♋	♌	♍	♎	♏	♐	♑	♒	♓
Moon	13	7	9	12	14	5	14	13	10	9	12	11
Sun	12	18	13	7	8	8	13	16	10	4	12	8
Mercury	9	10	14	6	10	13	10	11	17	10	10	9
Venus	16	20	11	10	8	9	8	10	10	6	11	10
Mars	17	4	16	11	8	13	10	16	6	7	9	12
Jupiter	8	16	9	16	13	10	8	12	4	7	15	11
Saturn	5	18	8	9	10	17	7	9	12	8	16	10
TOTAL	80	93	80	71	71	75	70	87	69	51	85	71

Houses	1st	2nd	3rd	4th	5th	6th	7th	8th	9th	10th	11th	12th
Moon	9	18	11	8	11	14	6	8	5	17	14	8
Sun	13	16	11	16	5	9	9	7	14	11	9	9
Mercury	17	12	9	15	11	8	7	6	12	13	7	12
Venus	11	11	16	8	6	9	12	14	8	11	10	13
Mars	10	14	11	13	9	6	15	14	7	13	11	6
Jupiter	18	8	15	10	14	7	8	9	10	4	13	13
Saturn	13	7	15	9	11	14	3	15	12	9	11	10
TOTAL	91	86	88	79	67	67	60	73	68	78	75	71
N. Node	16	13	10	13	9	5	10	16	9	5	16	7

Planets

Aspects	Parallels	4 Axis	Direction	☉/☽ Midpoint
Sun-Jupiter	Moon-Saturn	Jupiter (Asc)	Chiron (D)	Chiron
				Moon's Nodes
				Jupiter (S)

NOVELISTS

Theme: Aquarius, Saturn

Aquarius:
Aquarians are imaginative and expressive people who are talented at articulating their creative thoughts through language.

Aries is the weakest sign in the 1st and 9th Harmonics, and the 1st House is weak:
This suggests that novelists lack impulsive will. Arians need a lot of freedom and are not interested in being tied down to demanding long-term projects and, therefore, often lack the patience to write books. A weak 1st house indicates that a person has little interest in projecting the self into their environment.

Saturn:
Saturnine people are conservative, serious people who are capable of long periods of work. They seek accomplishments that are enduring and purposeful.

Signs

Size: 331	1st Harmonic	9th Harmonic	Houses	Rising Sign
Moon	Pisces	Aquarius		Aquarius
Sun				Lunar Phase
Mercury			12th	Virgo
Venus	Taurus		3rd	☉/☽ Midpoint
Mars			4th	Aries
Jupiter		Capricorn, Aquarius		Moon's Nodes
Saturn		Gemini	10th	Aquarius
TOTAL		Aquarius		Pisces
(weak)		Aries	1st	4th, 8th

1H	♈	♉	♊	♋	♌	♍	♎	♏	♐	♑	♒	♓
Moon	21	27	30	28	26	32	26	25	24	28	27	37
Sun	27	20	29	27	32	30	30	21	32	23	27	33
Mercury	24	20	30	21	27	30	29	29	28	29	36	28
Venus	25	37	16	35	25	37	18	22	36	25	31	24
Mars	20	25	26	27	36	32	40	30	30	26	22	17
Jupiter	27	26	29	30	27	24	29	32	28	29	24	26
Saturn	26	29	24	23	26	21	32	26	31	38	29	26
TOTAL	170	184	184	191	199	206	204	185	209	198	196	191
Rising	18	17	22	32	33	39	40	42	33	19	26	10
N. Node	30	24	25	31	26	30	30	16	29	20	35	35

9H	♈	♉	♊	♋	♌	♍	♎	♏	♐	♑	♒	♓
Moon	31	23	17	29	34	25	18	28	20	36	40	30
Sun	25	28	28	29	19	32	35	30	28	20	29	28
Mercury	21	28	25	25	23	19	32	32	33	33	28	32
Venus	22	32	14	27	31	31	23	35	30	23	30	33
Mars	21	25	34	29	32	30	23	28	26	28	26	29
Jupiter	17	31	30	23	25	20	29	28	23	39	36	30
Saturn	29	31	37	18	21	30	30	32	19	23	35	26
TOTAL	166	198	185	180	185	187	190	213	179	202	224	208

Houses	1st	2nd	3rd	4th	5th	6th	7th	8th	9th	10th	11th	12th
Moon	21	21	25	30	30	32	31	26	25	32	24	34
Sun	33	28	28	27	21	23	24	32	19	25	38	33
Mercury	25	35	33	20	23	28	22	26	21	27	29	42
Venus	26	31	40	22	21	17	29	26	25	25	35	34
Mars	28	29	20	36	32	22	28	27	27	30	30	22
Jupiter	20	28	31	32	28	19	34	26	29	29	24	31
Saturn	31	16	29	25	23	33	24	22	35	36	28	29
TOTAL	184	188	206	192	178	174	192	185	181	204	208	225
N. Node	31	31	20	38	21	23	24	39	31	28	29	16

Planets

Aspects	Parallels	4 Axis	Direction	☉/☽ Midpoint
Mercury-Saturn	Sun-Saturn		Saturn (D)	
	Moon-Saturn			
				Moon's Nodes
				Venus (N)
				Mars (S)
				Saturn (N)

OBESITY

Theme: Moon/Jupiter

This group is composed entirely of people born in the Southern Hemisphere. Therefore, reversing the signs should be considered since the seasons are reversed.

Virgo is strong & Pisces is weak:
Virgo is the most frequent sign for health problems. Pisces represents health and well-being since it is the strongest sign for people living long lives. (If we consider reversing the signs for this study group, Pisces would be strongest and Virgo would be weakest. While this goes against the majority of the findings, it does fit with the conventional idea of Virgos being thin and Pisceans being more rounded figures.)

Mars in Cancer (Capricorn) in the 9th Harmonic (1 in 15,193):
This indicates struggles (Mars) with eating for comfort (Cancer). (Mars in the opposite sign, Capricorn, would indicate a strong repression of aggressive impulses, which could be sublimated towards food consumption.)

Sun in the 4th House (1 in 39,301) & Sun on the Imum Coeli (1 in 6,712):
Obese people spend much of their lives at home and veer away from activities that receive public attention.

Moon & Jupiter:
Lunar people often seek emotional comfort from food. Jupiter, being the planet of abundance and excess, inclines people toward overdoing that which they enjoy. Jupiter aspects are the most frequent (1 in 345).

Venus aspects are infrequent:
This indicates that they lack attractiveness and sensuality. Venus aspects are common for people in long-term relationships, which many obese people have a difficult time finding.

Signs

Size: 161	1st Harmonic	9th Harmonic	Houses	Rising Sign
Moon			1st, 5th	Scorpio
Sun	Gemini		4th	
Mercury		Aries, Virgo	12th	**Lunar Phase**
Venus				
Mars	Pisces	Cancer, Scorpio	5th	☉/☽ Midpoint
Jupiter	Libra			Cancer
Saturn		Capricorn		Moon's Nodes
TOTAL	Virgo, Libra		4th, 6th	Aquarius
(weak)	Scorpio, Pisces		8th	5th

1H	♈	♉	♊	♋	♌	♍	♎	♏	♐	♑	♒	♓
Moon	11	8	19	18	11	12	14	10	17	15	13	13
Sun	19	10	20	15	12	18	17	12	5	14	16	3
Mercury	18	9	14	14	14	19	15	14	8	18	8	10
Venus	16	16	17	17	10	20	16	9	8	13	13	6
Mars	12	9	12	14	16	19	17	17	11	6	11	17
Jupiter	12	16	14	13	12	15	23	7	16	14	11	8
Saturn	11	16	7	14	11	15	13	13	19	13	16	13
TOTAL	99	84	103	105	86	118	115	82	84	93	88	70
Rising	12	16	11	11	14	4	15	21	10	14	18	15
N. Node	17	15	12	17	6	13	17	17	11	12	19	5

9H	♈	♉	♊	♋	♌	♍	♎	♏	♐	♑	♒	♓
Moon	15	14	14	8	12	11	11	11	13	18	17	17
Sun	11	14	9	19	14	14	8	13	14	14	17	14
Mercury	21	9	14	14	19	20	13	8	9	12	8	14
Venus	11	11	13	15	12	17	11	17	9	13	16	16
Mars	16	9	9	27	11	14	9	20	18	9	10	9
Jupiter	14	16	12	13	13	15	15	12	12	17	12	10
Saturn	13	13	12	12	17	13	13	10	17	21	9	11
TOTAL	101	86	83	108	98	104	80	91	92	104	89	91

Houses	1st	2nd	3rd	4th	5th	6th	7th	8th	9th	10th	11th	12th
Moon	21	10	11	18	23	11	13	10	10	10	10	14
Sun	17	8	13	28	10	17	7	12	7	16	13	13
Mercury	12	8	17	18	16	17	6	12	11	10	13	21
Venus	11	12	16	17	14	18	12	7	11	13	16	14
Mars	11	13	18	10	20	11	15	9	17	12	16	9
Jupiter	14	15	16	9	8	17	16	6	19	12	16	13
Saturn	14	12	16	14	12	19	12	10	17	6	13	16
TOTAL	100	78	107	114	103	110	81	66	92	79	97	100
N. Node	14	8	19	7	21	11	17	14	8	17	15	10

Planets

Aspects	Parallels	4 Axis	Direction	☉/☽ Midpoint
Jupiter	Venus-Pluto	Moon (Asc)	Jupiter (Rx)	Jupiter
Moon-Sun		Sun (IC)	Pluto (D)	Chiron
Moon-Jupiter				
				Moon's Nodes
				Moon (S)

ONLY CHILD

Theme: Aquarius, Venus

Aquarius:
Aquarius is strong in the 1st harmonic (1 in 8,158) and the 9th harmonic (1 in 369). Aquarians enjoy their space, yet relationships are very important to them. Being an only child causes a person to simultaneously wish for more relationships and to develop a close relationship with one's parents. Mars is the most frequent planet in Aquarius (1 in 3,639). Aquarius rising is also very frequent (1 in 1,189).

Sun & Moon in Leo:
Only children typically receive an abundance of attention and are often treated as special by their parents.

Mercury in Pisces (1 in 1,096):
This indicates imaginative thinking in solitude.

Jupiter in Virgo (1 in 11,514):
This position indicates a critical attitude towards growth and expansion.

7th House:
Planets in this house show an emphasis on relationships. Venus is the most frequent planet (1 in 342). This suggests a good ability to have rewarding relationships.

Venus aspects:
Relationships and being loved are a major theme in their lives.

Signs

Size: 144	1st Harmonic	9th Harmonic	Houses	Rising Sign
Moon	Leo, Scorpio		8th	Aquarius
Sun	Leo, Aquarius	Aquarius		
Mercury	Pisces		11th	Lunar Phase
Venus	Capricorn	Taurus	7th	
Mars	Aquarius	Aquarius	7th	☉/☽ Midpoint
Jupiter	Virgo	Scorpio		Capricorn
Saturn		Aquarius	9th	Moon's Nodes
TOTAL	Virgo, Aquarius	Taurus, Aquarius	7th, 11th	
(weak)	Aries, Taurus	Cancer, Libra		

1H	♈	♉	♊	♋	♌	♍	♎	♏	♐	♑	♒	♓
Moon	4	11	5	12	19	17	6	20	17	7	15	11
Sun	10	6	10	13	19	8	11	10	11	13	17	16
Mercury	7	5	10	11	12	13	12	11	14	10	17	22
Venus	11	8	12	12	10	16	12	14	7	18	13	11
Mars	9	7	16	10	11	19	11	8	12	12	20	9
Jupiter	12	9	9	9	9	26	10	11	11	12	16	10
Saturn	9	9	14	10	10	14	12	10	12	13	17	14
TOTAL	62	55	76	77	90	113	74	84	84	85	115	93
Rising	9	8	7	15	20	15	13	15	4	12	17	9
N. Node	15	16	10	14	13	15	11	13	8	14	6	9

9H	♈	♉	♊	♋	♌	♍	♎	♏	♐	♑	♒	♓
Moon	12	16	9	9	8	9	14	17	8	17	14	11
Sun	6	16	9	16	4	13	10	13	11	10	22	14
Mercury	8	11	16	13	12	10	14	8	14	9	15	14
Venus	16	18	11	5	17	9	10	6	17	15	9	11
Mars	12	14	11	6	10	15	9	9	15	13	19	11
Jupiter	16	14	11	12	10	7	5	18	13	14	11	13
Saturn	12	10	8	8	16	11	7	11	10	16	19	16
TOTAL	82	99	75	69	77	74	69	82	88	94	109	90

Houses	1st	2nd	3rd	4th	5th	6th	7th	8th	9th	10th	11th	12th
Moon	11	14	9	12	12	12	9	18	8	16	14	9
Sun	14	6	12	10	11	15	14	10	8	10	17	17
Mercury	17	9	10	10	10	11	15	14	11	8	20	9
Venus	9	11	9	9	16	11	19	6	3	15	17	19
Mars	12	12	15	11	10	11	17	14	7	10	15	10
Jupiter	11	10	11	9	17	12	15	5	12	15	10	17
Saturn	12	14	11	15	8	10	10	5	18	12	14	15
TOTAL	86	76	77	76	84	82	99	72	67	86	107	96
N. Node	7	11	10	8	13	13	15	14	17	13	8	15

Planets

Aspects	Parallels	4 Axis	Direction	☉/☽ Midpoint
Moon-Venus	Moon-Chiron	Chiron (Asc)	Mercury (Rx)	
Venus-Neptune	Venus-Uranus	Neptune (Dsc)	Uranus (Rx)	
Mars-Chiron				
				Moon's Nodes
				Venus (N)

Out-of-Body & Near Death Experiences

Theme: Virgo/Capricorn, Venus

Virgo:
Virgo is strong in the 1st and 9th Harmonics (1 in 72,676 - 1H).
The Near Death Experience is typically caused by serious bodily harm, which is common with Virgo. The result of a NDE is an out-of-body experience, which often involves an encounter with the divine or other after-life conditions. Out-of-Body experiences can also occur in a non-violent manner without a NDE.

Capricorn:
Capricorn is frequent in the 9th harmonic. Capricorn is the sign of the cave, being the lowest sign in the annual sun cycle. It is the place where solitary work is done. Leaving the body is withdrawing within in order to gain valuable insight.

Venus:
Venus aspects are frequent (1 in 148), especially Venus-Saturn aspects (1 in 2,235). Venus relates to all aspects of beauty and pleasure. Many Out-of-Body experiences contain great revelations and beauty. However, when combined with such a strong Virgo influence, the result is often being critical towards beauty. The Saturn-Venus aspect, specifically, suggests the restriction of pleasure.

Moon-Neptune parallels (1 in 252)
This is an indication of an abundant imagination, but also an insecure attachment to the body and physical environment.

Signs

Size: 188	1st Harmonic	9th Harmonic	Houses	Rising Sign
Moon	Gemini, Virgo		10th	
Sun		Leo, Pisces		Lunar Phase
Mercury		Cancer	3rd, 5th	
Venus		Virgo, Capricorn		☉/☽ Midpoint
Mars	Virgo			Gemini
Jupiter	Virgo, Sag, Capricorn	Gemini	8th	Virgo
Saturn	Cancer, Virgo	Virgo, Capricorn		Moon's Nodes
TOTAL	Virgo	Virgo, Capricorn	5th	Capricorn (S)
(weak)	Aries			5th

1H	♈	♉	♊	♋	♌	♍	♎	♏	♐	♑	♒	♓
Moon	11	17	22	11	15	22	17	12	19	11	17	14
Sun	14	15	18	20	16	17	16	11	19	16	13	13
Mercury	13	18	9	17	17	20	16	12	18	18	13	17
Venus	9	14	15	19	17	22	13	15	14	13	19	18
Mars	13	19	10	18	19	28	21	17	13	12	6	12
Jupiter	14	8	8	10	14	26	15	17	24	24	13	15
Saturn	12	12	13	21	15	24	12	16	21	15	13	14
TOTAL	86	103	95	116	113	159	110	100	128	109	94	103
Rising	10	10	16	12	22	23	22	16	24	14	11	8
N. Node	16	15	20	22	14	12	17	16	15	13	10	18

9H	♈	♉	♊	♋	♌	♍	♎	♏	♐	♑	♒	♓
Moon	18	20	19	8	19	14	15	14	17	21	10	13
Sun	14	7	14	15	22	21	18	19	7	15	12	24
Mercury	20	14	11	22	16	17	14	13	19	17	15	10
Venus	15	12	13	14	14	23	11	18	20	22	17	9
Mars	15	20	15	9	12	16	17	15	20	18	16	15
Jupiter	15	17	25	15	18	17	15	7	16	17	17	9
Saturn	16	15	17	14	14	26	13	13	11	24	12	13
TOTAL	113	105	114	97	115	134	103	99	110	134	99	93

Houses	1st	2nd	3rd	4th	5th	6th	7th	8th	9th	10th	11th	12th
Moon	15	14	11	15	21	13	16	18	11	22	13	19
Sun	18	15	19	18	11	19	12	13	16	20	7	20
Mercury	14	20	24	10	22	13	10	12	17	15	16	15
Venus	8	20	21	16	15	16	14	13	14	13	18	20
Mars	24	11	11	8	20	16	14	20	7	20	22	15
Jupiter	15	12	18	19	16	10	20	22	8	11	16	21
Saturn	19	20	13	14	19	14	17	16	11	10	15	20
TOTAL	113	112	117	100	124	101	103	114	84	111	107	130
N. Node	13	15	9	17	22	10	16	18	18	12	20	18

Planets

Aspects	Parallels	4 Axis	Direction	☉/☽ Midpoint
Venus	Moon-Mercury	Neptune (MC)		Mars
Sun-Venus	Moon-Neptune	Uranus (IC)		
Venus-Jupiter				
Venus-Saturn				**Moon's Nodes**
				Mercury (S)

Philosophers

Theme: Scorpio

Scorpio:
Scorpios are deep-thinking people who often seek to understand the mysteries of existence. Mars is especially frequent in Scorpio (1 in 3,567).

Saturn in the 1st House (1 in 247):
This shows a person who is serious and responsible, and projects authority.

Mercury-Saturn aspects (1 in 464) and parallels:
This planetary combination represents a serious and authoritative thinker. They have excellent concentration and memory. Mercury is also strong in Capricorn (1 in 311).

Sun on the Moon's South Node (1 in 1,957):
This indicates that a person has an innate sense of being special and a desire to achieve prominence.

Signs

Size: 110	1st Harmonic	9th Harmonic	Houses	Rising Sign
Moon	Aries	Scorpio	5th	Aquarius
Sun	Cancer, Aquarius	Scorpio, Capricorn	6th	
Mercury	Capricorn	Scorpio	7th	Lunar Phase
Venus	Leo	Virgo		Leo
Mars	Scorpio	Capricorn	7th	☉/☽ Midpoint
Jupiter	Taurus	Aquarius		Taurus
Saturn		Aries	1st, 11th	Moon's Nodes
TOTAL			2nd	Scorpio
(weak)	Sagittarius	Pisces		8th (S)

1H	♈	♉	♊	♋	♌	♍	♎	♏	♐	♑	♒	♓
Moon	14	12	6	7	7	7	9	12	6	8	12	10
Sun	8	9	4	16	5	6	12	9	8	10	15	8
Mercury	9	5	8	12	6	8	9	10	8	18	8	9
Venus	7	11	6	11	13	8	9	6	7	10	13	9
Mars	9	8	10	13	8	11	13	20	8	2	5	3
Jupiter	10	17	7	10	13	8	7	10	3	8	7	10
Saturn	7	11	9	5	11	13	4	9	9	11	12	9
TOTAL	64	73	50	74	63	61	63	76	49	67	72	58
Rising	6	2	9	12	12	11	10	11	15	5	12	5
N. Node	13	6	8	11	10	13	7	15	10	7	5	5

9H	♈	♉	♊	♋	♌	♍	♎	♏	♐	♑	♒	♓
Moon	7	11	11	8	6	8	12	14	9	9	7	8
Sun	8	9	10	4	10	10	8	14	9	15	6	7
Mercury	4	13	11	8	9	9	7	15	10	9	9	6
Venus	10	13	8	8	11	16	2	7	11	7	11	6
Mars	6	11	9	7	12	9	12	8	7	14	9	6
Jupiter	9	10	8	12	8	8	3	11	8	10	16	7
Saturn	14	5	13	13	5	10	9	5	10	4	14	8
TOTAL	58	72	70	60	61	70	53	74	64	68	72	48

Houses	1st	2nd	3rd	4th	5th	6th	7th	8th	9th	10th	11th	12th
Moon	8	11	11	4	14	7	11	6	9	12	11	6
Sun	11	14	9	6	5	13	11	6	8	10	8	9
Mercury	7	14	11	5	6	12	14	7	4	9	10	11
Venus	11	12	12	9	11	10	3	10	11	1	9	11
Mars	8	10	9	6	13	5	15	9	10	10	7	8
Jupiter	8	8	9	10	8	8	6	10	9	12	11	11
Saturn	17	12	10	12	7	7	3	5	11	7	14	5
TOTAL	70	81	71	52	64	62	63	53	62	61	70	61
N. Node	11	17	8	8	9	11	8	10	11	4	6	7

Planets

Aspects	Parallels	4 Axis	Direction	☉/☽ Midpoint
Moon-Uranus	Mercury-Saturn	Uranus (Dsc)	Chiron (D)	
Mercury-Saturn				
				Moon's Nodes
				Sun (S)

POETS

Theme: Pisces

Pisces:
Pisceans are the most contemplative and creative people. The Moon, representing the subconscious and emotions, is the planet most frequent in Pisces (1 in 206). People with the Moon in Pisces tend to be sensitive and withdrawn, preferring privacy and few people in their personal lives. The Moon's South Node is also frequent in Pisces (1 in 363).

Jupiter on the Moon's North Node (1 in 177):
This indicates having a life goal of growth directed towards wisdom.

Signs

Size: 203	1st Harmonic	9th Harmonic	Houses	Rising Sign
Moon	Gemini, Pisces	Taurus	2nd, 12th	Taurus
Sun			1st, 4th, 5th	Lunar Phase
Mercury	Pisces		3rd, 6th	Virgo
Venus	Cancer	Virgo	3rd	☉/☽ Midpoint
Mars	Pisces	Gemini	2nd	Cancer
Jupiter		Capricorn	8th	Aquarius
Saturn		Pisces	8th	Moon's Nodes
TOTAL	Pisces			Pisces (S)
(weak)		Libra	9th, 10th	

1H	♈	♉	♊	♋	♌	♍	♎	♏	♐	♑	♒	♓
Moon	17	15	25	17	18	12	18	9	11	16	18	27
Sun	22	18	17	11	23	19	15	9	15	21	11	22
Mercury	14	17	15	15	16	20	12	17	18	20	15	24
Venus	22	20	11	27	9	25	13	10	23	12	20	11
Mars	16	11	19	22	19	18	17	18	13	14	15	21
Jupiter	13	14	15	20	16	18	18	25	16	16	16	16
Saturn	12	15	15	17	14	22	15	24	21	22	12	14
TOTAL	116	110	117	129	115	134	108	112	117	121	107	135
Rising	11	18	10	10	25	24	24	23	18	18	11	11
N. Node	11	21	14	17	22	28	10	11	21	14	20	14

9H	♈	♉	♊	♋	♌	♍	♎	♏	♐	♑	♒	♓
Moon	13	24	12	17	16	14	15	21	14	19	21	17
Sun	22	22	18	21	13	18	14	13	10	18	19	15
Mercury	14	22	12	16	16	18	15	22	15	22	17	14
Venus	13	16	18	17	21	25	13	20	20	16	11	13
Mars	13	16	25	20	21	9	14	20	17	12	22	14
Jupiter	15	15	19	16	21	14	10	14	14	24	20	21
Saturn	16	16	13	17	8	17	12	20	22	18	18	26
TOTAL	106	131	117	124	116	115	93	130	112	129	128	120

Houses	1st	2nd	3rd	4th	5th	6th	7th	8th	9th	10th	11th	12th
Moon	16	24	12	15	16	19	15	18	12	18	14	24
Sun	28	17	18	24	23	11	10	15	9	12	19	17
Mercury	25	19	27	19	12	21	10	8	15	13	13	21
Venus	19	23	24	18	16	12	15	13	17	12	14	20
Mars	20	26	19	19	17	20	16	14	8	16	17	11
Jupiter	18	20	17	19	16	18	14	24	14	14	13	16
Saturn	7	12	18	11	17	20	15	28	17	17	22	19
TOTAL	133	141	135	125	117	121	95	120	92	102	112	128
N. Node	21	19	15	13	19	17	17	23	15	12	18	14

Planets

Aspects	Parallels	4 Axis	Direction	☉/☽ Midpoint
	Mars-Jupiter	Sun (Asc)		Uranus
		Mars (Dsc)		Neptune
		Saturn (MC)		
		Venus (IC)		**Moon's Nodes**
				Sun (N)
				Mercury (N)
				Jupiter (N)
				Chiron (S)

Police Officers

Theme: Leo/Scorpio (8th), Neptune

Scorpio & the 8th House:
Scorpios have an exceptional awareness of the dark side of human nature. Police officers make it their business to prevent the more dangerous elements of society from overwhelming us. The Sun (1 in 204) and Pluto (1 in 363) are also frequent in the 8th house, which represents the hidden and taboo aspects of our nature.

Leo:
Saturn is especially frequent in Leo (1 in 2,449). Police officers have confidence in their ability to be both authoritative and punitive.

Moon's North Node in the 5th House (1 in >1,000,000):
This surprising finding indicates that they aspire towards a life of joy, freedom, romance, and creativity.

Neptune on the Midheaven & Imum Coeli:
They experience much loss, confusion, and deception in their home and career.

Jupiter Retrograde is frequent (1 in 2,113), and Jupiter aspects are infrequent (1 in 338):
Police officers have the highest frequency of Jupiter retrograde of all the study groups. This shows a repression of growth, optimism, and expansion as well as the tendency to restrict the freedom of others.

Signs

Size: 88	1st Harmonic	9th Harmonic	Houses	Rising Sign
Moon	Cancer	Capricorn		Sagittarius
Sun	Aries, Scorpio		8th	
Mercury			7th	**Lunar Phase**
Venus	Aquarius	Scorpio, Capricorn	4th	
Mars		Leo, Virgo		☉/☽ Midpoint
Jupiter	Gemini, Leo			Virgo
Saturn	Leo	Gemini, Leo	3rd	**Moon's Nodes**
TOTAL	Scorpio		8th	5th
(weak)		Cancer, Libra	11th	

1H	♈	♉	♊	♋	♌	♍	♎	♏	♐	♑	♒	♓
Moon	6	4	9	11	9	11	10	9	5	5	3	6
Sun	12	4	8	5	4	7	7	12	8	7	8	6
Mercury	6	4	3	9	4	4	10	9	10	9	9	11
Venus	6	11	4	8	4	6	10	8	4	8	13	6
Mars	9	6	9	5	9	12	8	8	9	5	4	4
Jupiter	3	2	11	4	12	11	6	11	10	5	6	7
Saturn	5	8	8	4	12	5	7	7	12	6	8	6
TOTAL	47	39	52	46	54	56	58	64	58	45	51	46
Rising	5	5	3	9	13	5	10	9	15	8	5	1
N. Node	9	8	6	5	6	6	8	8	9	10	6	7

9H	♈	♉	♊	♋	♌	♍	♎	♏	♐	♑	♒	♓
Moon	3	10	8	5	6	8	7	5	7	12	11	6
Sun	10	10	10	4	2	8	4	7	8	11	6	8
Mercury	9	6	7	8	11	4	5	10	8	8	8	4
Venus	10	6	10	4	5	5	6	12	4	12	5	9
Mars	5	8	8	7	12	13	8	6	5	3	8	5
Jupiter	5	8	6	4	7	10	4	6	10	9	10	9
Saturn	10	5	12	6	16	4	5	9	10	1	5	5
TOTAL	52	53	61	38	59	52	39	55	52	56	53	46

Houses	1st	2nd	3rd	4th	5th	6th	7th	8th	9th	10th	11th	12th
Moon	9	7	8	8	6	6	10	7	6	8	4	9
Sun	5	5	11	9	9	9	8	13	2	5	4	8
Mercury	7	6	10	10	7	9	12	5	8	4	5	5
Venus	3	7	7	12	7	9	7	10	5	6	6	9
Mars	6	10	7	4	8	10	5	7	7	10	7	7
Jupiter	7	9	8	8	9	4	6	10	7	10	7	3
Saturn	9	9	12	6	7	8	9	8	4	6	6	4
TOTAL	46	53	63	57	53	55	57	60	39	49	39	45
N. Node	6	5	2	9	20	8	8	6	3	8	7	6

Planets

Aspects	Parallels	4 Axis	Direction	☉/☽ Midpoint
Venus-Uranus	Moon-Jupiter	Uranus (Asc)	Jupiter (Rx)	
		Sun (Dsc)	Neptune (Rx)	
		Mercury (Dsc)	Pluto (Rx)	
		Neptune (MC)		
		Neptune (IC)		Moon's Nodes
				Pluto (S)
				Chiron (S)

POLITICIANS

Theme: Aries/Pisces, Mars

Fire signs:
Leo in the 1st harmonic, Aries in the 9th harmonic, and the 9th house are frequent. Leo and Aries are the two most personally oriented Fire signs. Leos are typically very talented and impressive people who are widely admired, while Arians are assertive leaders. The 9th house shows aptitude for promoting a vision or philosophy.

Pisces:
Pisceans are often servers who are concerned with healing humanitarian issues. Mercury is especially frequent in Pisces (1 in 877), which indicates empathic communication and open-mindedness.

Mercury in Aquarius in the 1st & 9th Harmonics:
This indicates a talent for public speaking and creating relationships through sharing ideas.

Moon's North Node in Gemini (1 in 90,761):
This shows a life path towards much communication and human interaction.

Scorpio is infrequent in the 1st & 9th Harmonics (1 in 419 - 9H):
Scorpios are very private, while Politicians are extremely public.

Sun aspects are infrequent:
The Sun functions more powerfully when it is alone in the sky and not in any aspect relationship with other planetary bodies. This indicates that they strive to be special and be in charge of their own lives. Directors, Salespeople, Sports Coaches, and the Wealthy also have a solitary Sun.

Signs

Size: 545	1st Harmonic	9th Harmonic	Houses	Rising Sign
Moon				
Sun	Aquarius	Aries, Taurus, Capr	10th	
Mercury	Aquarius, Pisces	Libra, Aquarius	9th	Lunar Phase
Venus				
Mars	Gemini, Pisces	Taurus	7th	☉/☽ Midpoint
Jupiter				Pisces
Saturn	Leo	Aries		Moon's Nodes
TOTAL	Leo, Pisces	Aries, Capricorn	9th	Gemini
(weak)	Virgo, Scorpio	Scorpio	6th	8th

1H	♈	♉	♊	♋	♌	♍	♎	♏	♐	♑	♒	♓
Moon	52	55	45	42	52	36	37	48	37	50	39	52
Sun	53	42	44	54	39	39	45	31	49	40	59	50
Mercury	40	34	38	51	41	44	35	44	52	44	58	64
Venus	55	50	39	50	45	48	29	52	42	39	56	40
Mars	33	42	59	51	64	52	49	38	42	37	32	46
Jupiter	36	48	34	52	50	46	59	50	48	35	40	47
Saturn	43	48	41	42	55	40	49	35	56	46	53	37
TOTAL	312	319	300	342	346	305	303	298	326	291	337	336
Rising	28	26	46	53	59	61	54	66	58	42	26	26
N. Node	50	39	73	49	36	42	42	39	36	40	51	48

9H	♈	♉	♊	♋	♌	♍	♎	♏	♐	♑	♒	♓
Moon	53	42	40	44	46	52	43	40	48	42	48	47
Sun	57	59	51	45	43	39	44	35	35	56	44	37
Mercury	44	41	42	39	48	46	60	46	34	46	57	42
Venus	47	40	39	54	54	36	42	47	47	48	45	46
Mars	45	63	40	44	41	54	36	29	54	52	45	42
Jupiter	45	45	42	51	37	37	44	43	49	52	47	53
Saturn	58	55	51	30	49	50	36	32	52	53	49	30
TOTAL	349	345	305	307	318	314	305	272	319	349	335	297

Houses	1st	2nd	3rd	4th	5th	6th	7th	8th	9th	10th	11th	12th
Moon	53	43	46	44	42	40	46	47	53	39	43	49
Sun	56	49	45	40	46	28	39	39	46	59	51	47
Mercury	61	48	43	43	43	33	34	34	53	51	47	55
Venus	59	41	46	51	47	26	39	44	44	53	42	53
Mars	41	45	37	41	40	38	55	42	49	43	56	58
Jupiter	42	48	43	51	45	45	45	50	39	45	47	45
Saturn	44	37	46	40	50	47	51	50	50	39	41	50
TOTAL	356	311	306	310	313	257	309	306	334	329	327	357
N. Node	44	44	48	47	46	42	39	56	47	46	39	47

Planets

Aspects	Parallels	4 Axis	Direction	☉/☽ Midpoint
Mercury-Chiron	Moon-Mars	Sun (MC)	Mars (Rx)	
Venus-Mars			Uranus (D)	
			Neptune (Rx)	
			Pluto (Rx)	**Moon's Nodes**
				Moon (N)

PRESIDENTS (LEADERS OF NATIONS)

Theme: Aquarius, Jupiter

Aquarius:
Aquarians are highly charismatic speakers who have a broad vision and humanitarian purpose.

Sagittarius Lunar Phase (1 in 175):
Presidents have purpose with a philosophical vision.

Jupiter:
Jupiter is especially frequent on the Ascendant (1 in 248). Jupiterian people express great enthusiasm and leadership towards high-minded goals. They inspire confidence in people to support their ideas.

Uranus on the Ascendant (1 in 611)
These people are high-minded, unique, and independent.

Mercury on the Descendant (1 in 276):
Presidents are highly communicative and skilled at negotiating with people.

Signs

Size: 183	1st Harmonic	9th Harmonic	Houses	Rising Sign
Moon			3rd	
Sun	Aquarius		2nd	Lunar Phase
Mercury	Pisces	Cancer	2nd	Virgo, Sagittar
Venus			8th	☉/☽ Midpoint
Mars			1st, 7th	Virgo, Pisces
Jupiter			3rd	Aquarius
Saturn	Gemini	Aquarius	10th, 12th	Moon's Nodes
TOTAL		Cancer		Aquarius
(weak)			5th	3rd

1H	♈	♉	♊	♋	♌	♍	♎	♏	♐	♑	♒	♓
Moon	17	20	13	13	14	15	12	19	13	20	12	15
Sun	19	19	11	13	14	11	15	14	13	14	21	19
Mercury	15	18	11	8	15	10	15	14	20	20	15	22
Venus	19	14	13	19	11	11	13	14	17	12	22	18
Mars	13	16	13	11	17	23	18	19	15	12	15	11
Jupiter	14	17	12	19	18	17	16	15	11	17	9	18
Saturn	13	14	20	16	19	14	12	18	14	14	17	12
TOTAL	110	118	93	99	108	101	101	113	103	109	111	115
Rising	11	12	14	16	22	21	23	18	12	16	12	6
N. Node	16	14	14	20	10	12	20	15	16	12	22	12

9H	♈	♉	♊	♋	♌	♍	♎	♏	♐	♑	♒	♓
Moon	10	15	10	14	13	17	18	15	20	20	13	18
Sun	15	8	9	16	13	15	19	20	20	17	18	13
Mercury	11	12	18	27	15	12	13	20	12	13	16	14
Venus	16	18	13	20	10	16	16	17	14	16	12	15
Mars	9	13	16	17	14	17	21	18	16	17	12	13
Jupiter	17	13	9	20	15	18	11	14	15	18	16	17
Saturn	13	14	18	13	15	13	12	11	19	16	25	14
TOTAL	91	93	93	127	95	108	110	115	116	117	112	104

Houses	1st	2nd	3rd	4th	5th	6th	7th	8th	9th	10th	11th	12th
Moon	21	15	22	11	14	12	11	16	20	15	18	8
Sun	13	23	15	14	12	10	15	13	14	18	20	16
Mercury	20	23	11	19	7	17	10	18	13	18	13	14
Venus	20	18	11	18	13	14	8	21	12	17	16	15
Mars	22	14	14	16	17	15	22	14	13	11	14	11
Jupiter	19	17	22	15	9	12	16	14	5	20	15	19
Saturn	13	17	10	14	11	15	10	14	14	24	19	22
TOTAL	128	127	105	107	83	95	92	110	91	123	115	105
N. Node	16	13	22	11	9	15	16	12	15	20	17	17

Planets

Aspects	Parallels	4 Axis	Direction	☉/☽ Midpoint
Moon-Venus		Jupiter (Asc)		Jupiter
Moon-Mars		Uranus (Asc)		
Sun-Mars		Mercury (Dsc)		
Mercury-Chiron		Neptune (MC)		Moon's Nodes
Venus-Pluto		Jupiter (IC)		Venus (N)
		Pluto (IC)		Jupiter (S)
				Uranus (N)

PRIESTS

Theme: Taurus/Pisces, 7th/8th Houses

Pisces:
Mercury is frequent in Pisces in the 1st and 9th harmonics (1 in 274 – 9H). Pisceans are highly reflective people who think about spiritual topics and seek to be of service to others.

Sun in Leo in the 9th Harmonic (1 in 1,571):
Leos like to have attention and to be of central importance. Priests have been looked up to as the individual representatives of holiness. The Sun being strong in the 9th harmonic indicates that this is an inner quality which is not readily observed.

7th House (1 in 934) & 8th House:
These houses show much involvement with people, in both a public and private manner. Pluto is especially frequent in the 8th house (1 in 1,125), showing that priests often involve themselves in the shadowy aspects of peoples lives.

Neptune in the 9th House (1 in 46,094):
This position indicates much interest in ideals, beliefs, and religions.

Venus-Chiron aspects:
Venus with Chiron represents damaged relationships. This accurately symbolizes the many sex abuse cases that have come to light in recent years.

Signs

Size: 216	1st Harmonic	9th Harmonic	Houses	Rising Sign
Moon	Gemini	Cancer	4th, 7th	Aquarius
Sun	Aries, Scorpio	Leo	1st	Lunar Phase
Mercury	Pisces	Pisces	7th	Aries
Venus	Taurus, Virgo		5th, 8th	☉/☽ Midpoint
Mars	Taurus	Aries	8th	Aries
Jupiter		Pisces	11th	Virgo
Saturn	Capricorn	Aquarius	7th	Moon's Nodes
TOTAL	Taurus, Virgo		7th, 8th	Taurus
(weak)	Cancer		3rd, 6th	Pisces

1H	♈	♉	♊	♋	♌	♍	♎	♏	♐	♑	♒	♓
Moon	17	18	25	12	21	17	18	20	20	14	19	15
Sun	27	19	12	17	15	21	15	26	11	19	13	21
Mercury	15	20	16	9	13	21	24	17	22	18	16	25
Venus	18	26	14	13	15	26	20	19	16	11	18	20
Mars	20	25	23	20	21	27	23	17	13	12	9	6
Jupiter	14	21	16	20	17	23	17	17	25	17	15	14
Saturn	15	13	19	16	15	16	20	14	21	30	19	18
TOTAL	126	142	125	107	117	151	137	130	128	121	109	119
Rising	12	8	9	16	18	24	25	27	25	22	21	9
N. Node	19	26	23	21	16	12	15	15	13	16	15	25

9H	♈	♉	♊	♋	♌	♍	♎	♏	♐	♑	♒	♓
Moon	17	25	20	26	17	18	9	22	19	10	18	15
Sun	19	19	12	18	31	23	19	13	16	16	16	14
Mercury	17	16	17	17	12	23	16	16	16	23	14	29
Venus	20	23	15	18	17	18	17	18	18	19	16	17
Mars	25	16	15	18	20	15	15	17	16	24	17	18
Jupiter	17	16	16	17	19	15	18	16	21	18	17	26
Saturn	21	19	19	10	19	19	19	18	15	14	27	16
TOTAL	136	134	114	124	135	131	113	120	121	124	125	135

Houses	1st	2nd	3rd	4th	5th	6th	7th	8th	9th	10th	11th	12th
Moon	18	20	20	27	18	12	28	14	12	12	22	13
Sun	28	17	17	20	9	13	19	20	17	21	18	17
Mercury	24	22	16	21	11	11	24	14	19	16	19	19
Venus	23	18	11	13	24	11	20	26	18	11	17	24
Mars	19	19	10	14	21	22	14	25	17	18	16	21
Jupiter	17	17	15	24	22	3	21	23	12	17	28	17
Saturn	20	24	18	23	12	13	24	19	13	17	20	13
TOTAL	149	137	107	142	117	85	150	141	108	112	140	124
N. Node	23	21	21	21	17	23	13	14	11	20	17	15

Planets

Aspects	Parallels	4 Axis	Direction	☉/☽ Midpoint
Saturn	Moon-Jupiter	Mercury (Asc)	Venus (Rx)	
Moon-Saturn				
Venus-Chiron				
				Moon's Nodes

Prison Inmates

Theme: Scorpio/Aquarius, Mars

Scorpio:
Scorpios often struggle with the dark side of life. They encounter conflict and difficulty through involvements with others.

Mars in Virgo (1 in 3,787):
This position indicates having problems with health, addictions, and injuries.

Mars in Aquarius (1 in 2,393):
Mars in Aquarius can show that a person is against the trends of the majority of culture.

Mars in the 7th House (1 in 13,419):
This placement shows much conflict in relationships.

Mars on the Sun/Moon midpoint (1 in 246):
Mars in the Sun-Moon midpoint indicates a person with great energy and aggression.

Saturn in the 12th House (1 in 681):
This placement indicates having one's freedom restricted. The 12th house is the place of liberation, where planets are first visible each day.

Chiron aspects:
Chiron aspects show an abundance of trauma in a person's psyche.

Signs

Size: 372	1st Harmonic	9th Harmonic	Houses	Rising Sign
Moon	Capricorn	Scorpio		Taurus
Sun		Scorpio, Aquarius		Libra
Mercury	Gemini, Sagittarius			Aquarius
Venus	Scorpio	Gemini	2nd	Lunar Phase
Mars	Virgo, Aquarius		7th	Aquarius
Jupiter		Virgo, Pisces	5th, 10th	☉/☽ Midpoint
Saturn	Sagittarius	Aries	12th	
TOTAL	Virgo, Scorpio	Scorpio, Aquarius		Moon's Nodes
(weak)	Cancer		11th	

1H	♈	♉	♊	♋	♌	♍	♎	♏	♐	♑	♒	♓
Moon	38	30	26	21	29	31	35	29	27	43	33	30
Sun	35	32	33	26	34	38	29	35	26	38	25	21
Mercury	32	23	38	24	33	37	27	34	44	30	27	23
Venus	31	32	29	36	22	38	29	41	32	17	37	28
Mars	25	25	22	27	35	58	25	34	32	16	41	32
Jupiter	27	34	34	30	33	36	41	39	27	26	18	27
Saturn	25	34	29	24	26	25	31	41	43	21	34	39
TOTAL	213	210	211	188	212	263	217	253	231	191	215	200
Rising	15	31	26	45	39	31	53	31	34	23	30	14
N. Node	24	33	37	33	36	38	22	27	34	33	29	26

9H	♈	♉	♊	♋	♌	♍	♎	♏	♐	♑	♒	♓
Moon	36	31	23	25	35	35	22	43	35	23	39	25
Sun	35	34	29	33	26	25	31	40	28	26	41	24
Mercury	27	33	28	30	33	39	39	34	18	36	31	24
Venus	26	34	41	25	24	28	36	31	35	30	33	29
Mars	27	37	28	30	24	31	28	33	31	30	35	38
Jupiter	23	27	23	28	29	41	30	36	30	33	31	41
Saturn	43	27	30	27	33	36	29	25	39	27	32	24
TOTAL	217	223	202	198	204	235	215	242	216	205	242	205

Houses	1st	2nd	3rd	4th	5th	6th	7th	8th	9th	10th	11th	12th
Moon	35	32	38	35	21	33	29	34	33	30	26	26
Sun	36	36	25	37	26	24	24	35	30	28	30	41
Mercury	34	34	31	34	26	29	29	27	36	28	25	39
Venus	34	44	26	24	36	24	24	36	29	28	30	37
Mars	39	28	37	31	26	22	50	25	27	21	34	32
Jupiter	29	36	38	24	40	20	29	23	26	46	31	30
Saturn	25	36	29	23	24	34	25	36	34	35	24	47
TOTAL	232	246	224	208	199	186	210	216	215	216	200	252
N. Node	27	32	28	35	36	41	23	28	39	31	28	24

Planets

Aspects	Parallels	4 Axis	Direction	☉/☽ Midpoint
Chiron		Pluto (Dsc)	Jupiter (D)	Mars
Moon-Neptune		Jupiter (IC)	Uranus (Rx)	Saturn
Sun-Pluto				
				Moon's Nodes
				Uranus (N)

PROFESSORS

Theme: Scorpio, Saturn

Scorpio:
Scorpios are deep, introspective thinkers who seek to understand the roots of ideas and the causes of phenomena.

Mars in Capricorn (1 in 759):
Professors have a strong control over their aggressive impulses, and sublimate their energies towards academic pursuits.

Saturn:
Saturn aspects are frequent (1 in 466). Professors are authorities in their fields of expertise. They tend to have serious, conservative personalities.

Moon-Saturn aspects (1 in 14,730) and parallels (1 in 113):
They are very emotionally reserved. Professors seem to disdain great shows of emotion, viewing this as an indication of a lack of seriousness or intellectual ability.

Neptune aspects are weak (1 in 356):
Professors prefer factual information and have a strong disdain towards the spiritual and imaginary realms of thought.

Signs

Size: 175	1st Harmonic	9th Harmonic	Houses	Rising Sign
Moon	Libra		10th	
Sun	Aries		10th	
Mercury	Pisces	Scorpio	9th	Lunar Phase
Venus				
Mars	Capricorn			☉/☽ Midpoint
Jupiter	Scorpio		8th	Scorpio
Saturn			3rd	Moon's Nodes
TOTAL	Aries			Scorpio
(weak)			1st, 7th	1st

1H	♈	♉	♊	♋	♌	♍	♎	♏	♐	♑	♒	♓
Moon	13	20	12	9	13	15	20	14	17	18	12	12
Sun	24	13	14	15	9	15	15	15	10	9	18	18
Mercury	18	10	15	11	12	13	13	20	9	14	19	21
Venus	19	20	14	10	14	9	16	17	13	13	14	16
Mars	15	12	16	15	18	10	16	13	18	23	10	9
Jupiter	13	16	7	14	14	19	12	23	14	15	17	11
Saturn	14	16	12	14	15	17	16	15	20	10	13	13
TOTAL	116	107	90	88	95	98	108	117	101	102	103	100
Rising	3	10	15	20	24	16	22	17	22	10	11	5
N. Node	13	17	17	15	15	15	8	21	14	16	11	13

9H	♈	♉	♊	♋	♌	♍	♎	♏	♐	♑	♒	♓
Moon	17	16	20	16	15	19	12	11	9	14	12	14
Sun	17	18	17	12	10	13	13	16	16	15	14	14
Mercury	12	11	11	15	14	17	19	22	11	17	14	12
Venus	15	15	12	14	12	10	15	13	16	17	20	16
Mars	13	19	13	17	17	12	11	13	14	17	16	13
Jupiter	13	14	15	17	14	14	15	19	10	12	17	15
Saturn	13	17	17	17	16	18	16	12	11	13	12	13
TOTAL	100	110	105	108	98	103	101	106	87	105	105	97

Houses	1st	2nd	3rd	4th	5th	6th	7th	8th	9th	10th	11th	12th
Moon	12	12	15	13	19	14	15	13	11	22	19	10
Sun	11	18	12	16	9	14	6	15	14	24	15	21
Mercury	19	16	9	15	14	11	7	12	20	12	20	20
Venus	14	18	15	16	14	12	11	13	12	14	16	20
Mars	6	12	20	15	18	14	14	12	10	20	19	15
Jupiter	11	19	17	15	17	17	11	21	15	12	9	11
Saturn	15	14	22	16	16	13	10	10	16	14	19	10
TOTAL	88	109	110	106	107	95	74	96	98	118	117	107
N. Node	21	19	10	15	14	9	10	18	13	14	20	12

Planets

Aspects	Parallels	4 Axis	Direction	☉/☽ Midpoint
Saturn	Moon-Saturn	Pluto (Asc)	Jupiter (D)	
Moon-Saturn		Sun (MC)		
Mars-Saturn				
				Moon's Nodes
				Saturn (N)

PSYCHIATRISTS-PSYCHOLOGISTS

Theme: Taurus (2nd), Scorpio (8th), Chiron

Scorpio & the 8th House:
Scorpios have penetrating minds and are adept at research and investigation. They are more comfortable than most people in dealing with traumatic circumstances and are skilled at analyzing the psychological condition of others in order to support healing. People with many planets in the 8th house are very attuned to the hidden side of life. The Sun is especially frequent in the 8th house (1 in 341).

Cancer:
Cancer is strong in the 9th harmonic (1 in 985), especially for Jupiter (1 in 1,441). Cancers have a well-developed ability to acknowledge and support the emotional needs of others. Psychiatric workers have a healing philosophy (Jupiter). They seek to promote growth through the understanding of their client's personal issues.

Taurus & the 2nd House:
Taureans are patient and calm, which creates a safe environment for clients to work through personal difficulties.

Extroverted signs and houses are weak:
Extroverts enjoy a variety of casual, social relationships with others, while psychiatrists work with people in private over long periods of time.

Chiron:
Chiron is the wounded healer. Due to their own experiences, people in the psychiatric profession are drawn to healing the trauma and illnesses of others.

Moon's North Node in the 3rd House (1 in 953):
This indicates that they seek to develop their ability to communicate and relate to others.

Size: 158	1st Harmonic	Signs 9th Harmonic	Houses	Rising Sign
Moon	Scorpio, Aquar		2nd, 12th	
Sun		Aries, Cancer	2nd, 8th	
Mercury	Leo, Pisces		8th	Lunar Phase
Venus	Virgo		6th	Gemini
Mars	Pisces	Capricorn, Aquarius		☉/☽ Midpoint
Jupiter	Capricorn	Cancer		
Saturn		Taurus		Moon's Nodes
TOTAL	Scorpio	Taurus, Cancer, Scorpio	1st, 2nd, 8th	3rd
(weak)	Libra	Gemini, Sagittarius	5th, 9th	

1H	♈	♉	♊	♋	♌	♍	♎	♏	♐	♑	♒	♓
Moon	12	17	9	15	6	13	2	21	15	16	20	12
Sun	18	15	13	17	19	11	11	11	12	10	6	15
Mercury	13	12	14	14	20	10	12	14	13	7	10	19
Venus	17	16	17	18	10	22	4	16	9	9	10	10
Mars	11	11	12	14	14	12	15	16	13	13	11	16
Jupiter	7	9	9	17	17	5	11	17	19	20	17	10
Saturn	14	9	13	8	13	16	10	18	11	15	15	16
TOTAL	92	89	87	103	99	89	65	113	92	90	89	98

Rising	5	11	11	14	19	16	17	15	22	10	10	8
N. Node	18	13	13	12	19	13	5	15	8	16	14	12

9H	♈	♉	♊	♋	♌	♍	♎	♏	♐	♑	♒	♓
Moon	9	15	10	15	12	16	14	11	11	15	13	17
Sun	22	10	10	19	12	12	16	17	9	9	12	10
Mercury	9	13	9	13	13	12	14	17	15	15	12	16
Venus	17	14	12	14	13	10	15	17	13	13	11	9
Mars	10	18	6	18	14	11	9	14	7	21	20	10
Jupiter	10	17	11	25	6	9	13	18	10	14	16	9
Saturn	12	21	11	17	15	11	10	15	10	18	10	8
TOTAL	89	108	69	121	85	81	91	109	75	105	94	79

Houses	1st	2nd	3rd	4th	5th	6th	7th	8th	9th	10th	11th	12th
Moon	16	19	15	12	12	13	8	9	9	12	10	23
Sun	14	21	8	12	7	9	14	21	9	15	10	18
Mercury	16	19	14	10	6	10	16	19	10	14	12	12
Venus	18	17	15	11	5	20	14	15	13	8	8	14
Mars	18	19	15	13	14	7	9	11	7	12	17	16
Jupiter	19	19	17	10	17	13	4	13	13	10	11	12
Saturn	15	12	16	15	11	17	13	16	7	12	13	11
TOTAL	116	126	100	83	72	89	78	104	68	83	81	106

N. Node	10	17	22	14	15	14	10	11	9	11	11	14

Planets

Aspects	Parallels	4 Axis	Direction	☉/☽ Midpoint
Mercury-Chiron		Chiron (Asc)	Neptune (D)	
Mars-Jupiter		Saturn (Dsc)	Pluto (D)	
		Neptune (MC)	Chiron (D)	
				Moon's Nodes
				Venus (S)

PSYCHICS

Theme: Pisces, 8th House, Mercury

Pisces:
The Sun in Pisces is especially frequent (1 in 647). Pisceans often have a deeply spiritual nature and keen intuition.

Saturn in Sagittarius in the 9th Harmonic (1 in 23,215):
This indicates being an authority in religious and philosophical knowledge.

8th House:
The 8th house is the place of mysticism and occult knowledge. People with this house strong are often aware of the hidden side of life.

Mercury:
An influential Mercury indicates a person of high intelligence and communication ability. Mercury-Venus aspects are especially frequent (1 in 187). Mercury in Aquarius is frequent in the 1st and 9th harmonics, showing that psychics are highly intelligent people who have the ability to appeal and relate to others.

Jupiter on the Ascendant (1 in 3,313):
Psychics present themselves as people of wisdom and spiritual knowledge.

Uranus Direct (1 in 317):
Uranus is the planet of intuition and awaking.

Signs

Size: 192	1st Harmonic	9th Harmonic	Houses	Rising Sign
Moon				
Sun	Pisces		8th	
Mercury	Aquarius	Aquarius		Lunar Phase
Venus			8th	Scorpio
Mars			3rd	☉/☽ Midpoint
Jupiter		Leo	1st, 12th	
Saturn		Sagittarius	9th	Moon's Nodes
TOTAL	Pisces		8th	Pisces (S)
(weak)	Capricorn		2nd, 11th	7th

1H	♈	♉	♊	♋	♌	♍	♎	♏	♐	♑	♒	♓
Moon	17	20	12	17	17	16	16	14	17	19	15	12
Sun	17	18	16	15	15	16	12	12	17	12	15	27
Mercury	19	13	18	13	11	16	15	15	18	11	23	20
Venus	19	17	16	18	13	13	15	18	15	10	23	15
Mars	12	14	16	20	23	16	18	15	16	16	12	14
Jupiter	12	19	11	16	19	12	22	18	17	14	11	21
Saturn	16	14	15	17	12	15	21	22	16	8	18	18
TOTAL	112	115	104	116	110	104	119	114	116	90	117	127
Rising	10	12	15	13	27	21	23	19	17	16	9	10
N. Node	12	20	20	13	12	25	16	13	13	15	17	16

9H	♈	♉	♊	♋	♌	♍	♎	♏	♐	♑	♒	♓
Moon	17	17	13	19	15	22	15	11	16	13	16	18
Sun	16	21	21	20	17	11	13	11	18	15	14	15
Mercury	10	17	10	21	14	15	14	22	14	11	24	20
Venus	16	15	20	17	19	13	12	18	14	14	19	15
Mars	16	17	11	12	20	19	13	20	14	19	13	18
Jupiter	12	10	15	15	25	15	21	16	15	18	13	17
Saturn	13	16	16	18	15	12	14	11	31	16	18	12
TOTAL	100	113	106	122	125	107	102	109	122	106	117	115

Houses	1st	2nd	3rd	4th	5th	6th	7th	8th	9th	10th	11th	12th
Moon	11	11	13	16	15	18	16	21	19	21	17	14
Sun	19	19	13	17	16	11	10	21	14	14	15	23
Mercury	22	15	19	14	18	12	12	19	13	12	15	21
Venus	18	17	21	17	14	11	17	21	9	17	18	12
Mars	19	16	23	10	11	16	21	14	13	20	9	20
Jupiter	26	10	10	15	15	19	16	12	18	18	11	22
Saturn	20	10	11	21	16	12	13	16	22	19	12	20
TOTAL	135	98	110	110	105	99	105	124	108	121	97	132
N. Node	20	10	5	17	19	21	24	13	17	10	14	22

Planets

Aspects	Parallels	4 Axis	Direction	☉/☽ Midpoint
Mercury	Sun-Neptune	Mercury (Asc)	Mercury (D)	
Venus & Mars		Jupiter (Asc)	Uranus (D)	
Moon-Mars		Saturn (MC)	Neptune (R)	
Sun-Jupiter		Pluto (MC)		
Mercury-Venus		Chiron (MC)		Moon's Nodes
Mercury-Pluto				
Venus-Chiron				
Mars-Saturn				

Racers – Cars & Bicycles

Theme: Aries/Taurus

Aries (1 in 2,809):
Arians are the most competitive, aggressive people. They have an insatiable desire to overcome obstacles and challenges. The Sun in Aries (1 in 227) is most frequent in the 1st harmonic, and Jupiter is the most frequent planet in the 9th harmonic (1 in 3,323). The Sun/Moon midpoint is also frequent in Aries (1 in 9,810).

Taurus:
Taureans are the most physically competent people. They are often powerful and skilled in their bodily pursuits. Venus is the most frequent planet in Taurus (1 in 4,280).

Signs

Size: 173	1st Harmonic	9th Harmonic	Houses	Rising Sign
Moon			4th	
Sun	Aries, Capricorn	Taurus	9th	
Mercury	Aries, Pisces		9th	Lunar Phase
Venus	Taurus, Aquarius	Aries, Taurus	8th	Aries
Mars				☉/☽ Midpoint
Jupiter	Scorpio	Aries, Cancer	10th	Aries
Saturn		Virgo		Moon's Nodes
TOTAL	Aries, Aquarius	Taurus		
(weak)	Leo, Libra, Sag	Aquarius	5th, 12th	

1H	♈	♉	♊	♋	♌	♍	♎	♏	♐	♑	♒	♓
Moon	19	13	15	20	16	11	14	9	9	19	15	13
Sun	24	8	19	17	9	11	14	9	6	21	16	19
Mercury	20	13	12	11	12	15	11	12	10	18	18	21
Venus	16	28	13	10	11	12	7	13	16	10	28	9
Mars	14	17	10	13	14	16	17	15	14	12	17	14
Jupiter	17	14	11	16	9	16	18	23	11	12	13	13
Saturn	21	18	12	12	7	16	6	13	18	16	14	20
TOTAL	131	111	92	99	78	97	87	94	84	108	121	109
Rising	7	9	15	18	25	17	14	22	20	15	5	6
N. Node	14	14	10	19	13	12	17	16	12	16	16	14

9H	♈	♉	♊	♋	♌	♍	♎	♏	♐	♑	♒	♓
Moon	9	20	14	19	18	15	13	16	16	12	9	12
Sun	15	21	12	11	13	16	11	14	16	17	8	19
Mercury	13	14	13	19	15	17	11	16	18	10	14	13
Venus	21	23	5	19	12	17	9	8	18	13	18	10
Mars	12	15	13	12	13	17	15	20	16	10	17	13
Jupiter	27	14	14	22	10	11	20	14	4	15	8	14
Saturn	14	11	15	12	18	23	16	14	16	16	11	7
TOTAL	111	118	86	114	99	116	95	102	104	93	85	88

Houses	1st	2nd	3rd	4th	5th	6th	7th	8th	9th	10th	11th	12th
Moon	16	14	13	21	9	13	11	16	15	17	14	14
Sun	18	16	13	13	5	16	13	13	21	13	20	12
Mercury	15	17	14	13	8	16	14	10	20	14	19	13
Venus	19	17	11	10	12	10	15	19	10	20	16	14
Mars	14	17	19	16	12	15	13	16	17	10	16	8
Jupiter	16	19	11	12	16	11	17	11	15	21	8	16
Saturn	10	19	15	17	8	10	16	16	9	20	18	15
TOTAL	108	119	96	102	70	91	99	101	107	115	111	92
N. Node	10	13	20	9	15	16	19	14	17	11	15	14

Planets

Aspects	Parallels	4 Axis	Direction	☉/☽ Midpoint
Saturn		Mercury (MC)	Uranus (D)	
Moon-Neptune			Neptune (R)	
Mars-Jupiter			Pluto (R)	
				Moon's Nodes
				Uranus (S)
				Chiron (N)

ROYALTY

Theme: Pisces (12th), Jupiter

Pisces & the 12th House:
Pisces is the sign of service. Politicians also have planets frequent in Pisces. The Sun, Venus, and Jupiter are most frequent in the 12th house, showing this placement to be beneficial.

Scorpio (1 in 345):
Scorpios can be devious, scheming, and powerful people. They are capable of both the greatest loyalty and the worst acts of betrayal.

Capricorn Rising (1 in 3,405):
They appear serious and authoritative.

Jupiter:
Jupiter is frequent with the Sun, Moon, and Ascendant. Jupiter in these positions are indications for achieving goals and having a successful life. Jupiter is also frequent on the Moon's South Node (1 in 355).

Sun on the Moon's North Node (1 in 426):
This indicates a life path leading towards individual success and specialness.

The fact that astrological signatures show up among members of Royalty is compelling evidence that there is a cosmic connection between parents and their children. In most categories, it could be explained that the astrological traits led them into their circumstances. But here, the circumstances are present at birth.

Signs

Size: 347	1st Harmonic	9th Harmonic	Houses	Rising Sign
Moon			4th	Leo
Sun	Cancer			Capricorn
Mercury		Libra		**Lunar Phase**
Venus	Taurus	Cancer		
Mars				☉/☽ **Midpoint**
Jupiter		Libra	10th, 12th	
Saturn	Aquarius	Pisces		**Moon's Nodes**
TOTAL	Scorpio	Pisces	12th	Pisces (S)
(weak)	Leo	Gemini	6th	

1H	♈	♉	♊	♋	♌	♍	♎	♏	♐	♑	♒	♓
Moon	30	27	25	25	30	24	31	37	23	31	32	32
Sun	23	35	35	40	21	27	37	30	27	25	24	23
Mercury	28	27	30	30	27	30	27	36	30	35	17	30
Venus	27	42	35	26	21	36	26	35	26	25	28	20
Mars	24	29	22	32	35	32	28	39	30	29	22	25
Jupiter	27	30	26	30	22	27	30	40	27	28	27	33
Saturn	26	24	26	20	20	22	34	32	34	37	43	29
TOTAL	185	214	199	203	176	198	213	249	197	210	193	192
Rising	14	13	25	34	49	46	36	26	25	44	20	15
N. Node	28	18	30	30	37	39	27	30	32	30	21	25

9H	♈	♉	♊	♋	♌	♍	♎	♏	♐	♑	♒	♓
Moon	25	32	28	28	28	35	30	23	28	30	29	31
Sun	34	29	27	35	20	23	25	32	29	31	28	34
Mercury	25	26	32	26	37	33	39	24	30	31	19	25
Venus	32	30	20	38	29	27	21	33	22	30	33	32
Mars	27	28	19	21	30	34	30	37	27	31	32	31
Jupiter	29	17	22	34	26	21	44	32	32	28	28	34
Saturn	30	36	22	17	30	22	29	32	29	27	33	40
TOTAL	202	198	170	199	200	195	218	213	197	208	202	227

Houses	1st	2nd	3rd	4th	5th	6th	7th	8th	9th	10th	11th	12th
Moon	26	30	29	37	24	28	36	31	29	19	24	34
Sun	29	28	30	34	25	20	20	25	23	39	35	39
Mercury	38	29	29	29	27	25	24	18	21	37	39	31
Venus	30	29	29	21	25	22	24	30	24	34	40	39
Mars	33	30	33	33	28	23	21	29	21	28	34	34
Jupiter	34	25	24	26	29	21	26	32	28	40	22	40
Saturn	30	35	23	21	34	24	39	26	32	28	27	28
TOTAL	220	206	197	201	192	163	190	191	178	225	221	245
N. Node	24	37	28	32	36	25	29	30	23	30	28	25

Planets

Aspects	Parallels	4 Axis	Direction	☉/☽ Midpoint
Moon-Jupiter	Sun-Pluto	Jupiter (Asc)	Saturn (D)	Saturn
Sun-Jupiter		Chiron (MC)	Uranus (Rx)	
				Moon's Nodes
				Moon (S)
				Sun (N)
				Mercury (N)
				Jupiter (S)

SALESPEOPLE

Theme: Aries, Saturn

Aries & Sagittarius:
Aries is strong in the 1st harmonic, and Sagittarius is strong in the 9th harmonic. Energetic fire sign people like to work independently and make their own decisions.

Aquarius:
Aquarians are sociable, intelligent people. Salespeople use their natural ease with people to gain trust and create interest in their products. The Moon is most frequent in Aquarius (1 in 232).

Cardinal mode & Cardinal sector are strong:
This is the only group that is strong in both the Cardinal mode and Cardinal sector (Aries to Cancer). It indicates a tremendous amount of initiative and drive.

Sun aspects are infrequent:
The Sun functions more powerfully when it is alone in the sky and not in any aspect relationship with other planetary bodies. This indicates that they strive to be special and be in charge of their own lives. Directors, Politicians, Sports Coaches, and the Wealthy also have a solitary Sun.

Saturn on the Ascendant (1 in 963):
People with Saturn on the Ascendant like to be in control of their environment. They appear responsible and authoritative.

Signs

Size: 103	1st Harmonic	9th Harmonic	Houses	Rising Sign
Moon	Aries, Aquarius	Sagittarius		Aries
Sun	Gemini	Cancer		Taurus
Mercury			3rd	Lunar Phase
Venus				Virgo
Mars	Libra	Scorpio, Sagittarius	8th	O/D Midpoint
Jupiter	Aries		7th, 12th	Aries, Taurus
Saturn	Cancer	Leo, Capricorn		Moon's Nodes
TOTAL	Aries, Cancer, Aquar	Sagittarius	3rd, 12th	10th
(weak)	Leo, Capricorn	Taurus	9th	

1H	♈	♉	♊	♋	♌	♍	♎	♏	♐	♑	♒	♓
Moon	14	2	10	8	7	7	12	5	5	7	16	10
Sun	11	10	15	12	4	6	10	6	8	4	10	7
Mercury	8	12	9	11	6	5	7	9	9	9	8	10
Venus	13	11	8	10	10	10	8	7	7	2	13	4
Mars	7	5	8	10	5	8	16	6	11	9	11	7
Jupiter	15	5	9	12	4	11	9	12	6	3	10	7
Saturn	7	8	5	13	9	11	8	5	7	13	8	9
TOTAL	75	53	64	76	45	58	70	50	53	47	76	54
Rising	10	10	6	13	11	13	10	13	1	7	5	4
N. Node	9	9	12	10	8	7	8	12	4	10	4	10

9H	♈	♉	♊	♋	♌	♍	♎	♏	♐	♑	♒	♓
Moon	10	4	13	10	4	6	6	13	14	5	9	9
Sun	10	4	7	14	6	5	9	11	13	5	9	10
Mercury	8	8	12	7	13	12	8	7	7	6	7	8
Venus	7	11	8	2	8	9	7	11	13	7	13	7
Mars	7	2	4	8	8	10	6	15	14	9	8	12
Jupiter	9	11	11	11	5	3	10	5	7	11	11	9
Saturn	5	5	7	9	16	12	4	6	5	17	12	5
TOTAL	56	45	62	61	60	57	50	68	73	60	69	60

Houses	1st	2nd	3rd	4th	5th	6th	7th	8th	9th	10th	11th	12th
Moon	10	11	12	10	4	9	4	9	6	8	10	10
Sun	10	9	12	5	7	6	11	4	4	8	13	14
Mercury	13	5	14	8	9	7	10	2	4	11	8	12
Venus	12	11	10	8	6	7	11	6	3	6	12	11
Mars	10	5	12	8	5	8	7	13	5	9	9	12
Jupiter	10	6	7	11	8	9	15	5	7	5	4	16
Saturn	9	6	11	13	9	7	5	9	7	7	9	11
TOTAL	74	53	78	63	48	53	63	48	36	54	65	86
N. Node	6	6	9	4	11	9	11	13	7	15	8	4

Planets

Aspects	Parallels	4 Axis	Direction	☉/☽ Midpoint
Saturn-Chiron	Mars-Saturn	Venus (Asc)	Mars (D)	Jupiter
		Saturn (Asc)	Uranus (D)	
		Jupiter (Dsc)		
		Mars (IC)		Moon's Nodes
		Chiron (IC)		

SAME JOB – 10 YEARS OR MORE

Theme: Taurus (2nd), Leo (5th)

Taurus:
Taurus has long been associated with money and a stubborn tenacity to pursue goals. Taureans are very reliable people, who like to stay in a comfort zone and will only change life circumstances under duress. The Moon (1 in 947), Venus (1 in 235), and Sun (1 in 222 - 9H) are most frequent in Taurus.

Moon in Fixed signs:
The Moon in fixed signs seeks stability in the home and finances. This suggests an economically stable (Taurus) and loving (Leo) home and family life (Moon). It also inclines a person to be comfortable in routines (Taurus) and emotionally (Moon) well adjusted. The Moon in Taurus is strongest in the 1st harmonic (1 in 947), and the Moon in Leo is strongest in the 9th harmonic (1 in 387).

Jupiter in Cancer (1 in 136,092):
This indicates positive experiences in the home and family. It also indicates buoyant feelings of well being and positive expectations in developing close connections with people.

Mars in 5th House (1 in 569), and Mars in Leo in the 9th Harmonic:
These positions indicate having much confidence expressing oneself assertively.

Libra is weak in both the 1st and 9th Harmonics:
Librans are strongly independent and stand up for their values. Stability, harmony, and security are less important for them.

Mercury-Venus aspects:
Mercury-Venus aspects incline a person towards congenial relations with others.

Signs

Size: 219	1st Harmonic	9th Harmonic	Houses	Rising Sign
Moon	Taurus, Scorpio, Aquar	Leo		
Sun	Aries	Taurus	3rd	
Mercury				Lunar Phase
Venus	Taurus			Leo
Mars	Gemini	Leo, Pisces	5th	☉/☽ Midpoint
Jupiter	Cancer	Virgo	2nd	Aries
Saturn	Gemini, Pisces			Moon's Nodes
TOTAL	Taurus		2nd, 5th	Cancer
(weak)	Libra	Libra	12th	1st, 5th (S)

1H	♈	♉	♊	♋	♌	♍	♎	♏	♐	♑	♒	♓
Moon	15	31	17	14	18	12	11	25	20	18	25	13
Sun	26	17	20	17	21	20	16	22	15	16	19	10
Mercury	21	15	19	11	22	18	19	24	18	14	20	18
Venus	15	30	18	21	18	16	21	13	18	16	21	12
Mars	15	22	26	23	21	20	21	22	11	19	11	8
Jupiter	16	13	16	35	24	18	11	14	22	22	12	16
Saturn	12	16	23	16	16	14	14	24	16	23	16	29
TOTAL	120	144	139	137	140	118	113	144	120	128	124	106
Rising	7	15	15	29	26	26	23	23	18	13	9	15
N. Node	23	21	17	27	23	22	15	11	17	14	12	17

9H	♈	♉	♊	♋	♌	♍	♎	♏	♐	♑	♒	♓
Moon	21	19	16	20	30	20	11	22	20	11	12	17
Sun	22	29	21	17	10	20	16	19	11	16	18	20
Mercury	19	15	20	20	17	14	16	18	18	18	24	20
Venus	9	20	22	24	14	21	14	18	21	20	22	14
Mars	23	19	22	19	25	14	15	12	15	16	13	26
Jupiter	16	18	18	23	14	26	15	18	17	15	21	18
Saturn	18	19	13	17	17	22	16	23	17	15	21	21
TOTAL	128	139	132	140	127	137	103	130	119	111	131	136

Houses	1st	2nd	3rd	4th	5th	6th	7th	8th	9th	10th	11th	12th
Moon	18	21	19	14	21	15	20	15	24	19	18	15
Sun	13	22	27	13	21	15	12	19	22	17	17	21
Mercury	22	21	24	23	16	16	18	16	19	18	18	8
Venus	20	24	18	17	18	17	20	21	17	13	18	16
Mars	17	20	17	19	29	12	14	22	15	19	17	18
Jupiter	16	28	14	16	16	23	12	21	18	20	20	15
Saturn	23	15	18	10	22	19	15	19	18	18	22	20
TOTAL	129	151	137	112	143	117	111	133	133	124	130	113
N. Node	28	15	18	11	20	14	15	23	14	19	26	16

Planets

Aspects	Parallels	4 Axis	Direction	☉/☽ Midpoint
Mercury-Venus	Venus-Pluto			
	Mars-Uranus			
				Moon's Nodes

SCIENTISTS

Theme: Sagittarius, 3rd House, Moon

Sagittarius (1 in 218):
Sagittarians are the seekers of the zodiac. They have a need to know how life functions, both physically and spiritually. Scientists have a combination of adventurous and conservative qualities (Capricorn). Their quest for knowledge mostly happens within the confines of the structured university setting following traditional methods for research. The Moon is most frequent in Sagittarius (1 in 1,267). Mercury in Sagittarius is frequent in both the 1st and 9th harmonics (1 in 452 – 9H).

Sun in Capricorn (1 in 6,155):
They embrace the identity of a hard worker capable of achieving difficult goals and receiving recognition amongst authoritative peers.

Sun in Taurus in the 9th Harmonic (1 in 7,722):
They have the goal of understanding physical nature. They are also practical, hard-working, and patient.

3rd House:
The 3rd house represents learning and knowledge.

Uranus on the Imum Coeli (1 in 431):
This indicates having the ability to develop unique insights while working in a private setting.

Signs

Size: 175	1st Harmonic	9th Harmonic	Houses	Rising Sign
Moon	Sagittarius			
Sun	Capricorn	Taurus	3rd, 4th	Lunar Phase
Mercury	Sagittarius, Capr, Pisces	Sagittarius	3rd	Libra
Venus			3rd	☉/☽ Midpoint
Mars			2nd	Sagittarius
Jupiter	Pisces			Pisces
Saturn	Sagittarius			Moon's Nodes
TOTAL	Sagittarius, Capricorn	Taurus, Sagittarius	3rd	
(weak)	Libra	Cancer	12th	

1H	♈	♉	♊	♋	♌	♍	♎	♏	♐	♑	♒	♓
Moon	15	8	11	11	17	18	7	15	26	17	20	10
Sun	16	15	13	7	12	18	12	9	10	27	16	20
Mercury	13	11	13	9	11	14	13	10	22	22	14	23
Venus	14	18	17	13	10	13	9	16	17	16	18	14
Mars	9	14	11	17	14	14	21	16	18	17	15	9
Jupiter	12	12	10	19	15	18	8	15	15	12	18	21
Saturn	13	11	14	14	18	12	16	18	22	14	11	12
TOTAL	92	89	89	90	97	107	86	99	130	125	112	109

	♈	♉	♊	♋	♌	♍	♎	♏	♐	♑	♒	♓
Rising	3	10	9	20	19	22	20	22	21	12	8	9
N. Node	12	8	13	19	12	17	16	11	19	13	15	20

9H	♈	♉	♊	♋	♌	♍	♎	♏	♐	♑	♒	♓
Moon	18	15	18	13	19	11	18	12	14	9	15	13
Sun	18	28	7	9	13	14	13	18	15	12	14	14
Mercury	12	19	16	7	9	18	15	15	25	13	13	13
Venus	17	18	18	15	6	9	16	14	13	19	15	15
Mars	20	18	15	8	17	12	12	13	17	12	14	17
Jupiter	10	15	16	17	14	12	17	15	16	14	15	14
Saturn	17	15	11	14	11	20	12	16	18	12	11	18
TOTAL	112	128	101	83	89	96	103	103	118	91	97	104

Houses	1st	2nd	3rd	4th	5th	6th	7th	8th	9th	10th	11th	12th
Moon	17	17	10	17	15	19	17	13	13	14	13	10
Sun	20	14	22	22	11	12	9	14	10	15	14	12
Mercury	22	15	23	19	12	10	15	7	11	15	14	12
Venus	19	14	21	18	8	17	11	10	16	11	22	8
Mars	16	23	14	14	14	12	16	13	12	18	13	10
Jupiter	14	13	15	16	10	15	15	15	12	13	17	20
Saturn	17	13	17	10	15	11	19	14	13	14	14	18
TOTAL	125	109	122	116	85	96	102	86	87	100	107	90

	1st	2nd	3rd	4th	5th	6th	7th	8th	9th	10th	11th	12th
N. Node	14	15	17	16	20	12	10	19	11	14	18	9

Planets

Aspects	Parallels	4 Axis	Direction	☉/☽ Midpoint
Moon-Saturn	Moon-Pluto	Mars (Asc)	Jupiter (D)	
Mars-Pluto		Uranus (IC)		
				Moon's Nodes
				Moon (S)

Sex Abuse Victims

Theme: Taurus (2nd), Saturn/Chiron

Taurus & the 2nd House:
Taureans are the most sensual, physical people. This shows that sex abuse victims have an allure that attracts abusers, who are Virgo dominant.

Saturn:
Saturn shows a person to be very serious or depressed, causing one's goals in life to be restricted or slow to develop. Sun-Saturn and Moon-Saturn aspects are the most frequent.

Chiron:
Chiron represents trauma that needs to be healed. Sun-Chiron and Mars-Chiron aspects are frequent. Chiron is also frequent on the Midheaven (1 in 120).

Aquarius is weak:
Aquarians are sociable, yet very non-physical people.

Signs

Size: 123	1st Harmonic	9th Harmonic	Houses	Rising Sign
Moon	Libra			Taurus
Sun	Gemini	Gemini	2nd	
Mercury		Leo		Lunar Phase
Venus	Leo, Virgo	Sagittarius	2nd, 3rd	Gemini
Mars	Taurus		4th, 8th	☉/☽ Midpoint
Jupiter		Taurus	5th	Virgo
Saturn	Cancer	Capricorn	7th	Moon's Nodes
TOTAL	Virgo	Taurus	2nd, 12th	Cancer
(weak)	Aries, Aquarius	Aquarius	6th	10th

1H	♈	♉	♊	♋	♌	♍	♎	♏	♐	♑	♒	♓
Moon	7	9	6	13	11	8	17	13	11	6	10	12
Sun	9	5	16	10	13	12	14	6	11	13	8	6
Mercury	7	10	8	11	11	13	12	10	16	10	7	8
Venus	9	12	3	10	15	17	3	15	13	3	13	10
Mars	7	14	11	9	11	16	15	10	7	7	9	7
Jupiter	13	6	9	11	9	13	10	15	16	11	5	5
Saturn	2	10	9	14	13	13	9	13	11	14	5	10
TOTAL	54	66	62	78	83	92	80	82	85	64	57	58
Rising	8	14	9	9	13	11	13	15	14	10	4	3
N. Node	7	10	14	15	15	13	9	14	7	8	5	6

9H	♈	♉	♊	♋	♌	♍	♎	♏	♐	♑	♒	♓
Moon	9	13	15	8	11	9	12	5	9	15	7	10
Sun	6	11	16	7	10	5	14	9	9	11	10	15
Mercury	9	11	7	12	20	13	9	10	9	11	5	7
Venus	12	11	14	8	11	7	7	8	19	11	7	8
Mars	7	12	11	15	14	7	7	12	9	5	14	10
Jupiter	9	17	12	15	6	12	10	14	4	11	5	8
Saturn	8	10	5	13	5	13	9	12	6	19	11	12
TOTAL	60	85	80	78	77	66	68	70	65	83	59	70

Houses	1st	2nd	3rd	4th	5th	6th	7th	8th	9th	10th	11th	12th
Moon	10	12	11	9	10	6	9	12	13	12	8	11
Sun	16	18	10	8	7	5	5	13	4	15	10	12
Mercury	15	14	15	6	4	8	9	8	6	14	10	14
Venus	11	17	16	5	7	8	9	5	11	8	10	16
Mars	8	9	11	17	7	10	9	17	4	10	9	12
Jupiter	12	5	13	6	17	9	10	12	7	8	10	14
Saturn	11	13	5	11	8	7	17	10	13	12	4	12
TOTAL	83	88	81	62	60	53	68	77	58	79	61	91
N. Node	14	8	4	13	10	9	9	12	13	16	3	12

Planets

Aspects	Parallels	4 Axis	Direction	☉/☽ Midpoint
Jupiter	Sun-Mars	Venus (Asc)	Chiron (D)	
Moon-Jupiter		Chiron (MC)		
Moon-Saturn		Mars (IC)		
Sun-Saturn				Moon's Nodes
Sun-Chiron				Saturn (N)
Mars-Chiron				

Sex Offenders

Theme: Virgo, 4th House

Virgo (1 in 227):
Virgo is the sign most often associated with physical injuries and difficult life circumstances. It is also the sign most common with the other sex-related study groups. Jupiter is especially frequent in Virgo, indicating a frustrated and critical attitude towards life.

4th House:
The 4th house is the most hidden placement for planets. It indicates that much of the person's life occurs in private.

Mercury-Mars & Mercury-Pluto aspects:
These aspects show a preponderance of sexual thoughts. Their minds are attracted to the hidden, taboo aspects of life, including sexuality. They can also be very secretive and manipulative.

Venus-Jupiter parallels (1 in 1,216):
This indicates the ability to seek frequent relationship partners.

Neptune on the Ascendant (1 in 837), and the Sun-Moon midpoint:
As the planet of illusion, Neptune can influence a person to be involved in deceptions and delusional thinking.

Signs

Size: 101	1st Harmonic	9th Harmonic	Houses	Rising Sign
Moon			8th	Pisces
Sun	Leo, Pisces	Aries, Aquarius	4th	
Mercury	Virgo		4th	Lunar Phase
Venus	Virgo, Aquarius		6th	
Mars		Scorpio	4th, 7th	☉/☽ Midpoint
Jupiter	Gemini, Virgo	Gemini, Virgo, Pisces		Virgo
Saturn			12th	Moon's Nodes
TOTAL	Virgo		4th	Cancer
(weak)	Capricorn			6th

1H	♈	♉	♊	♋	♌	♍	♎	♏	♐	♑	♒	♓
Moon	9	7	10	9	5	11	6	9	11	9	7	8
Sun	5	5	9	6	17	9	7	7	7	9	7	13
Mercury	7	5	6	8	12	13	5	13	8	5	10	9
Venus	10	5	10	11	3	14	8	11	5	4	15	5
Mars	7	8	8	13	6	14	5	9	9	4	8	10
Jupiter	8	10	14	3	7	16	12	11	6	6	0	8
Saturn	5	9	6	10	9	5	7	9	12	7	13	9
TOTAL	51	49	63	60	59	82	50	69	58	44	60	62
Rising	4	4	10	6	11	14	12	13	7	6	3	11
N. Node	3	5	12	13	8	12	11	9	8	7	7	6

9H	♈	♉	♊	♋	♌	♍	♎	♏	♐	♑	♒	♓
Moon	11	5	5	10	9	11	8	8	8	6	11	9
Sun	13	9	10	7	9	9	9	6	3	8	13	5
Mercury	4	11	7	11	9	8	9	9	9	6	9	9
Venus	7	9	11	4	9	8	10	7	12	10	9	5
Mars	5	9	5	7	6	5	11	15	9	9	10	10
Jupiter	7	4	14	8	7	13	5	8	5	7	9	14
Saturn	11	8	10	7	8	9	9	9	3	8	9	10
TOTAL	58	55	62	54	57	63	61	62	49	54	70	62

Houses	1st	2nd	3rd	4th	5th	6th	7th	8th	9th	10th	11th	12th
Moon	8	8	10	5	7	9	13	13	9	8	6	5
Sun	10	6	12	14	6	7	8	8	8	6	10	6
Mercury	6	9	9	15	8	6	11	6	8	8	8	7
Venus	9	11	8	9	8	13	5	7	10	7	6	8
Mars	10	9	5	14	4	5	13	7	5	10	9	10
Jupiter	10	11	6	5	9	10	10	9	12	7	5	7
Saturn	13	5	8	9	6	10	5	6	7	7	12	13
TOTAL	66	59	58	71	48	60	65	56	59	53	56	56
N. Node	5	9	10	10	11	13	6	6	11	6	8	6

Planets

Aspects	Parallels	4 Axis	Direction	☉/☽ Midpoint
Sun-Venus	Venus-Jupiter	Neptune (Asc)		Jupiter
Mercury-Mars		Uranus (Dsc)		Neptune
Mercury-Pluto		Pluto (MC)		
				Moon's Nodes

Sex Workers – Pornography and Prostitution

Theme: Virgo/Libra, Jupiter

Virgo (1 in 2,775):
Virgo is very strong for all of the sex-related groups in this study. While Scorpio is strong amongst Sex Symbols, who are magnetic and alluring, Virgo is strong with Sex Workers due to their more physical approach to sexuality. Venus is the most frequent planet in Virgo (1 in 9,401), an indication of sexual difficulties and guilt.

Libra:
Libras are very skilled at working with people closely. They are able to associate with many people in a sociable manner, yet still keep an emotional distance.

Mars in Scorpio in the 9th Harmonic (1 in 942):
This indicates a strong inclination towards the shadowy side of sexuality.

Moon-Jupiter aspects and parallels (1 in 7,387):
Moon-Jupiter contacts indicate emotional detachment. They often grow apart from their place of origin.

Signs

Size: 100	1st Harmonic	9th Harmonic	Houses	Rising Sign
Moon	Libra, Aquarius		11th	Leo
Sun	Libra	Libra	4th	Pisces
Mercury	Virgo, Libra	Aries, Pisces	2nd	Lunar Phase
Venus	Virgo		1st, 5th	Aquarius
Mars	Leo, Virgo	Scorpio		☉/☽ Midpoint
Jupiter		Aries		Libra
Saturn		Cancer, Leo	4th	Moon's Nodes
TOTAL	Virgo, Libra		1st, 2nd	Virgo (S)
(weak)	Gemini, Capricorn			

1H	♈	♉	♊	♋	♌	♍	♎	♏	♐	♑	♒	♓
Moon	10	7	6	6	4	9	13	11	6	7	13	8
Sun	7	12	3	12	12	11	14	6	5	5	5	8
Mercury	8	11	5	11	5	16	15	5	8	3	6	7
Venus	10	5	12	11	7	20	5	12	6	1	6	5
Mars	4	7	4	11	16	16	11	5	8	3	10	5
Jupiter	11	10	4	11	8	10	7	11	6	9	6	7
Saturn	9	6	9	8	6	5	9	10	11	10	8	9
TOTAL	59	58	43	70	58	87	74	60	50	38	54	49
Rising	5	3	6	4	16	8	8	12	9	11	9	9
N. Node	9	8	12	7	9	10	4	7	7	5	8	14

9H	♈	♉	♊	♋	♌	♍	♎	♏	♐	♑	♒	♓
Moon	7	8	9	12	11	2	10	9	7	8	7	10
Sun	8	6	11	2	8	8	15	6	7	8	10	11
Mercury	15	9	4	11	2	8	9	5	7	12	5	13
Venus	6	12	9	9	5	7	7	9	11	6	11	8
Mars	5	8	6	10	10	11	6	17	6	7	8	6
Jupiter	14	10	7	12	4	7	6	11	7	7	7	8
Saturn	9	6	1	13	13	8	7	7	11	11	10	4
TOTAL	64	59	47	69	53	51	60	64	56	59	58	60

Houses	1st	2nd	3rd	4th	5th	6th	7th	8th	9th	10th	11th	12th
Moon	11	11	7	11	5	10	8	7	7	4	13	6
Sun	14	11	12	9	7	4	5	5	8	8	10	7
Mercury	12	17	11	8	3	8	6	3	11	10	4	7
Venus	14	13	7	7	12	5	3	4	8	9	8	10
Mars	10	6	8	10	7	9	11	12	5	9	7	6
Jupiter	11	10	9	5	12	6	9	6	9	4	7	12
Saturn	11	9	5	16	5	6	11	8	8	10	7	4
TOTAL	83	77	59	66	51	48	53	45	56	54	56	52
N. Node	12	8	8	11	6	6	10	5	9	12	4	9

Planets

Aspects	Parallels	4 Axis	Direction	☉/☽ Midpoint
Moon-Jupiter	Moon-Jupiter	Jupiter (Asc)	Neptune (D)	
Sun-Pluto	Venus-Uranus	Mercury (IC)	Pluto (D)	
		Chiron (IC)		
				Moon's Nodes
				Mercury (S)

Singers

Theme: Taurus/Gemini

Taurus:
Taureans are known for having deep, strong, sensual voices. They enjoy the arts and their senses. Taurus is especially frequent for the Sun (1 in 387) and Mercury (1 in 171). Cancer is also strong in the 9th harmonic, showing singers' soulfulness.

Gemini:
Geminis love using their voices to share their ideas. Mars in Gemini (1 in 415) gives singers added force and energy to their voices. Saturn in Gemini is also frequent, indicating a career involving communication.

Mercury Direct (1 in 123):
This indicates being able to think and communicate clearly.

Venus-Neptune aspects:
They achieve personal dreams through the arts. They have highly artistic imaginations and are very sensitive to the tastes of their audience.

Signs

Size: 443	1st Harmonic	9th Harmonic	Houses	Rising Sign
Moon		Cancer	2nd, 3rd, 12th	Virgo
Sun	Taurus, Capricorn		1st, 9th	
Mercury	Taurus		9th	Lunar Phase
Venus	Aquarius			Leo
Mars	Gemini	Capricorn	2nd, 10th	☉/☽ Midpoint
Jupiter	Sagittarius		4th, 5th	Pisces
Saturn	Gemini	Gemini		Moon's Nodes
TOTAL	Taurus, Gemini	Cancer	12th	Cancer, Leo
(weak)	Libra		7th	3rd (S)

1H	♈	♉	♊	♋	♌	♍	♎	♏	♐	♑	♒	♓
Moon	38	30	31	40	40	45	29	41	46	38	33	32
Sun	32	54	42	29	38	29	32	27	36	45	36	43
Mercury	28	48	34	33	37	26	30	40	34	45	43	45
Venus	45	43	35	35	33	47	21	29	33	30	49	43
Mars	31	37	55	45	41	49	35	41	37	20	23	29
Jupiter	26	35	31	34	38	35	39	47	51	39	30	38
Saturn	28	38	51	37	42	33	39	21	35	48	35	36
TOTAL	228	285	279	253	269	264	225	246	272	265	249	266
Rising	22	27	37	53	43	59	38	38	41	37	34	14
N. Node	42	36	40	49	54	38	41	27	29	22	39	26

9H	♈	♉	♊	♋	♌	♍	♎	♏	♐	♑	♒	♓
Moon	29	36	39	46	38	40	40	34	39	35	36	31
Sun	40	30	32	46	36	42	42	33	34	35	32	41
Mercury	40	28	31	41	41	40	42	39	46	30	29	36
Venus	39	34	38	45	36	29	41	24	39	35	38	45
Mars	35	37	37	43	33	32	30	40	35	46	37	38
Jupiter	30	34	30	38	44	37	35	45	40	31	37	42
Saturn	42	43	47	30	31	40	24	25	36	44	38	43
TOTAL	255	242	254	289	259	260	254	240	269	256	247	276

Houses	1st	2nd	3rd	4th	5th	6th	7th	8th	9th	10th	11th	12th
Moon	40	48	50	32	39	31	26	27	37	34	31	48
Sun	53	42	34	33	34	22	24	37	42	31	49	42
Mercury	49	40	37	35	30	29	20	33	45	36	42	47
Venus	37	47	40	30	27	36	22	40	43	33	38	50
Mars	22	48	40	36	33	31	33	34	35	48	41	42
Jupiter	33	26	36	48	48	37	36	38	34	36	33	38
Saturn	36	32	36	43	38	39	36	36	27	35	42	43
TOTAL	270	283	273	257	249	225	197	245	263	253	276	310
N. Node	37	31	34	36	43	25	36	42	38	39	40	42

Planets

Aspects	Parallels	4 Axis	Direction	☉/☽ Midpoint
Sun-Uranus		Chiron (Asc)	Mercury (D)	
Venus-Neptune		Uranus (MC)	Neptune (Rx)	
		Chiron (MC)		
				Moon's Nodes
				Mercury (S)
				Pluto (S)

Songwriters – Jazz, Pop, Rock

Theme: Leo (5th), Sun

5th House (1 in 826,114):
The 5th house represents creativity, fun, and romance. Songwriting is a powerful expression of these activities. Venus is especially frequent in the 5th house (1 in 1,611).

Moon in Sagittarius in the 1st & 9th Harmonics:
Songwriters express their emotions enthusiastically through creative expressions. Sagittarians are often known as the bards.

Sun aspects:
They are confident people who enjoy the creative process and like publicly displaying their work. This reveals creative mental abilities. Sun-Mercury aspects are especially frequent (1 in 145). They have confidence in their writing and communication skills and seek adulation for their work.

Saturn is infrequent on the 4 Axis Points (1 in 832):
Songwriters' lives are not bound by rules and discipline. Instead, they are typically creative, spontaneous people who follow their inspiration.

Signs

Size: 216	1st Harmonic	9th Harmonic	Houses	Rising Sign
Moon	Sagittarius	Sagittarius	5th	
Sun	Libra, Capricorn	Pisces		
Mercury	Capricorn	Leo	5th, 6th	Lunar Phase
Venus		Cancer	5th	Taurus
Mars		Aquarius		☉/☽ Midpoint
Jupiter	Sagittarius			
Saturn	Taurus, Leo	Virgo	11th	Moon's Nodes
TOTAL			5th	
(weak)	Aries		8th, 9th	

1H	♈	♉	♊	♋	♌	♍	♎	♏	♐	♑	♒	♓
Moon	13	12	14	20	23	15	18	17	25	17	22	20
Sun	15	18	20	17	13	14	26	15	14	25	19	20
Mercury	15	16	20	18	11	12	19	24	15	26	23	17
Venus	18	22	17	18	8	26	13	17	17	19	21	20
Mars	14	16	19	17	20	28	23	18	15	16	10	20
Jupiter	15	17	16	18	12	11	21	25	27	22	20	12
Saturn	13	28	20	17	32	16	16	13	9	12	18	22
TOTAL	103	129	126	125	119	122	136	129	122	137	133	131
Rising	14	16	14	24	27	26	24	23	15	10	13	10
N. Node	20	21	22	19	24	21	14	18	15	16	14	12

9H	♈	♉	♊	♋	♌	♍	♎	♏	♐	♑	♒	♓
Moon	17	11	10	21	16	21	19	19	26	17	20	19
Sun	22	19	18	19	14	15	18	14	20	13	19	25
Mercury	18	13	16	15	26	17	22	16	18	15	23	17
Venus	18	20	19	26	19	6	20	16	11	19	23	19
Mars	21	21	16	14	22	18	15	16	16	16	26	15
Jupiter	15	22	13	13	20	22	20	21	23	17	11	19
Saturn	19	19	21	14	14	29	11	15	20	23	12	19
TOTAL	130	125	113	122	131	128	125	117	134	120	134	133

Houses	1st	2nd	3rd	4th	5th	6th	7th	8th	9th	10th	11th	12th
Moon	18	25	18	12	28	18	12	13	12	17	21	22
Sun	16	21	17	24	21	20	15	12	14	18	17	21
Mercury	23	12	24	19	23	24	14	8	17	18	14	20
Venus	16	17	19	19	29	20	14	15	20	14	17	16
Mars	14	15	20	18	23	20	13	16	19	17	24	17
Jupiter	21	17	12	23	21	15	21	17	16	18	17	18
Saturn	20	15	16	17	23	13	15	16	14	16	31	20
TOTAL	128	122	126	132	168	130	104	97	112	118	141	134
N. Node	17	18	13	20	20	16	21	15	21	20	16	19

Planets

Aspects	Parallels	4 Axis	Direction	☉/☽ Midpoint
Sun & Uranus	Moon-Sun		Chiron (D)	Pluto
Moon-Saturn				
Sun-Mercury				
Sun-Jupiter				**Moon's Nodes**
Sun-Neptune				Mercury (S)
				Neptune (S)
				Chiron (S)

SPIRITUAL TEACHERS

Theme: Scorpio/Pisces, Jupiter

Scorpio & Pisces:
Scorpio is strong in the 1st harmonic, and Pisces is strong in the 9th harmonic. Spiritual teachers have deep insight into the metaphysical and spiritual nature of existence. The Scorpio Lunar Phase is especially frequent (1 in 821).

1st House:
People with a strong 1st house project their personalities powerfully onto their environment. The Sun (1 in 450) and Mars (1 in 6,170) are especially frequent.

Jupiter in Leo in the 9th Harmonic (1 in 2,634):
This shows confidence and enjoyment in teaching philosophy.

Jupiter aspects and parallels:
This indicates a highly intellectual nature as well as a positive outlook on life.

Jupiter on the Midheaven & the Moon's North Node:
Their goals revolve around matters of personal growth, exploration, and philosophy. Jupiterians often seek to influence the thinking of large groups of people.

Signs

Size: 80	1st Harmonic	9th Harmonic	Houses	Rising Sign
Moon			7th	
Sun	Libra	Capricorn, Pisces	1st	
Mercury			12th	Lunar Phase
Venus	Scorpio	Virgo		Scorpio
Mars	Capricorn	Sagittarius	1st	☉/☽ Midpoint
Jupiter		Leo	5th	
Saturn	Libra, Scorpio, Pisces	Pisces		Moon's Nodes
TOTAL	Scorpio		1st	Aquarius
(weak)	Leo, Virgo	Gemini	11th	3rd

1H	♈	♉	♊	♋	♌	♍	♎	♏	♐	♑	♒	♓
Moon	6	9	4	10	5	8	8	5	10	6	3	6
Sun	6	8	6	10	4	5	11	8	3	5	8	6
Mercury	5	7	7	9	3	7	9	6	9	5	5	8
Venus	7	7	6	9	3	6	7	12	6	4	5	8
Mars	3	7	8	8	6	4	7	10	6	13	5	3
Jupiter	3	8	7	9	10	4	6	9	3	7	6	8
Saturn	7	4	6	1	4	3	11	13	6	5	8	12
TOTAL	37	50	44	56	35	37	59	63	43	45	40	51
Rising	5	1	6	7	13	6	9	8	6	9	5	5
N. Node	10	8	2	7	8	4	5	5	9	6	12	4

9H	♈	♉	♊	♋	♌	♍	♎	♏	♐	♑	♒	♓
Moon	7	9	4	6	7	7	9	7	8	9	2	5
Sun	4	8	7	6	4	7	6	7	5	11	2	13
Mercury	6	6	5	10	7	8	5	10	5	6	8	4
Venus	8	5	2	1	3	12	9	10	6	10	8	6
Mars	9	10	6	7	3	2	10	3	11	5	6	8
Jupiter	4	10	4	6	15	6	4	8	3	8	6	6
Saturn	4	8	5	9	1	4	9	8	9	5	6	12
TOTAL	42	56	33	45	40	46	52	53	47	54	38	54

Houses	1st	2nd	3rd	4th	5th	6th	7th	8th	9th	10th	11th	12th
Moon	1	6	6	4	9	6	11	9	10	8	4	6
Sun	15	5	3	7	5	4	6	7	4	8	7	9
Mercury	9	7	5	6	5	7	5	6	3	7	6	14
Venus	12	5	9	6	8	6	2	6	7	10	6	3
Mars	16	8	4	8	3	7	6	4	6	3	5	10
Jupiter	5	6	5	10	12	6	5	7	8	4	5	7
Saturn	5	8	9	10	2	7	3	7	9	8	2	10
TOTAL	63	45	41	51	44	43	38	46	47	48	35	59
N. Node	6	10	8	8	7	3	7	5	10	7	3	6

Planets

Aspects	Parallels	4 Axis	Direction	☉/☽ Midpoint
Jupiter & Pluto	Moon-Jupiter	Sun (Dsc)	Venus (Rx)	
Sun-Chiron	Mercury-Jupiter	Moon (MC)		
Mercury-Pluto	Mars-Jupiter	Jupiter (MC)		
Venus-Pluto		Venus (IC)		**Moon's Nodes**
Jupiter-Saturn		Saturn (IC)		Moon (N)
				Sun (N)
				Mercury (N)
				Jupiter (N)

Sports – Professional and Olympics

Theme: Aries/Gemini

The 1st Quadrant is strong & the 3rd Quadrant is weak:
Aries, Taurus and Gemini form the 1st quadrant, which is focused on individual experience and developing personal abilities. The opposite 3rd quadrant is the more psychologically oriented and focused on societal issues.

Aries:
Aries is the sign of the sports champion. They are aggressive competitors who focus on personal achievements and serve as examples of leadership to others. Saturn in Aries indicates a career involving competition and achievement. The Moon's North Node is especially frequent in Aries (1 in 7,486). The Sun/Moon midpoint is also frequent in Aries (1 in 3,779).

Jupiter in Scorpio (1 in 123) & Aries in the 9th Harmonic (1 in 168):
This indicates a great deal of power and confidence when competing with others.

9th & 10th Houses:
These houses are at the top of the sky. Athletes receive a great amount of attention for their achievements. The 9th house is focused on striving towards goals with a positive vision, while the 10th house is focused on determination and hard work.

Signs

Size: 1,214	1st Harmonic	9th Harmonic	Houses	Rising Sign
Moon	Gemini			Scorpio
Sun		Virgo	9th	
Mercury	Gemini, Pisces		10th	Lunar Phase
Venus	Gemini, Aquarius		10th	Scorpio
Mars	Leo			☉/☽ Midpoint
Jupiter	Virgo, Scorpio	Aries, Aquarius	2nd	Aries
Saturn	♈ ♉ ♊ ♓	Aries, Virgo, Scorp	11th	Moon's Nodes
TOTAL	♈ ♉ ♊ ♓		9th, 10th	Aries
(weak)	Virgo, Libra, Sagitt			Pisces

1H	♈	♉	♊	♋	♌	♍	♎	♏	♐	♑	♒	♓
Moon	100	97	132	99	112	98	103	97	93	86	94	103
Sun	108	105	109	112	112	75	97	92	78	111	101	114
Mercury	99	89	109	92	104	96	94	101	95	111	108	116
Venus	115	115	113	120	84	113	73	89	99	77	134	82
Mars	97	100	107	117	127	101	123	94	98	76	85	89
Jupiter	95	101	88	101	92	130	117	137	101	92	81	79
Saturn	135	153	117	88	69	67	52	78	92	120	110	133
TOTAL	749	760	775	729	700	680	659	688	656	673	713	716
Rising	49	68	84	127	139	114	137	154	125	94	64	59
N. Node	136	99	102	107	82	101	95	78	88	90	113	123

9H	♈	♉	♊	♋	♌	♍	♎	♏	♐	♑	♒	♓
Moon	89	116	102	105	96	93	109	109	103	102	86	104
Sun	89	107	109	90	100	122	104	95	100	106	93	99
Mercury	115	98	115	111	88	99	90	95	107	101	112	83
Venus	102	106	75	92	106	101	93	101	114	114	109	101
Mars	107	97	98	112	111	99	91	111	100	96	86	106
Jupiter	126	97	87	103	101	107	95	106	82	95	124	91
Saturn	117	89	106	110	76	127	91	118	78	96	102	104
TOTAL	745	710	692	723	678	748	673	735	684	710	712	688

Houses	1st	2nd	3rd	4th	5th	6th	7th	8th	9th	10th	11th	12th
Moon	107	113	94	110	88	95	101	94	112	103	99	98
Sun	105	107	118	80	80	92	85	89	112	105	123	118
Mercury	105	122	104	87	87	98	89	70	105	120	120	107
Venus	115	99	113	100	79	86	84	89	109	119	103	118
Mars	104	102	98	92	100	93	94	103	95	112	108	113
Jupiter	93	124	112	99	109	92	100	94	106	104	92	89
Saturn	104	108	97	117	93	92	97	103	82	98	120	103
TOTAL	733	775	736	685	636	648	650	642	721	761	765	746
N. Node	115	74	107	81	112	108	103	106	109	101	108	90

Planets

Aspects	Parallels	4 Axis	Direction	☉/☽ Midpoint
Moon-Pluto		Uranus (Dsc)		
Venus-Saturn		Pluto (MC)		
		Moon (IC)		
				Moon's Nodes
				Jupiter (S)
				Uranus (S)
				Chiron (D)

Sports Coaches

Theme: Aries (1st)

Aries:
Aries is the sign of the sports champion. They are aggressive competitors who focus on personal achievement and serve as examples of leadership to others. Saturn in Aries (1 in 288) indicates a leadership involving competition and achievement. The Aries lunar phase (1 in 450,690) is especially frequent.

1st House:
People with a strong 1st house project their personalities powerfully onto their environment. The Sun (1 in 4,441) and Saturn (1 in 237) are most common.

Sun aspects are infrequent:
The Sun functions more powerfully when it is alone in the sky and not in any aspect relationship with other planetary bodies. This indicates that they strive to be special and be in charge of their own lives. Directors, Politicians, Salespeople, and the Wealthy also have a solitary Sun.

Signs

Size: 102	1st Harmonic	9th Harmonic	Houses	Rising Sign
Moon			6th	Aquarius
Sun	Pisces	Leo	1st	Lunar Phase
Mercury	Pisces		12th	Aries
Venus	Aries	Aries, Leo	12th	☉/☽ Midpoint
Mars	Aries			Aries
Jupiter				Libra
Saturn	Leo	Aries, Virgo	1st	Moon's Nodes
TOTAL		Aries	1st, 7th	5th (S)
(weak)		Pisces		

1H	♈	♉	♊	♋	♌	♍	♎	♏	♐	♑	♒	♓
Moon	5	9	5	9	11	6	7	12	7	11	13	7
Sun	8	9	13	6	10	6	10	6	3	9	8	14
Mercury	12	6	10	8	5	10	11	5	5	7	10	13
Venus	14	8	5	11	6	13	4	4	11	6	13	7
Mars	13	3	9	13	9	7	10	7	7	8	8	8
Jupiter	6	9	3	8	4	7	13	12	11	12	6	11
Saturn	11	12	5	6	16	9	6	6	6	10	9	6
TOTAL	69	56	50	61	61	58	61	52	50	63	67	66
Rising	6	4	12	9	7	7	15	11	6	10	10	5
N. Node	10	10	11	9	8	6	10	10	8	6	7	7

9H	♈	♉	♊	♋	♌	♍	♎	♏	♐	♑	♒	♓
Moon	11	5	7	10	10	10	8	7	7	7	11	9
Sun	6	6	11	11	13	7	8	9	6	6	11	8
Mercury	11	10	8	7	9	5	10	7	8	12	10	5
Venus	14	5	9	10	13	7	8	3	9	12	5	7
Mars	6	12	11	11	4	11	2	8	10	9	12	6
Jupiter	11	13	9	12	7	6	5	10	8	4	11	6
Saturn	16	11	5	6	7	14	10	8	4	9	8	4
TOTAL	75	62	60	67	63	60	51	52	52	59	68	45

Houses	1st	2nd	3rd	4th	5th	6th	7th	8th	9th	10th	11th	12th
Moon	7	5	11	4	8	15	12	5	10	12	6	7
Sun	20	8	12	2	5	4	10	5	9	8	6	13
Mercury	12	13	4	7	5	5	10	4	6	12	9	15
Venus	11	15	3	6	4	6	7	11	6	7	11	15
Mars	11	12	8	6	5	6	10	7	7	9	10	11
Jupiter	6	11	11	13	9	8	12	11	5	5	7	4
Saturn	16	4	8	11	8	9	8	10	8	4	9	7
TOTAL	83	68	57	49	44	53	69	53	51	57	58	72
N. Node	11	6	4	6	11	3	8	6	12	12	13	10

Planets

Aspects	Parallels	4 Axis	Direction	☉/☽ Midpoint
Mercury-Venus	Venus-Neptune	Mercury (Asc)	Mars (D)	
	Mars-Pluto	Moon (Dsc)	Chiron (D)	
		Saturn (Dsc)		
		Pluto (MC)		**Moon's Nodes**
				Saturn (N)

STROKE VICTIMS

Theme: Libra (7th), Chiron

A stroke is caused by a lack of oxygen to the brain, and is associated with the Air element.

Gemini (1 in 588) & Sagittarius:
These signs represent the Axis of Communication. People strong in these signs are highly mentally active and talkative. It is possible that all this thinking puts a strain on the brain.

Jupiter in Aquarius (1 in 828) & Jupiter in Libra in the 9th Harmonic (1 in 593):
Jupiter in Libra and Aquarius indicates a strong propensity for thinking and intellectual engagement with others.

7th House (1 in 224):
The 7th house depicts how we relate to people. Mars is especially frequent in the 7th house (1 in 248), indicating conflicts in relationships with others.

Moon-Uranus aspects:
The Moon represents emotions and the subconscious mind. Moon-Uranus aspects are indicative of having unpredictable emotions and surprising thoughts.

Mercury-Chiron parallels (1 in 1,668):
This can be an indication of brain damage.

Signs

Size: 106	1st Harmonic	9th Harmonic	Houses	Rising Sign
Moon	Cancer			Aquarius
Sun		Capricorn	1st, 3rd	
Mercury	Taurus, Leo			Lunar Phase
Venus	Sagittarius	Aquarius	2nd, 11th	Capricorn
Mars	Gemini	Leo	2nd, 7th	☉/☽ Midpoint
Jupiter	Aquarius	Libra	4th	Capricorn
Saturn	Scorpio		10th	Moon's Nodes
TOTAL	Gemini, Sagittarius	Libra	7th	Cancer
(weak)	Libra	Cancer	6th	7th

1H	♈	♉	♊	♋	♌	♍	♎	♏	♐	♑	♒	♓
Moon	6	10	13	14	9	6	9	6	11	6	6	10
Sun	8	9	13	7	10	10	5	10	10	9	6	9
Mercury	10	15	5	7	15	6	7	6	13	12	5	5
Venus	10	8	11	14	6	8	5	8	17	4	11	4
Mars	9	9	16	8	9	10	4	8	11	11	6	5
Jupiter	5	6	11	12	5	6	7	11	8	10	17	8
Saturn	7	9	12	5	5	7	9	17	8	9	9	9
TOTAL	55	66	81	67	59	53	46	66	78	61	60	50
Rising	6	3	7	10	8	13	14	11	10	8	10	6
N. Node	11	8	5	16	11	11	9	8	7	8	5	7

9H	♈	♉	♊	♋	♌	♍	♎	♏	♐	♑	♒	♓
Moon	7	7	6	8	11	8	12	8	10	8	10	11
Sun	9	7	6	3	10	6	12	11	9	14	10	9
Mercury	11	6	13	8	9	9	11	12	8	7	6	6
Venus	8	3	11	3	11	9	8	11	11	10	14	7
Mars	7	10	12	5	14	11	8	7	7	4	10	11
Jupiter	6	12	8	8	3	7	17	11	8	6	9	11
Saturn	10	6	6	7	4	6	10	10	9	12	13	13
TOTAL	58	51	62	42	62	56	78	70	62	61	72	68

Houses	1st	2nd	3rd	4th	5th	6th	7th	8th	9th	10th	11th	12th
Moon	9	3	13	12	10	5	11	13	7	9	7	7
Sun	15	10	15	7	6	4	12	4	5	7	8	13
Mercury	12	9	12	10	5	9	7	7	6	5	10	14
Venus	10	16	11	7	6	5	9	7	7	4	16	8
Mars	10	14	8	6	5	4	16	9	8	10	7	9
Jupiter	8	12	4	14	10	5	10	8	6	6	11	12
Saturn	9	4	8	10	9	10	12	5	11	14	7	7
TOTAL	73	68	71	66	51	42	77	53	50	55	66	70
N. Node	7	5	7	10	8	5	16	13	6	6	10	13

Planets

Aspects	Parallels	4 Axis	Direction	☉/☽ Midpoint
Jupiter	Mercury-Chiron	Sun (Asc)	Mercury (Rx)	Pluto
Moon-Uranus		Jupiter (Asc)	Uranus (Rx)	Chiron
Mercury-Jupiter		Saturn (MC)		**Moon's Nodes**
		Moon (IC)		Pluto (N)
				Chiron (S)

Suicide Victims

Theme: Aquarius, Mars/Uranus

Aquarius & Taurus:
Aquarius is strong in the 1st harmonic, and Taurus is strong in the 9th harmonic. These are the two signs most associated with Venus. This suggests that many suicides are due to problems with love and money. The Sun is strong in Aquarius in both the 1st and 9th harmonics, indicating that their life goals often revolve around relationships, since they are such likable people.

Capricorn Lunar Phase (1 in 446)
This starts the last quadrant of the lunar cycle and triggers endings. During this phase, there is a mood of seriousness that can incline a person towards despondency. This phase is also frequent for murderers.

Libra is weak:
Librans are very intellectual and independent in relationships with others.

Mars:
Mars-Uranus aspects are frequent (1 in 135). These aspects can indicate sudden violence. In the personality, it shows an impulsive way of asserting one's desires. Mars is also frequent on the Descendant (1 in 284), indicating that they often experience other people as hostile towards them. Mars-Chiron parallels (1 in 3,738) are a common indication of suffering physical injuries.

Uranus aspects (1 in 299):
This shows the likelihood of having many sudden shocks and surprises in one's life.

Signs

Size: 251	1st Harmonic	9th Harmonic	Houses	Rising Sign
Moon	Aries			Leo
Sun	Capricorn, Aquarius	Aquarius	5th	Lunar Phase
Mercury	Aquarius	Cancer, Capricorn	4th	Capricorn
Venus	Capricorn, Aquarius			☉/☽ Midpoint
Mars	Virgo, Pisces	Taurus		Aquarius
Jupiter		Gemini	4th	Pisces
Saturn		Taurus	7th	Moon's Nodes
TOTAL	Aquarius	Taurus	6th	Sagittarius
(weak)	Libra	Libra		12th

1H	♈	♉	♊	♋	♌	♍	♎	♏	♐	♑	♒	♓
Moon	28	23	24	12	22	17	16	24	18	21	25	21
Sun	20	25	20	22	21	16	17	18	16	29	29	18
Mercury	21	18	16	23	24	15	14	23	21	24	31	21
Venus	20	13	26	25	19	18	12	21	18	28	30	21
Mars	22	18	18	15	19	35	21	23	26	11	18	25
Jupiter	24	16	16	20	25	20	28	29	25	20	14	14
Saturn	20	23	18	13	25	20	20	27	26	22	24	13
TOTAL	155	136	138	130	155	141	128	165	150	155	171	133
Rising	10	8	23	26	39	32	31	18	23	13	18	10
N. Node	22	16	25	20	23	18	21	17	31	16	15	27

9H	♈	♉	♊	♋	♌	♍	♎	♏	♐	♑	♒	♓
Moon	26	19	23	24	23	19	15	22	18	20	25	17
Sun	24	27	17	23	16	21	20	20	19	14	29	21
Mercury	15	26	16	31	18	21	23	23	19	29	17	13
Venus	17	20	27	14	23	28	19	18	19	21	20	25
Mars	14	29	20	23	17	16	19	23	28	19	18	25
Jupiter	23	17	28	15	12	21	22	21	26	25	24	17
Saturn	18	32	17	15	21	27	10	23	28	21	18	21
TOTAL	137	170	148	145	130	153	128	150	157	149	151	139

Houses	1st	2nd	3rd	4th	5th	6th	7th	8th	9th	10th	11th	12th
Moon	17	27	25	22	19	23	20	12	23	19	18	26
Sun	25	18	24	17	25	22	15	25	16	24	19	21
Mercury	20	22	18	30	21	22	20	12	21	21	20	24
Venus	25	19	21	23	22	21	21	15	20	17	23	24
Mars	16	16	24	22	25	25	26	20	15	16	23	23
Jupiter	22	21	18	30	23	26	15	13	20	20	18	25
Saturn	19	23	17	18	18	22	28	17	26	23	18	22
TOTAL	144	146	147	162	153	161	145	114	141	140	139	165
N. Node	26	19	20	18	15	24	19	16	22	19	24	29

Planets

Aspects	Parallels	4 Axis	Direction	☉/☽ Midpoint
Uranus	Mars-Chiron	Mars (Dsc)	Mars (D)	
Moon-Uranus		Chiron (Dsc)	Jupiter (D)	
Sun-Pluto				
Mercury-Uranus				Moon's Nodes
Mars-Uranus				

TEACHERS – KINDERGARTEN THROUGH 12TH GRADES

Theme: Cancer/Libra

Cancer:
Cancers are the most nurturing, affectionate people. They have a natural affinity for children and they love to see them grow. Teaching is also an effective way for Cancers to use their leadership abilities (Cardinal sign).

Moon in Libra & the 7th House:
This position represents nurturing through the intellect. Librans are known for their idealism and truthfulness. The Moon seeks to express this idealism through the development of children's minds.

Pisces Lunar Phase (1 in 6,266):
This is a phase of service towards humanity.

Saturn in the 3rd House (1 in 289):
This indicates being an authority in the field of learning and communication.

Signs

Size: 92	1st Harmonic	9th Harmonic	Houses	Rising Sign
Moon	Libra	Libra	7th	Taurus
Sun		Aries		
Mercury	Cancer	Aries	4th	Lunar Phase
Venus	Cancer		6th, 10th	Pisces
Mars		Taurus		☉/☽ Midpoint
Jupiter		Pisces	6th	Libra
Saturn		Aries	3rd, 5th	Moon's Nodes
TOTAL	Cancer			9th
(weak)	Scorpio		1st, 2nd	

1H	♈	♉	♊	♋	♌	♍	♎	♏	♐	♑	♒	♓
Moon	5	4	11	10	6	3	14	7	8	7	11	6
Sun	11	4	12	11	5	8	9	5	6	6	11	4
Mercury	3	10	5	12	6	6	10	7	7	5	12	9
Venus	9	5	9	14	7	12	2	2	9	7	7	9
Mars	10	7	8	6	10	11	8	7	9	6	4	6
Jupiter	3	8	8	6	9	9	10	9	11	6	7	6
Saturn	7	9	7	7	9	9	8	5	6	12	7	6
TOTAL	48	47	60	66	52	58	61	42	56	49	59	46
Rising	5	10	10	11	8	12	9	5	7	7	4	4
N. Node	9	10	10	8	8	11	6	5	5	6	6	8

9H	♈	♉	♊	♋	♌	♍	♎	♏	♐	♑	♒	♓
Moon	9	6	6	2	11	5	13	9	7	4	9	11
Sun	12	9	5	10	4	9	9	8	8	5	10	3
Mercury	13	5	8	5	7	8	8	6	10	5	10	7
Venus	7	6	8	4	6	6	11	7	9	8	9	11
Mars	4	15	8	9	5	5	5	11	10	8	7	5
Jupiter	6	4	8	6	6	10	9	5	8	9	8	13
Saturn	12	7	10	11	7	10	4	5	9	9	2	6
TOTAL	63	52	53	47	46	53	59	51	61	48	55	56

Houses	1st	2nd	3rd	4th	5th	6th	7th	8th	9th	10th	11th	12th
Moon	5	9	6	10	8	4	13	10	9	4	6	8
Sun	3	6	10	10	5	8	5	9	10	8	11	7
Mercury	7	5	6	13	8	3	10	9	9	11	8	3
Venus	8	3	10	6	9	11	4	7	5	13	6	10
Mars	8	6	6	10	8	8	3	11	11	6	9	6
Jupiter	6	9	8	6	8	12	6	8	3	9	8	9
Saturn	6	5	15	1	13	9	6	6	8	10	7	6
TOTAL	43	43	61	56	59	55	47	60	55	61	55	49
N. Node	7	9	10	10	6	3	8	6	12	7	6	8

Planets

Aspects	Parallels	4 Axis	Direction	
Sun-Neptune	Venus-Mars		Mars (D)	
Mercury-Venus	Mercury-Chiron			
				Moon's Nodes
				Moon (N)

TECHNICAL PROFESSIONS

Theme: Capricorn, Mercury

This group is composed of Computer Programmers, Engineers, Mathematicians, Physicists, and Scientists.

Capricorn (1 in 853):
Capricorns are able to commit long periods of time to accomplish goals. They are patient and practical and want tangible, enduring results from their work.

Scorpio:
Scorpios have good analytical ability and are able to examine problems deeply.

Moon's North Node in the 6th House (1 in 544) & the Moon's South Node in Virgo:
This can indicate that a person's life path involves analytical work.

5th House is weak:
The 5th house represents fun, creativity, and romance.

Mercury:
Mercury represents logic and reasoning. Mercurial people are good at solving intricate problems.

Signs

Size: 493	1st Harmonic	9th Harmonic	Houses	Rising Sign
Moon	Sagittarius	Leo		
Sun	Aries, Capricorn	Taurus	4th	Lunar Phase
Mercury	Capricorn, Pisces	Scorpio	4th	
Venus	Taurus	Scorpio	1st	☉/☽ Midpoint
Mars	Capricorn		10th	Capricorn
Jupiter				Pisces
Saturn		Taurus, Virgo	12th	Moon's Nodes
TOTAL	Aries, Capricorn	Scorpio	11th	Virgo (S)
(weak)	Gemini, Cancer, Libra		5th	6th

1H	♈	♉	♊	♋	♌	♍	♎	♏	♐	♑	♒	♓
Moon	35	35	35	32	41	48	38	36	53	45	51	44
Sun	55	43	36	28	35	40	33	36	37	54	47	49
Mercury	47	38	30	31	31	39	32	37	54	55	48	51
Venus	51	58	35	37	29	40	25	44	44	40	45	45
Mars	40	34	42	43	40	42	51	51	40	45	36	29
Jupiter	40	37	33	42	45	43	38	42	36	46	43	48
Saturn	40	49	37	36	41	37	46	44	46	47	40	30
TOTAL	308	294	248	249	262	289	263	290	310	332	310	296
Rising	23	35	41	50	61	60	47	52	40	37	21	26
N. Node	45	27	33	48	46	49	35	40	42	31	44	53

9H	♈	♉	♊	♋	♌	♍	♎	♏	♐	♑	♒	♓
Moon	45	33	43	35	53	35	39	45	40	40	42	43
Sun	40	53	31	36	35	51	37	50	46	42	37	35
Mercury	38	40	44	38	27	45	50	51	41	32	45	42
Venus	33	39	38	46	39	43	32	54	42	48	45	34
Mars	50	46	37	40	49	33	39	43	37	32	46	41
Jupiter	38	44	37	45	39	45	39	39	37	41	46	43
Saturn	44	56	39	35	34	53	40	40	43	35	34	40
TOTAL	288	311	269	275	276	305	276	322	286	270	295	278

Houses	1st	2nd	3rd	4th	5th	6th	7th	8th	9th	10th	11th	12th
Moon	33	43	42	50	41	49	40	35	43	41	48	28
Sun	49	48	46	56	26	32	22	37	34	41	49	53
Mercury	57	47	47	51	30	31	32	23	37	48	46	44
Venus	60	47	47	41	26	39	30	31	38	45	51	38
Mars	44	49	36	37	25	40	42	41	34	56	48	41
Jupiter	34	45	50	47	30	37	40	42	37	41	44	46
Saturn	42	36	46	24	34	43	42	44	38	44	45	55
TOTAL	319	315	314	306	212	271	248	253	261	316	331	305
N. Node	38	34	36	41	49	59	39	40	35	38	41	43

Planets

Aspects	Parallels	4 Axis	Direction	☉/☽ Midpoint
Moon-Sun	Jupiter-Saturn	Mars (Asc)		Mercury
Mercury-Uranus				
Mercury-Neptune				
Mercury-Chiron				Moon's Nodes
Mars-Pluto				Moon (S)

TRANSVESTITES

Theme: Scorpio, Moon/Mars

Scorpio:
Transvestites are often burdened with sex identity and emotional struggles. Scorpios are deep, mysterious people. Venus in Scorpio is frequent in the 1st and 9th harmonics (1 in 279 – 1H, 1 in 898 - 9H). Their relationships are intense, complex, and shadowy. Venus also represents the female persona.

Aries is weak in the 1st & 9th Harmonics and the 1st House:
Scorpio and Aries both represent sexuality. Scorpio is secretive, while Aries is open and direct. Transvestites tend to keep private and do not project themselves into their environment.

Sun-Mars aspects (1 in 189) and parallels (1 in 1,287):
Sun-Mars relations indicate that a person's identity revolves around sexuality. Their lives contain themes of aggression, desire, and conflict.

Moon:
The Moon is frequently in aspect with Venus, Mars, and the Descendant, which are all relational. Lunar relationships are emotional, moody, nurturing, and reclusive. Moon-Saturn aspects show that nurturing was restricted in the early life, which tends to depress our emotional state.

Signs

Size: 85	1st Harmonic	9th Harmonic	Houses	Rising Sign
Moon		Scorpio	12th	Leo
Sun	Virgo			
Mercury		Virgo		Lunar Phase
Venus	Gemini, Scorpio	Scorpio	10th	Cancer
Mars			5th	☉/☽ Midpoint
Jupiter		Sagittarius		Cancer
Saturn	Scorpio, Capricorn	Virgo	4th	Moon's Nodes
TOTAL	Scorpio		3rd, 5th, 6th	Leo
(weak)	Aries	Aries	1st	

1H	♈	♉	♊	♋	♌	♍	♎	♏	♐	♑	♒	♓
Moon	7	11	6	11	9	3	5	7	4	7	6	9
Sun	4	9	11	4	7	13	5	9	6	8	7	2
Mercury	3	8	8	8	7	8	9	10	9	6	3	6
Venus	5	3	15	7	6	4	8	14	6	4	8	5
Mars	5	7	6	5	6	7	11	5	7	8	9	9
Jupiter	6	3	6	9	6	5	12	7	9	9	10	3
Saturn	4	6	3	4	4	8	9	15	6	16	6	4
TOTAL	34	47	55	48	45	48	59	67	47	58	49	38
Rising	2	4	10	8	15	9	8	10	10	5	2	2
N. Node	7	4	1	8	12	7	4	7	9	10	6	10

9H	♈	♉	♊	♋	♌	♍	♎	♏	♐	♑	♒	♓
Moon	4	5	8	9	6	8	7	12	8	5	5	8
Sun	6	11	7	3	6	8	4	7	7	7	8	11
Mercury	4	5	4	9	3	14	7	6	8	8	10	7
Venus	8	9	6	6	4	3	6	15	6	6	7	9
Mars	5	7	10	5	7	6	7	9	9	7	9	4
Jupiter	5	6	7	10	6	4	9	5	13	6	6	8
Saturn	4	11	10	0	8	15	7	5	5	6	7	7
TOTAL	36	54	52	42	40	58	47	59	56	45	52	54

Houses	1st	2nd	3rd	4th	5th	6th	7th	8th	9th	10th	11th	12th
Moon	8	7	7	3	7	8	9	8	5	6	5	12
Sun	4	6	10	9	7	9	5	5	5	10	8	7
Mercury	7	5	12	10	5	7	5	8	8	7	5	6
Venus	5	8	9	8	8	8	8	1	3	13	9	5
Mars	8	5	11	6	12	9	5	5	9	7	5	3
Jupiter	2	8	11	7	10	8	5	9	5	9	5	6
Saturn	5	7	7	13	11	9	7	5	6	5	4	6
TOTAL	39	46	67	56	60	58	44	41	41	57	41	45
N. Node	3	7	11	6	8	9	3	10	8	5	8	7

Planets

Aspects	Parallels	4 Axis	Direction	☉/☽ Midpoint
Moon	Sun-Mars	Moon (Dsc)	Chiron (D)	
Moon-Venus	Mars-Saturn	Venus (Dsc)		
Moon-Mars		Neptune (MC)		
Moon-Saturn				**Moon's Nodes**
Sun-Mars				Mercury (S)
Sun-Uranus				Venus (S)
Mercury-Mars				

TWINS

Theme: Sagittarius, 11th House

Sagittarius:
Sagittarians are highly communicative people.

Gemini Lunar Phase (1 in 343):
Gemini is the sign of the twins. They enjoy frequent social contact in their environment.

11th House:
The 11th house represents higher goals of brotherly love and concern for humanity.

Mercury in Air Houses (1 in 264):
Mercury is very communicative in Air houses.

Signs

Size: 132	1st Harmonic	9th Harmonic	Houses	Rising Sign
Moon		Gemini	2nd, 4th	Gemini
Sun	Capricorn		11th	Lunar Phase
Mercury	Capricorn		5th, 11th	Gemini
Venus	Capricorn, Pisces	Leo	6th	☉/☽ Midpoint
Mars	Sagittarius			Aquarius
Jupiter	Sagittarius	Sagittarius		Moon's Nodes
Saturn			11th	Sagittarius
TOTAL	Sagittarius, Capricorn		11th	Aquarius
(weak)				7th

1H	♈	♉	♊	♋	♌	♍	♎	♏	♐	♑	♒	♓
Moon	14	10	13	8	9	8	13	13	7	15	12	10
Sun	7	7	13	11	5	8	13	11	15	16	11	15
Mercury	6	10	10	10	9	4	12	16	16	18	15	6
Venus	6	6	14	9	7	10	9	15	11	16	13	16
Mars	10	9	9	18	10	13	13	14	18	6	5	7
Jupiter	14	8	8	9	7	11	16	12	19	8	10	10
Saturn	9	8	10	10	8	12	15	12	13	12	13	10
TOTAL	66	58	77	75	55	66	91	93	99	91	79	74
Rising	6	6	15	12	13	15	13	10	11	13	11	7
N. Node	12	4	10	6	10	8	11	11	16	11	18	15

9H	♈	♉	♊	♋	♌	♍	♎	♏	♐	♑	♒	♓
Moon	14	9	17	15	13	9	13	9	12	7	6	8
Sun	13	7	15	10	8	11	13	12	11	12	7	13
Mercury	15	11	10	12	8	11	10	12	7	16	12	8
Venus	11	12	9	9	17	10	8	7	11	9	16	13
Mars	9	11	9	15	13	11	15	12	11	4	11	11
Jupiter	14	12	6	15	13	6	14	8	16	7	12	9
Saturn	12	8	11	6	13	10	10	11	13	13	13	12
TOTAL	88	70	77	82	85	68	83	71	81	68	77	74

Houses	1st	2nd	3rd	4th	5th	6th	7th	8th	9th	10th	11th	12th
Moon	9	17	11	16	8	12	7	7	15	10	9	11
Sun	12	10	8	12	11	12	14	7	11	6	18	11
Mercury	6	6	16	10	16	8	12	12	8	8	19	11
Venus	11	12	10	6	12	15	14	10	5	15	12	10
Mars	16	6	9	8	9	14	8	12	10	12	14	14
Jupiter	14	11	6	7	15	13	13	7	12	9	12	13
Saturn	8	9	6	15	11	8	14	12	12	16	17	4
TOTAL	76	71	66	74	82	82	82	67	73	76	101	74
N. Node	15	12	13	11	12	7	16	9	10	13	9	5

Planets

Aspects	Parallels	4 Axis	Direction	☉/☽ Midpoint
	Moon-Chiron	Uranus (Asc)		Uranus
		Sun (Dsc)		
		Pluto (Dsc)		
		Saturn (MC)		Moon's Nodes

Wealthy

Theme: Capricorn, Mars/Jupiter

Moon in Capricorn (1 in 1,209):
Their emotions are well controlled, and they are comfortable working long, hard hours. This also indicates that many wealthy people come from conservative family backgrounds that value career success.

Mars:
Mars is the god of war. Wealthy people overcome all kinds of challenges using diverse tactics. Mars-Pluto aspects are especially frequent (1 in 465), showing them to be powerful people who often use shadowy methods to achieve goals.

Jupiter Direct (1 in 1,549):
This indicates having great confidence and enthusiasm to achieve goals.

Sun aspects are infrequent:
The Sun functions more powerfully when it is alone in the sky and not in any aspect relationship with other planetary bodies. This indicates that they strive to be special and be in charge of their own lives. Directors, Politicians, Salespeople, and Sports Coaches also have a solitary Sun.

Signs

Size: 419	1st Harmonic	9th Harmonic	Houses	Rising Sign
Moon	Capricorn	Leo	5th	
Sun	Aquarius	Virgo		
Mercury	Pisces			Lunar Phase
Venus	Leo, Aquarius	Capricorn		Taurus
Mars	Gemini	Capricorn	6th	☉/☽ Midpoint
Jupiter	Scorpio			Capricorn
Saturn	Scorpio		3rd, 9th	Moon's Nodes
TOTAL	Pisces	Sagittarius	9th	10th
(weak)	Sagittarius			

1H	♈	♉	♊	♋	♌	♍	♎	♏	♐	♑	♒	♓
Moon	28	29	24	33	31	36	40	37	37	53	29	42
Sun	37	34	29	42	35	36	34	33	29	25	48	37
Mercury	28	32	32	30	35	37	37	36	33	31	40	48
Venus	34	31	31	36	43	41	34	26	24	30	49	40
Mars	33	37	46	36	38	37	42	41	25	35	25	24
Jupiter	28	26	37	31	34	40	38	51	38	34	31	31
Saturn	39	31	39	31	34	26	34	48	33	31	38	35
TOTAL	227	220	238	239	250	253	259	272	219	239	260	257
Rising	22	29	30	35	38	49	51	45	49	35	19	17
N. Node	34	41	34	40	44	40	36	22	31	37	33	27

9H	♈	♉	♊	♋	♌	♍	♎	♏	♐	♑	♒	♓
Moon	21	28	37	38	49	26	34	37	35	38	35	41
Sun	32	36	29	31	32	45	38	39	42	31	35	29
Mercury	34	33	29	39	34	35	26	35	44	27	43	40
Venus	28	34	41	29	33	42	38	33	39	45	25	32
Mars	31	36	24	38	29	38	27	34	41	45	36	40
Jupiter	39	39	24	34	43	36	34	32	41	35	27	35
Saturn	36	32	38	37	36	45	33	33	40	25	34	30
TOTAL	221	238	222	246	256	267	230	243	282	246	235	247

Houses	1st	2nd	3rd	4th	5th	6th	7th	8th	9th	10th	11th	12th
Moon	36	39	36	27	46	31	44	29	38	36	26	31
Sun	48	33	40	36	22	30	37	25	32	40	35	41
Mercury	42	46	36	34	29	26	30	30	38	37	37	34
Venus	43	36	28	42	36	25	30	36	35	35	33	40
Mars	43	30	32	36	38	42	30	39	27	35	35	32
Jupiter	39	33	34	35	35	34	32	37	42	40	32	26
Saturn	32	48	36	39	31	26	28	32	45	23	41	38
TOTAL	283	265	242	249	237	214	231	228	257	246	239	242
N. Node	32	33	29	36	27	32	29	37	38	45	40	41

Planets

Aspects	Parallels	4 Axis	Direction	☉/☽ Midpoint
Mars	Moon-Mercury	Moon (Asc)	Jupiter (D)	Jupiter
Moon-Mars			Chiron (D)	
Venus-Saturn				
Mars-Jupiter				Moon's Nodes
Mars-Saturn				
Mars-Pluto				

Widowed

Theme: Moon/Chiron

Venus in Aries (1 in 702):
This indicates a need for freedom in relationships.

Venus in Scorpio in the 9th Harmonic (1 in 806):
This can indicate death and transformation within relationships.

Moon in Mutable Houses (1 in 1,198):
This indicates that the home life undergoes fluctuation.

Moon aspects:
Moon-Mars aspects show a strong temper and frequent conflicts in the home. Moon-Uranus aspects indicate a propensity towards sudden changes in one's home environment. These people are forced to adjust themselves emotionally due to unpredictable circumstances.

Chiron:
Chiron indicates that a wound is a significant issue for a person. Chiron is especially frequent on the Moon's North Node (1 in 1,794), showing a life path of healing trauma.

Signs

Size: 376	1st Harmonic	9th Harmonic	Houses	Rising Sign
Moon		Sagittarius	3rd, 12th	Aries, Libra
Sun	Taurus	Libra	5th	Scorpio
Mercury		Leo	10th	**Lunar Phase**
Venus	Aries	Scorpio		Cancer
Mars				**☉/☽ Midpoint**
Jupiter	Gemini		9th, 10th, 12th	
Saturn	Cancer	Pisces		**Moon's Nodes**
TOTAL			9th, 10th	Cancer
(weak)	Capricorn			

1H	♈	♉	♊	♋	♌	♍	♎	♏	♐	♑	♒	♓
Moon	37	19	37	31	32	35	21	34	38	34	21	37
Sun	28	44	33	33	27	29	32	38	28	26	31	27
Mercury	36	32	28	33	24	28	36	35	31	34	25	34
Venus	48	35	26	34	27	37	24	36	29	22	38	20
Mars	18	27	34	39	37	44	35	31	30	26	24	31
Jupiter	25	26	40	27	33	38	34	41	39	22	29	22
Saturn	29	28	20	36	36	24	39	30	40	30	36	28
TOTAL	221	211	218	233	216	235	221	245	235	194	204	199
Rising	26	16	32	36	38	25	55	52	36	25	19	16
N. Node	23	33	33	40	29	40	31	24	24	33	31	35

9H	♈	♉	♊	♋	♌	♍	♎	♏	♐	♑	♒	♓
Moon	30	34	28	30	34	32	30	39	45	23	25	26
Sun	40	24	30	29	34	34	41	29	22	22	38	33
Mercury	35	24	36	38	42	32	35	22	25	35	26	26
Venus	24	33	36	32	25	31	27	48	32	31	24	33
Mars	39	37	24	31	28	30	29	31	31	36	28	32
Jupiter	29	30	38	29	31	26	30	32	37	34	33	27
Saturn	31	38	28	30	21	30	30	34	30	28	36	40
TOTAL	228	220	220	219	215	215	222	235	222	209	210	217

Houses	1st	2nd	3rd	4th	5th	6th	7th	8th	9th	10th	11th	12th
Moon	28	37	41	19	22	35	37	26	38	25	28	40
Sun	27	32	36	34	38	18	21	29	27	40	35	39
Mercury	36	37	32	39	28	25	20	23	35	43	27	31
Venus	30	37	27	24	33	24	34	34	31	31	37	34
Mars	23	37	27	27	32	37	29	36	36	29	29	34
Jupiter	35	26	28	27	30	30	22	25	43	45	24	41
Saturn	33	29	35	30	27	24	28	33	29	37	32	39
TOTAL	212	235	226	200	210	193	191	206	239	250	212	258
N. Node	27	28	42	33	35	28	29	34	36	23	29	32

Planets

Aspects	Parallels	4 Axis	Direction	☉/☽ Midpoint
Moon & Chiron	Moon-Neptune	Mercury (Asc)	Mars (Rx)	
Uranus		Mars (Dsc)	Saturn (D)	
Moon-Mars		Mercury (MC)	Chiron (D)	
Moon-Uranus		Jupiter (MC)		Moon's Nodes
Sun-Chiron				Mercury (N)
Mercury-Venus				Mars (S)
				Pluto (S)
				Chiron (N)

WRITERS

Theme: Aquarius

This group contains writers of Journalism, Literature, Novels, and Poetry.

4th Quadrant:
Capricorn, Aquarius, and Pisces are the signs of the Universal quadrant of the zodiac. This quadrant represents the reception of intuitive ideas that serve a purpose for humanity.

Aquarius:
Aquarians enjoy reflecting on ideas and putting them into words.

Sun on the Imum Coeli (1 in 181):
This indicates that writers are most productive in their homes working in private.

Signs

Size: 851	1st Harmonic	9th Harmonic	Houses	Rising Sign
Moon	Gemini	Capricorn, Aquarius	12th	Aquarius
Sun		Cancer, Virgo	4th	Lunar Phase
Mercury	Aquarius	Capricorn, Pisces	2nd, 3rd, 6th	Capricorn
Venus		Scorpio	3rd	☉/☽ Midpoint
Mars	Pisces		4th	Aries
Jupiter		Aquarius		Moon's Nodes
Saturn			6th	Aquarius
TOTAL	Capricorn, Pisces	Aquarius		Pisces (S)
(weak)		Aries, Libra, Sagitt	9th, 11th	8th

1H	♈	♉	♊	♋	♌	♍	♎	♏	♐	♑	♒	♓
Moon	64	67	86	62	73	71	80	59	63	73	71	82
Sun	79	68	71	59	78	79	66	51	69	75	75	81
Mercury	63	62	65	58	60	76	74	67	73	84	88	81
Venus	83	86	44	87	59	92	48	60	86	64	79	63
Mars	55	63	63	87	85	78	90	70	73	62	56	69
Jupiter	58	64	70	77	74	70	83	90	64	76	65	60
Saturn	54	64	60	59	66	72	80	86	78	89	70	73
TOTAL	456	474	459	489	495	538	521	483	506	523	504	509
Rising	45	53	48	72	87	99	105	102	86	59	61	34
N. Node	62	69	62	74	78	85	63	54	79	72	83	70

9H	♈	♉	♊	♋	♌	♍	♎	♏	♐	♑	♒	♓
Moon	74	74	54	66	72	62	64	76	62	88	88	71
Sun	63	73	72	87	58	86	71	70	62	65	74	70
Mercury	64	77	61	67	63	60	64	82	76	87	64	86
Venus	60	77	61	79	75	83	56	85	65	62	78	70
Mars	67	55	81	73	82	81	63	65	79	71	71	63
Jupiter	57	80	72	72	66	60	72	74	48	82	87	81
Saturn	69	82	78	62	62	78	55	78	64	62	83	78
TOTAL	454	518	479	506	478	510	445	530	456	517	545	519

Houses	1st	2nd	3rd	4th	5th	6th	7th	8th	9th	10th	11th	12th
Moon	61	76	71	70	70	78	73	76	56	76	60	84
Sun	85	84	80	89	61	59	61	74	45	61	73	79
Mercury	69	92	94	82	53	80	58	55	56	54	71	87
Venus	73	78	90	73	74	57	68	65	61	63	71	78
Mars	75	80	71	83	69	71	74	66	54	77	70	61
Jupiter	63	78	72	74	69	62	76	81	74	71	64	67
Saturn	75	67	68	54	71	88	61	68	75	82	68	74
TOTAL	501	555	546	525	467	495	471	485	421	484	477	530
N. Node	79	73	68	72	68	66	70	90	67	62	75	61

Planets

Aspects	Parallels	4 Axis	Direction	☉/☽ Midpoint
		Sun (IC)	Mercury (D)	Saturn
			Chiron (D)	
				Moon's Nodes
				Mercury (N)

Personality Trait Study Groups

ACTIVE

Theme: Aries (1st), Leo (5th)

Aries & the 1st House:
Arians are the most assertive people. A strong 1st house indicates that a person projects themself effectively in their environment.

Leo & the 5th House:
Leos are highly confident and expressive people. Mercury is the most frequent in Leo (1 in 191). A strong 5th house shows creative self-expression.

11th House:
A strong 11th house shows a person who is highly involved with people in their community. Mars is the most frequent planet in the 11th House (1 in 489).

Sun (1 in 434) & Mercury in the Fire Houses (1 in 236):
The Fire houses emphasize confident self-expression.

Venus in the 3rd House (1 in 676):
This indicates enjoyment from communicating and relating to people in the immediate environment.

Mercury on the Moon's South Node (1 in 4,744):
This indicates that a person comes into life with an active mind and a need to connect with others.

Signs

Size: 117	1st Harmonic	9th Harmonic	Houses	Rising Sign
Moon		Virgo	11th	
Sun	Leo, Aquarius	Aries, Sagittarius	9th	
Mercury	Leo, Pisces		5th	Lunar Phase
Venus			3rd	
Mars	Aries, Gemini, Pisces		5th, 11th	☉/☽ Midpoint
Jupiter		Virgo	1st, 8th	
Saturn	Virgo	Aries, Gemini	12th	Moon's Nodes
TOTAL		Virgo	1st, 11th, 12th	5th
(weak)			7th	

1H	♈	♉	♊	♋	♌	♍	♎	♏	♐	♑	♒	♓
Moon	11	9	9	12	5	8	11	11	11	12	10	8
Sun	14	7	10	9	15	8	7	8	5	8	16	10
Mercury	9	8	9	7	17	7	7	9	6	11	12	15
Venus	12	10	9	14	8	8	11	7	10	6	14	8
Mars	13	13	15	6	10	7	10	7	5	9	8	14
Jupiter	10	7	13	9	13	14	9	12	6	6	10	8
Saturn	3	13	11	9	9	18	9	12	14	5	8	6
TOTAL	72	67	76	66	77	70	64	66	57	57	78	69
Rising	6	4	10	16	12	8	16	12	14	7	8	4
N. Node	9	8	10	10	17	11	10	10	9	8	8	7

9H	♈	♉	♊	♋	♌	♍	♎	♏	♐	♑	♒	♓
Moon	2	9	14	9	8	17	10	8	11	9	10	10
Sun	15	13	4	6	4	10	9	8	15	6	14	13
Mercury	12	14	8	7	10	12	10	14	7	7	11	5
Venus	10	10	6	12	10	9	7	10	10	13	13	7
Mars	10	8	10	11	10	10	11	11	5	10	10	11
Jupiter	13	9	10	12	4	15	6	8	14	8	6	12
Saturn	15	7	16	12	8	12	8	6	5	11	8	9
TOTAL	77	70	68	69	54	85	61	65	67	64	72	67

Houses	1st	2nd	3rd	4th	5th	6th	7th	8th	9th	10th	11th	12th
Moon	13	14	11	10	10	6	5	6	8	10	15	9
Sun	15	12	7	7	12	5	5	7	14	9	9	15
Mercury	16	11	9	5	14	6	4	9	11	10	10	12
Venus	8	9	19	7	6	7	6	8	7	14	12	14
Mars	12	7	4	7	14	7	8	11	11	6	19	11
Jupiter	16	8	12	9	5	11	5	16	10	7	8	10
Saturn	7	5	9	3	9	11	11	10	12	12	12	16
TOTAL	87	66	71	48	70	53	44	67	73	68	85	87
N. Node	14	9	11	12	6	9	4	9	8	12	11	12

Planets

Aspects	Parallels	4 Axis	Direction	☉/☽ Midpoint
Moon-Venus	Moon-Chiron	Sun (Asc)		Mars
Sun-Saturn		Mars (Asc)		
Mercury-Neptune		Uranus (Dsc)		
		Moon (IC)		**Moon's Nodes**
				Mercury (S)

AGGRESSIVE-BRASH

Theme: Aries/Libra

Aries Rising (1 in 2,915):
Aries as the rising sign shows a person's persona to be influenced by the need to express oneself directly and assertively.

Jupiter on the Descendant (1 in 1,260):
This shows a person who is highly enthusiastic and confident relating to people.

Venus-Pluto aspects (1 in 2,337):
This indicates that a person interacts with others in an intense and coercive manner.

Signs

Size: 140	1st Harmonic	9th Harmonic	Houses	Rising Sign
Moon		Leo		Aries
Sun	Aries	Virgo, Libra, Pisces	3rd, 9th	
Mercury	Gemini	Capricorn		Lunar Phase
Venus	Cancer		3rd, 7th	
Mars				☉/☽ Midpoint
Jupiter	Cancer		6th	Aries
Saturn	Libra		2nd	Moon's Nodes
TOTAL	Aquarius		2nd	Aries
(weak)				Libra

1H	♈	♉	♊	♋	♌	♍	♎	♏	♐	♑	♒	♓
Moon	10	11	11	6	13	14	13	15	9	11	13	14
Sun	18	8	17	16	11	6	11	9	6	12	14	12
Mercury	9	6	17	13	11	11	10	9	9	14	15	16
Venus	11	15	11	19	11	9	6	9	13	8	15	13
Mars	10	11	13	10	18	15	13	13	7	7	12	11
Jupiter	14	11	11	19	5	8	18	9	15	8	13	9
Saturn	9	15	9	6	14	10	17	9	11	18	14	8
TOTAL	81	77	89	89	83	73	88	73	70	78	96	83
Rising	15	10	10	10	19	12	13	13	14	13	7	4
N. Node	17	13	14	8	11	14	19	7	9	7	10	11

9H	♈	♉	♊	♋	♌	♍	♎	♏	♐	♑	♒	♓
Moon	15	6	12	8	19	9	14	5	15	14	10	13
Sun	11	8	11	10	8	17	17	15	5	11	10	17
Mercury	9	11	14	11	14	13	12	9	6	18	13	10
Venus	11	11	11	12	12	13	10	9	16	13	10	12
Mars	14	10	13	15	15	8	9	11	9	9	13	14
Jupiter	11	15	12	8	12	15	13	11	9	11	12	11
Saturn	15	15	11	8	4	17	11	13	14	14	9	9
TOTAL	86	76	84	72	84	92	86	73	74	90	77	86

Houses	1st	2nd	3rd	4th	5th	6th	7th	8th	9th	10th	11th	12th
Moon	12	16	14	11	10	10	13	11	6	14	10	13
Sun	13	17	19	9	11	7	9	10	16	4	12	13
Mercury	11	16	18	10	15	8	7	8	14	8	10	15
Venus	13	15	18	13	9	5	16	5	9	14	12	11
Mars	12	15	7	8	12	9	11	15	10	12	12	17
Jupiter	12	10	10	10	9	20	11	10	13	12	10	13
Saturn	12	18	9	14	9	13	10	8	10	12	12	13
TOTAL	85	107	95	75	75	72	77	67	78	76	78	95
N. Node	16	14	11	16	10	13	11	8	13	11	8	9

Planets

Aspects	Parallels	4 Axis	Direction	☉/☽ Midpoint
Moon-Saturn	Moon-Uranus	Jupiter (Dsc)	Mercury (R)	
Mercury-Venus		Uranus (MC)		
Venus-Pluto		Neptune (MC)		
				Moon's Nodes
				Chiron (S)

BRILLIANT MIND

Theme: Aquarius/Pisces, 3rd/11th Houses

Aquarius & Pisces (1 in 708):
Aquarius, as a Fixed Air sign, represents the love of ideas and communication. It shows that brilliant people have a deep love for learning and sharing ideas. Mercury is most frequent in Aquarius in the 9th harmonic (1 in 1,316). Pisceans have the greatest capacity for reflective thinking in solitude. These are the last two signs of the zodiac and are the most concerned with uplifting humanity.

3rd & 11th Houses:
Air houses describe aptitude for thinking, learning, and communicating.

Mercury-Saturn aspects (1 in 140):
This is an indication of a serious, mature, well thought-out mind.

Mercury Direct (1 in 822):
Mercury is the planet associated with logic, reason, communication, learning ability, and intelligence.

Sun on the Midheaven (1 in 175):
The Sun on the Midheaven gives a strong personality that is highly regarded in public. They like to shine in the public eye and, therefore, have strong ambitions to succeed in their careers.

Size: 156	Signs		Houses	Rising Sign
	1st Harmonic	9th Harmonic		
Moon	Libra, Sagittarius		3rd	
Sun	Pisces	Aquarius	6th, 10th	
Mercury	Pisces	Aquarius	11th	Lunar Phase
Venus	Aries, Pisces	Scorpio	4th, 8th	
Mars	Capricorn			☉/☽ Midpoint
Jupiter	Aries	Taurus, Leo	6th	
Saturn		Aquarius	11th	Moon's Nodes
TOTAL	Aries, Pisces	Aquarius	3rd, 8th, 11th	3rd
(weak)	Leo, Virgo		1st, 9th, 12th	

1H	♈	♉	♊	♋	♌	♍	♎	♏	♐	♑	♒	♓
Moon	14	14	13	10	13	14	21	7	20	7	9	14
Sun	17	15	12	12	9	8	12	9	10	17	16	19
Mercury	17	12	14	9	9	12	9	8	17	18	10	21
Venus	19	13	9	13	10	9	7	15	15	13	14	19
Mars	14	15	10	16	14	13	12	14	11	17	15	5
Jupiter	18	12	8	13	6	13	14	16	11	15	13	17
Saturn	16	15	8	11	7	8	8	17	17	12	19	18
TOTAL	115	96	74	84	68	77	83	86	101	99	96	113
Rising	6	9	14	16	22	13	13	20	15	9	11	8
N. Node	16	12	12	9	18	12	11	14	12	15	12	13

9H	♈	♉	♊	♋	♌	♍	♎	♏	♐	♑	♒	♓
Moon	13	14	13	12	16	8	17	18	12	11	15	7
Sun	16	17	15	8	11	14	10	10	9	13	21	12
Mercury	8	11	17	12	11	15	12	14	10	13	24	9
Venus	14	12	16	10	18	11	6	21	14	10	11	13
Mars	11	10	18	10	9	16	14	13	15	13	12	15
Jupiter	9	19	12	13	22	7	13	9	13	14	12	13
Saturn	12	14	10	16	8	10	16	15	11	8	19	17
TOTAL	83	97	101	81	95	81	88	100	84	82	114	86

Houses	1st	2nd	3rd	4th	5th	6th	7th	8th	9th	10th	11th	12th
Moon	8	9	21	14	7	14	13	17	9	17	14	13
Sun	13	13	19	12	5	17	11	11	10	19	13	13
Mercury	8	18	17	8	14	9	15	15	7	13	20	12
Venus	14	13	15	20	10	5	15	19	15	8	12	10
Mars	9	18	12	11	11	13	16	14	10	12	17	13
Jupiter	11	12	17	12	12	19	8	18	9	14	15	9
Saturn	18	9	13	11	16	13	12	15	9	11	20	9
TOTAL	81	92	114	88	75	90	90	109	69	94	111	79
N. Node	13	13	21	13	11	14	12	15	11	12	13	8

Planets

Aspects	Parallels	4 Axis	Direction	☉/☽ Midpoint
Pluto	Venus-Mars	Sun (MC)	Mercury (D)	
Mercury-Saturn			Saturn (D)	
Venus-Jupiter				
				Moon's Nodes
				Moon (N)

Eccentric

Theme: 6th House, Mercury

Moon in Taurus (1 in 25,395):
This is a counter-intuitive finding since this position typically shows a very stable secure person.

6th House (1 in 25,360):
A strong 6th house indicates that a person is concerned with their health and well-being. They tend to be critical of themselves and their environment. Mercury is the most frequent planet in the 6th house (1 in 231).

Mercury aspects (1 in 159):
Mercurian people are clever and often difficult for others to figure out.

Chiron on the Moon's South Node (1 in 78,287):
This indicates that a person is carrying a wound from early life.

Signs

Size: 127	1st Harmonic	9th Harmonic	Houses	Rising Sign
Moon	Taurus, Virgo	Sagittarius, Capr	6th, 12th	Taurus
Sun	Cancer, Pisces		4th, 6th	Pisces
Mercury	Taurus		6th	**Lunar Phase**
Venus			5th	Virgo
Mars	Capricorn			**☉/☽ Midpoint**
Jupiter			10th	
Saturn	Sagittarius	Aries, Scorpio		**Moon's Nodes**
TOTAL	Capricorn		6th	Gemini
(weak)	Gemini, Scorpio	Libra	9th	8th

1H	♈	♉	♊	♋	♌	♍	♎	♏	♐	♑	♒	♓
Moon	14	23	9	7	7	16	13	10	6	8	9	5
Sun	7	12	6	17	12	6	9	8	9	13	12	16
Mercury	5	15	10	7	16	5	9	7	11	11	17	14
Venus	12	7	7	14	13	13	8	7	8	13	12	13
Mars	13	8	15	9	14	10	8	11	8	15	11	5
Jupiter	13	10	7	8	13	8	15	11	10	14	11	7
Saturn	9	8	3	12	10	13	7	8	17	16	9	15
TOTAL	73	83	57	74	85	71	69	62	69	90	81	75
Rising	9	12	7	9	15	14	13	13	11	9	4	11
N. Node	9	12	17	7	11	9	11	15	10	8	7	11

9H	♈	♉	♊	♋	♌	♍	♎	♏	♐	♑	♒	♓
Moon	15	10	2	13	13	7	5	7	16	16	12	11
Sun	6	8	8	14	13	15	9	9	13	12	9	11
Mercury	13	9	8	8	12	10	9	12	10	11	12	13
Venus	10	8	9	11	15	13	7	12	8	9	11	14
Mars	7	12	12	14	14	8	12	9	10	12	10	7
Jupiter	11	15	13	12	8	11	11	6	14	10	7	9
Saturn	16	9	11	9	9	15	7	17	12	9	4	9
TOTAL	78	71	63	81	84	79	60	72	83	79	65	74

Houses	1st	2nd	3rd	4th	5th	6th	7th	8th	9th	10th	11th	12th
Moon	12	12	9	14	9	16	7	9	6	11	5	17
Sun	14	8	7	18	10	16	7	8	6	10	13	10
Mercury	14	11	8	13	14	17	7	5	7	10	12	9
Venus	10	17	11	7	17	12	11	7	9	9	9	8
Mars	13	12	9	11	11	14	7	11	13	11	6	9
Jupiter	11	7	12	8	13	13	9	8	9	18	11	8
Saturn	9	15	11	10	9	13	13	12	6	6	11	12
TOTAL	83	82	67	81	83	101	61	60	56	75	67	73
N. Node	9	13	11	9	8	9	11	16	8	11	10	12

Planets

Aspects	Parallels	4 Axis	Direction	☉/☽ Midpoint
Mercury	Moon-Uranus	Uranus (Asc)		
Sun-Pluto				
Mercury-Venus				
Mercury-Jupiter				**Moon's Nodes**
				Saturn (N)
				Neptune (S)
				Chiron (S)

EMOTIONAL

Theme: Leo (5th), Venus/Mars

Leo & the 5th House:
Leos are the most self-expressive, passionate people. Mars is the most frequent planet in Leo in the 9th harmonic (1 in 17,111).

Mars in the 1st House (1 in 227):
In the 1st house, Mars shows an aggressive projection of the self.

Mars on the Moon's North Node (1 in 376):
This indicates that a person is learning to become more assertive and develop leadership ability.

Signs

Size: 141	1st Harmonic	9th Harmonic	Houses	Rising Sign
Moon		Taurus		Aquarius
Sun	Cancer			
Mercury		Leo, Virgo		Lunar Phase
Venus		Capricorn		
Mars		Leo	1st	☉/☽ Midpoint
Jupiter			5th	
Saturn	Leo	Leo		Moon's Nodes
TOTAL	Libra	Leo	5th	Leo (S)
(weak)		Sagittarius	6th	5th (S)

1H	♈	♉	♊	♋	♌	♍	♎	♏	♐	♑	♒	♓
Moon	12	9	11	11	11	11	15	13	11	13	14	10
Sun	7	15	13	19	11	13	13	10	11	5	14	10
Mercury	12	13	8	14	15	10	17	8	12	9	13	10
Venus	12	14	16	15	11	12	11	15	9	10	8	8
Mars	14	7	13	10	12	17	11	12	11	13	12	9
Jupiter	12	15	9	15	9	9	17	18	8	9	8	12
Saturn	14	11	14	9	16	4	17	9	14	10	14	9
TOTAL	83	84	84	93	85	76	101	85	76	69	83	68
Rising	4	5	10	16	17	19	12	12	12	14	14	6
N. Node	6	16	14	12	14	10	11	14	10	7	18	9

9H	♈	♉	♊	♋	♌	♍	♎	♏	♐	♑	♒	♓
Moon	9	20	8	8	11	12	10	16	7	15	13	12
Sun	9	11	11	10	10	10	14	14	12	9	16	15
Mercury	10	11	11	11	18	20	9	10	13	9	11	8
Venus	16	13	8	12	11	9	15	9	7	22	9	10
Mars	9	10	13	8	24	14	9	9	5	13	13	14
Jupiter	16	13	8	14	11	8	13	14	14	10	10	10
Saturn	15	11	13	6	19	14	14	12	9	13	6	9
TOTAL	84	89	72	69	104	87	84	84	67	91	78	78

Houses	1st	2nd	3rd	4th	5th	6th	7th	8th	9th	10th	11th	12th
Moon	10	15	11	11	13	11	14	14	11	13	10	8
Sun	15	9	14	12	13	10	13	6	9	12	13	15
Mercury	18	10	14	14	12	8	13	7	6	16	9	14
Venus	16	12	15	11	13	8	13	10	10	11	11	11
Mars	21	10	3	12	14	7	11	12	8	14	13	16
Jupiter	13	11	11	10	22	10	10	11	10	12	10	11
Saturn	7	16	12	12	14	6	11	15	13	12	14	9
TOTAL	100	83	80	82	101	60	85	75	67	90	80	84
N. Node	11	12	11	12	14	8	16	8	10	14	17	8

Planets

Aspects	Parallels	4 Axis	Direction	☉/☽ Midpoint
Chiron	Moon-Neptune	Pluto (Dsc)		Venus
Moon-Saturn	Venus-Pluto			Neptune
Sun-Mars				
Mercury-Mars				**Moon's Nodes**
Venus-Chiron				Mars (N)
				Pluto (S)

EXTROVERT

Theme: Aries/Aquarius, Mercury

Aries, Libra, & Aquarius:
These are extroverted Air and Fire signs.

Mercury:
Mercury is the planet of communication. Mercury-Uranus aspects (1 in 260), Mercury on the Midheaven (1 in 1,409), and Mercury on the Moon's North Node (1 in 607) are the most prominent findings for extroverts.

Signs

Size: 227	1st Harmonic	9th Harmonic	Houses	Rising Sign
Moon			3rd	
Sun		Aquarius	3rd	Lunar Phase
Mercury			9th	Scorpio
Venus	Aries, Taurus	Aquarius	4th, 9th	☉/☽ Midpoint
Mars	Aries, Taurus	Taurus, Gemini	12th	Aries
Jupiter	Libra, Virgo			Sagittarius
Saturn	Libra			Moon's Nodes
TOTAL	Aries	Aquarius		Virgo
(weak)	Sagittarius		6th	Libra

1H	♈	♉	♊	♋	♌	♍	♎	♏	♐	♑	♒	♓
Moon	20	15	15	21	14	23	19	21	21	17	23	18
Sun	24	21	23	15	20	17	22	9	11	23	21	21
Mercury	24	15	22	14	15	20	16	24	9	25	19	24
Venus	26	31	18	20	18	27	12	12	15	9	24	15
Mars	23	24	21	18	20	29	20	17	15	15	12	13
Jupiter	16	15	18	19	17	18	31	29	22	24	10	8
Saturn	16	16	18	19	22	20	28	11	18	19	19	21
TOTAL	149	137	135	126	126	154	148	123	111	132	128	120
Rising	9	12	15	18	22	30	26	22	27	19	17	10
N. Node	12	20	15	20	11	27	30	11	19	24	20	18

9H	♈	♉	♊	♋	♌	♍	♎	♏	♐	♑	♒	♓
Moon	15	19	14	21	23	21	15	23	13	21	22	20
Sun	22	13	18	17	20	23	11	22	14	23	28	16
Mercury	17	20	17	19	25	22	22	19	15	17	17	17
Venus	9	16	17	15	22	18	16	22	23	24	26	19
Mars	24	26	26	21	16	14	18	12	19	21	15	15
Jupiter	21	21	18	14	25	15	20	15	24	18	17	19
Saturn	15	19	18	19	14	23	22	14	17	22	26	18
TOTAL	123	134	128	126	145	136	124	127	125	146	151	124

Houses	1st	2nd	3rd	4th	5th	6th	7th	8th	9th	10th	11th	12th
Moon	17	23	27	16	16	20	25	16	19	15	16	17
Sun	21	24	28	17	17	6	16	12	19	26	23	18
Mercury	20	25	22	25	14	11	13	9	24	19	21	24
Venus	16	21	16	28	15	14	8	16	25	21	24	23
Mars	20	16	16	20	19	16	15	25	6	24	21	29
Jupiter	20	19	17	18	23	21	20	23	14	16	21	15
Saturn	21	16	14	25	18	17	13	16	21	23	18	25
TOTAL	135	144	140	149	122	105	110	117	128	144	144	151
N. Node	24	15	24	16	20	20	16	14	12	21	23	22

Planets

Aspects	Parallels	4 Axis	Direction	☉/☽ Midpoint
Mercury	Sun-Jupiter	Mercury (MC)		
Uranus	Mars-Jupiter			
Moon-Mars	Venus-Neptune			
Mercury-Uranus				Moon's Nodes
Venus-Saturn				Mercury (N)

GRACIOUS-SOCIABLE

Theme: Aquarius (11th)

Aquarius & the 11th House:
Aquarians are the most gifted socialites. They are attractive, intelligent, and well-spoken. Saturn is the most frequent planet in Aquarius in the 9th harmonic (1 in 12,360), and the Moon's South Node is most frequent in the 11th house (1 in 2,752).

Jupiter Direct (1 in 411):
This indicates having a positive, confident attitude towards life.

Moon on the Imum Coeli (1 in 345):
The Moon in this hidden position shows that a person does not demonstrate shy, protective characteristics.

Signs

Size: 167	1st Harmonic	9th Harmonic	Houses	Rising Sign
Moon	Aquarius	Leo	6th	Scorpio
Sun	Aquarius	Virgo	10th	**Lunar Phase**
Mercury	Taurus		11th	Aries
Venus	Aries	Capr, Aquarius		☉/☽ Midpoint
Mars	Aries	Taurus	11th	Aries
Jupiter		Leo	8th	Sagittarius
Saturn	Libra	Libra, Scorp, Aquar		**Moon's Nodes**
TOTAL	Gemini, Libra, Aquar	Aquarius		Virgo
(weak)			9th	11th (S)

1H	♈	♉	♊	♋	♌	♍	♎	♏	♐	♑	♒	♓
Moon	10	9	14	12	15	14	15	10	16	16	22	14
Sun	13	14	19	12	14	13	15	10	9	15	20	13
Mercury	15	20	12	7	12	12	19	12	12	16	17	13
Venus	20	12	17	12	14	19	10	9	14	9	16	15
Mars	20	14	19	12	21	18	17	10	11	5	12	8
Jupiter	13	13	11	14	11	13	19	18	15	17	10	13
Saturn	15	8	17	14	11	12	28	10	15	9	14	14
TOTAL	106	90	109	83	98	101	123	79	92	87	111	90
Rising	9	10	14	11	14	21	17	26	16	11	8	10
N. Node	10	14	8	11	14	21	18	10	17	14	13	17

9H	♈	♉	♊	♋	♌	♍	♎	♏	♐	♑	♒	♓
Moon	10	13	10	15	20	18	13	15	7	13	17	16
Sun	15	16	11	12	16	20	14	11	8	17	16	11
Mercury	17	13	13	17	14	13	12	10	19	9	15	15
Venus	9	11	16	13	13	16	11	15	12	21	21	9
Mars	19	20	18	15	12	8	13	17	9	11	17	8
Jupiter	13	17	15	12	23	8	8	16	20	12	11	12
Saturn	11	11	8	7	9	13	20	20	15	15	28	10
TOTAL	94	101	91	91	107	96	91	104	90	98	125	81

Houses	1st	2nd	3rd	4th	5th	6th	7th	8th	9th	10th	11th	12th
Moon	11	18	18	13	15	20	12	11	13	9	12	15
Sun	13	16	17	13	13	6	8	12	12	22	18	17
Mercury	15	18	13	14	11	10	6	10	17	14	21	18
Venus	15	13	16	18	13	7	10	13	14	16	18	14
Mars	16	11	7	18	12	7	16	17	6	19	22	16
Jupiter	11	16	13	16	18	14	19	20	9	11	8	12
Saturn	18	12	6	19	12	13	13	15	13	11	16	19
TOTAL	99	104	90	111	94	77	84	98	84	102	115	111
N. Node	12	9	15	8	26	17	13	10	8	17	19	13

Planets

Aspects	Parallels	4 Axis	Direction	☉/☽ Midpoint
Moon-Mars	Mars-Chiron	Moon (IC)	Jupiter (D)	Venus
Venus-Chiron				
Mars-Saturn				
				Moon's Nodes
				Chiron (N)

HARD WORKER

Theme: Virgo/Capricorn/Aquarius, Saturn

Virgo & Capricorn:
These are the two most diligent Earth signs. They are capable of burdening great responsibilities.

Venus in Cancer is weak (1 in 11,451):
This indicates that these people are not self-indulgent or in need of comforts or intimacy.

Ascendant:
Many planets are frequent on the Ascendant, showing great personal energy.

Jupiter on the Sun/Moon midpoint (1 in 4,667):
Their central energies are highly expansive and forward moving.

Signs

Size: 158	1st Harmonic	9th Harmonic	Houses	Rising Sign
Moon	Virgo, Aquarius	Cancer		Taurus
Sun	Virgo, Capr, Aquar	Scorpio, Aquar	9th, 12th	
Mercury	Capricorn	Aquarius		Lunar Phase
Venus	Leo, Capricorn			Capricorn
Mars	Taurus	Leo		☉/☽ Midpoint
Jupiter			8th	Aries
Saturn		Virgo	10th	Moon's Nodes
TOTAL	Virgo, Aquarius		6th	
(weak)	Gemini, Cancer		4th	

1H	♈	♉	♊	♋	♌	♍	♎	♏	♐	♑	♒	♓
Moon	10	7	16	13	14	19	10	10	10	13	22	14
Sun	11	12	7	16	9	19	11	11	13	19	19	11
Mercury	9	13	7	11	13	15	16	8	18	22	13	13
Venus	13	13	7	1	18	19	12	14	9	17	19	16
Mars	12	20	10	16	13	14	16	8	13	12	13	11
Jupiter	13	7	11	11	13	16	13	19	16	9	12	18
Saturn	9	9	12	10	11	17	17	17	19	10	14	13
TOTAL	77	81	70	78	91	119	95	87	98	102	112	96
Rising	8	16	12	11	15	21	18	19	15	12	4	7
N. Node	14	6	10	17	12	15	9	13	16	15	16	15

9H	♈	♉	♊	♋	♌	♍	♎	♏	♐	♑	♒	♓
Moon	13	10	8	19	11	11	12	12	17	14	15	16
Sun	12	11	11	13	11	11	13	20	11	10	19	16
Mercury	18	11	14	12	16	13	15	14	9	10	20	6
Venus	16	14	14	11	11	12	13	13	14	12	11	17
Mars	10	12	12	16	20	14	15	10	14	10	11	14
Jupiter	16	17	7	12	16	11	14	11	14	11	14	15
Saturn	13	12	17	12	11	21	15	11	10	11	11	14
TOTAL	98	87	83	95	96	93	97	91	89	78	101	98

Houses	1st	2nd	3rd	4th	5th	6th	7th	8th	9th	10th	11th	12th
Moon	9	15	12	18	18	14	9	14	13	10	11	15
Sun	12	15	15	7	12	13	11	11	17	9	13	23
Mercury	19	19	12	9	11	16	11	8	14	11	14	14
Venus	20	12	14	14	9	15	13	10	11	10	16	14
Mars	18	11	9	6	15	16	10	12	16	15	13	17
Jupiter	16	14	12	13	6	16	7	19	15	12	11	17
Saturn	13	11	15	9	10	14	13	16	14	20	9	14
TOTAL	107	97	89	76	81	104	74	90	100	87	87	114
N. Node	15	9	11	13	15	12	12	12	13	17	15	14

Planets

Aspects	Parallels	4 Axis	Direction	☉/☽ Midpoint
Mercury-Neptune	Saturn-Neptune	Venus (Asc)	Pluto (R)	Jupiter
Venus-Pluto		Mars (Asc)		
		Jupiter (Asc)		
		Saturn (Asc)		Moon's Nodes
		Saturn (MC)		

HUMOROUS-WITTY

Theme: Taurus/Cancer/Leo, Venus

Cancer & Leo:
These signs are the most personable and engaging with others.

Air signs:
Air signs represent the intellect and sharing of ideas. Air people are often clever and original. Gemini, Libra, and Aquarius are all strong. Venus in Aquarius is especially frequent in the 9th harmonic (1 in 223,301).

Moon in Taurus in the 9th Harmonic (1 in 1,150):
People with this placement are often very relaxed and at ease with themselves.

Uranus in the 9th House (1 in 1,158):
This indicates that a person has a unique philosophy of life.

Signs

Size: 132	1st Harmonic	9th Harmonic	Houses	Rising Sign
Moon	Sagittarius	Taurus, Leo	5th	Leo
Sun	Cancer	Taurus, Cancer	9th	
Mercury		Leo		Lunar Phase
Venus	Cancer	Aquarius	6th	
Mars	Pisces	Aquarius	4th	☉/☽ Midpoint
Jupiter	Sagitt, Capricorn	Taurus	8th	Sagittarius
Saturn	Gem, Scorp, Capr			Moon's Nodes
TOTAL	Gemini, Cancer	Taurus, Libra, Aquar		Leo
(weak)				

1H	♈	♉	♊	♋	♌	♍	♎	♏	♐	♑	♒	♓
Moon	11	9	11	9	9	14	12	7	16	12	12	10
Sun	9	11	16	17	13	9	10	7	13	7	10	10
Mercury	5	13	13	13	14	11	8	12	12	9	10	12
Venus	8	15	13	19	5	9	13	10	9	10	9	12
Mars	12	12	12	13	12	11	7	10	7	11	8	17
Jupiter	11	7	7	13	13	11	9	7	17	19	9	9
Saturn	7	4	16	8	9	10	13	17	10	18	11	9
TOTAL	63	71	88	92	75	75	72	70	84	86	69	79
Rising	5	6	8	11	21	15	12	17	12	9	9	7
N. Node	10	11	13	14	17	11	9	8	10	6	11	12

9H	♈	♉	♊	♋	♌	♍	♎	♏	♐	♑	♒	♓
Moon	12	21	6	6	17	11	15	8	9	8	8	11
Sun	7	19	5	18	12	10	12	5	13	10	11	10
Mercury	6	7	12	3	17	12	14	6	11	16	15	13
Venus	6	9	14	14	6	7	9	11	8	8	25	15
Mars	11	14	7	13	9	8	15	10	6	15	17	7
Jupiter	11	19	13	8	10	11	11	12	14	10	8	5
Saturn	16	10	9	7	3	14	16	14	12	15	9	7
TOTAL	69	99	66	69	74	73	92	66	73	82	93	68

Houses	1st	2nd	3rd	4th	5th	6th	7th	8th	9th	10th	11th	12th
Moon	14	13	12	11	16	12	7	12	5	13	7	10
Sun	14	11	6	15	7	9	9	13	16	10	11	11
Mercury	8	11	14	10	12	8	12	8	10	15	12	12
Venus	9	9	14	11	9	15	10	11	10	10	16	8
Mars	9	11	11	18	13	7	8	12	10	15	8	10
Jupiter	12	13	17	5	11	9	7	16	6	8	14	14
Saturn	15	7	8	14	8	16	8	11	12	16	7	10
TOTAL	81	75	82	84	76	76	61	83	69	87	75	75
N. Node	11	13	7	8	13	10	10	12	9	12	15	12

Planets

Aspects	Parallels	4 Axis	Direction	☉/☽ Midpoint
Moon-Jupiter	Sun-Neptune	Venus (Dsc)		
Moon-Saturn	Mars-Jupiter	Venus (IC)		
Mercury-Pluto	Mars-Pluto			
Venus-Uranus				Moon's Nodes
				Chiron (S)

INTROVERT

Theme: Capricorn, 12th House, Saturn

Capricorn:
Capricorn, the beginning of Winter, is the lowest position of the Sun's annual cycle. They are the most capable of living without attention, while pursuing goals. The extroverted signs Aries, Gemini, and Libra are all weak.

12th House:
This house has traditionally been known as the place of hidden activities.

Saturn on the Ascendant (1 in 8,099):
Saturn represses outward expressions of behavior, causing a person to appear cautious and controlled.

Mars aspects are infrequent (1 in 306):
This indicates a lack of outward aggression.

Mercury on the Moon's South Node (1 in 39,223):
This indicates that a person has an inborn and early life inclination to be thoughtful and intelligent.

Signs

Size: 134	1st Harmonic	9th Harmonic	Houses	Rising Sign
Moon	Taurus, Capricorn	Leo, Capricorn	12th	
Sun				
Mercury		Cancer, Aquarius	1st, 12th	Lunar Phase
Venus	Scorpio			
Mars	Leo, Sagittarius	Cancer		☉/☽ Midpoint
Jupiter	Gemini, Aquarius		7th	
Saturn	Capricorn	Aries		Moon's Nodes
TOTAL	Sagittarius		1st, 10th, 12th	10th
(weak)	Aries	Gemini, Libra	3rd, 8th	

1H	♈	♉	♊	♋	♌	♍	♎	♏	♐	♑	♒	♓
Moon	6	14	14	13	7	9	11	8	13	14	15	10
Sun	7	17	7	10	11	16	10	8	10	16	11	11
Mercury	7	12	10	9	5	16	16	11	12	13	11	12
Venus	10	7	12	15	4	16	11	18	13	7	11	10
Mars	5	8	12	11	20	11	11	9	19	11	7	10
Jupiter	13	7	15	7	11	9	11	14	16	11	16	4
Saturn	4	14	8	10	13	10	5	13	11	18	17	11
TOTAL	52	79	78	75	71	87	75	81	94	90	88	68
Rising	8	10	12	13	12	15	13	17	13	10	7	4
N. Node	9	13	10	13	14	15	12	9	5	11	10	13

9H	♈	♉	♊	♋	♌	♍	♎	♏	♐	♑	♒	♓
Moon	11	12	8	11	18	4	10	9	9	17	10	15
Sun	7	9	4	15	10	15	11	16	9	14	10	14
Mercury	13	5	9	17	14	6	10	7	13	10	21	9
Venus	13	11	12	9	12	15	9	8	15	14	5	11
Mars	4	12	13	18	14	15	6	13	12	5	11	11
Jupiter	13	12	7	11	10	11	10	15	14	8	13	10
Saturn	18	16	9	10	11	11	8	5	11	10	11	14
TOTAL	79	77	62	91	89	77	64	73	83	78	81	84

Houses	1st	2nd	3rd	4th	5th	6th	7th	8th	9th	10th	11th	12th
Moon	12	15	9	11	10	11	8	5	11	13	12	17
Sun	18	11	8	10	7	11	5	8	12	12	15	17
Mercury	19	13	6	13	4	13	5	7	12	13	10	19
Venus	16	10	8	11	8	7	10	8	8	16	17	15
Mars	16	13	11	8	15	10	10	9	6	13	17	6
Jupiter	12	8	7	12	7	11	17	11	9	16	15	9
Saturn	12	16	11	4	13	11	9	7	14	12	10	15
TOTAL	105	86	60	69	64	74	64	55	72	95	96	98
N. Node	7	8	16	13	13	9	9	8	11	20	9	11

Planets

Aspects	Parallels	4 Axis	Direction	☉/☽ Midpoint
Venus-Neptune	Sun-Saturn	Saturn (Asc)	Mars (D)	
Jupiter-Saturn		Sun (MC)		
		Mercury (MC)		
				Moon's Nodes
				Mercury (S)
				Chiron (N)

PIONEERS

Theme: Capricorn/Aquarius

Capricorn, Aquarius, & Pisces:
These are the 4th quadrant, Winter signs. This quadrant represent having Universal purpose. Information is often received by them through the intuition. The Sun in Capricorn (1 in 331), the Sun in Pisces (1 in 227), and the Sun in Aquarius in the 9th harmonic (1 in 299) are most frequent.

Sun in the 1st House (1 in 193):
This position indicates having a strong projection of self-will and confidence.

Signs

Size: 280	1st Harmonic	9th Harmonic	Houses	Rising Sign
Moon			7th	Capricorn
Sun	Capricorn, Pisces	Aquarius	1st	
Mercury	Aquarius	Aquarius		Lunar Phase
Venus	Taurus, Aquarius	Capricorn, Aquarius		
Mars	Capricorn	Aries, Capr, Aquar	2nd	☉/☽ Midpoint
Jupiter	Virgo, Scorpio	Capricorn	10th	
Saturn		Aries, Cancer	8th	Moon's Nodes
TOTAL	Pisces	Aquarius	1st	Aquarius (S)
(weak)		Cancer, Leo		

1H	♈	♉	♊	♋	♌	♍	♎	♏	♐	♑	♒	♓
Moon	19	21	28	21	19	27	29	17	25	23	23	28
Sun	26	20	18	31	19	16	18	15	23	35	24	35
Mercury	17	17	26	17	21	20	19	16	31	31	35	30
Venus	16	33	21	25	16	21	20	25	24	16	35	28
Mars	25	23	19	28	23	23	26	28	20	28	18	19
Jupiter	22	18	17	19	24	34	25	36	27	19	18	21
Saturn	23	19	24	20	22	28	27	31	28	18	18	22
TOTAL	148	151	153	161	144	169	164	168	178	170	171	183

	♈	♉	♊	♋	♌	♍	♎	♏	♐	♑	♒	♓
Rising	10	12	26	29	28	30	26	29	27	31	16	16
N. Node	27	19	25	12	33	25	23	27	18	21	29	21

9H	♈	♉	♊	♋	♌	♍	♎	♏	♐	♑	♒	♓
Moon	21	23	25	20	22	16	27	23	29	23	21	30
Sun	18	24	24	23	25	28	18	25	25	22	36	12
Mercury	25	26	32	12	19	22	22	23	25	18	31	25
Venus	19	24	19	16	18	24	22	24	23	35	31	25
Mars	32	27	15	16	12	21	21	23	27	31	36	19
Jupiter	19	30	23	24	24	20	15	26	26	31	21	21
Saturn	34	19	27	32	22	28	20	21	16	16	20	25
TOTAL	168	173	165	143	142	159	145	165	171	176	196	157

Houses	1st	2nd	3rd	4th	5th	6th	7th	8th	9th	10th	11th	12th
Moon	24	26	15	22	22	23	32	22	28	28	25	13
Sun	39	26	27	19	23	18	19	25	14	23	20	27
Mercury	34	30	33	21	19	20	19	19	18	22	17	28
Venus	24	28	23	19	21	22	23	19	23	20	28	30
Mars	26	35	23	20	21	22	23	27	25	18	23	17
Jupiter	25	16	20	21	19	29	22	20	26	32	24	26
Saturn	27	15	17	23	23	21	21	32	24	18	29	30
TOTAL	199	176	158	145	148	155	159	164	158	161	166	171
N. Node	25	20	18	23	24	20	28	27	21	24	28	22

Planets

Aspects	Parallels	4 Axis	Direction	☉/☽ Midpoint
Moon-Pluto	Moon-Jupiter	Sun (Asc)		
	Venus-Neptune	Mercury (Asc)		
				Moon's Nodes

PRINCIPLED

Theme: Taurus/Leo/Capricorn

Taurus & Capricorn:
Earth sign people are the most reliable and consistent in their behavior. They prefer an ordered pattern to their lives.

Leo:
Leos are proud people who like to be admired.

Jupiter in the 1st House (1 in 1,213):
Jupiterian people show outward exuberance and a positive philosophy of life.

Saturn in the 9th House (1 in 252):
This indicates having a strict value system and abiding by the laws of a community.

Signs

Size: 130	1st Harmonic	9th Harmonic	Houses	Rising Sign
Moon	Aries, Leo	Sagittarius	12th	Aries
Sun				
Mercury		Virgo, Capricorn		Lunar Phase
Venus	Leo	Taurus	9th	
Mars	Taurus		7th, 10th	☉/☽ Midpoint
Jupiter	Aquarius	Taurus, Capricorn	1st	
Saturn		Cancer	3rd, 9th	Moon's Nodes
TOTAL	Leo	Taurus, Capricorn		Leo (S)
(weak)		Aries, Scorpio	6th	

1H	♈	♉	♊	♋	♌	♍	♎	♏	♐	♑	♒	♓
Moon	16	11	16	11	16	10	13	5	9	11	6	6
Sun	10	13	9	10	12	15	13	11	4	11	10	12
Mercury	9	11	13	7	11	15	13	9	10	11	10	11
Venus	14	12	8	9	15	17	6	9	9	8	12	11
Mars	7	15	12	15	15	12	18	11	7	5	6	7
Jupiter	8	10	6	10	12	9	15	15	12	7	18	8
Saturn	7	8	11	10	12	12	11	8	13	12	13	13
TOTAL	71	80	75	72	93	90	89	68	64	65	75	68
Rising	10	3	11	12	14	17	17	13	13	8	8	4
N. Node	15	6	13	13	10	12	10	11	9	9	18	4

9H	♈	♉	♊	♋	♌	♍	♎	♏	♐	♑	♒	♓
Moon	6	12	13	13	11	12	10	9	18	8	8	10
Sun	4	8	15	15	8	10	8	15	8	14	13	12
Mercury	8	11	14	9	10	19	11	9	7	17	7	8
Venus	5	16	11	7	12	9	10	8	13	15	13	11
Mars	12	13	11	9	15	10	13	5	13	7	9	13
Jupiter	12	16	7	8	12	7	8	8	15	17	8	12
Saturn	14	14	14	17	4	14	8	6	11	13	8	7
TOTAL	61	90	85	78	72	81	68	60	85	91	66	73

Houses	1st	2nd	3rd	4th	5th	6th	7th	8th	9th	10th	11th	12th
Moon	13	9	11	10	8	12	13	9	7	15	6	17
Sun	8	12	11	13	7	8	11	14	4	15	12	15
Mercury	12	15	9	13	9	8	11	8	13	9	10	13
Venus	11	13	14	7	13	6	10	4	16	7	14	15
Mars	12	14	8	12	5	7	17	5	6	18	14	12
Jupiter	21	9	15	11	15	6	8	8	11	12	5	9
Saturn	13	6	17	9	6	10	8	14	19	6	13	9
TOTAL	90	78	85	75	63	57	78	62	76	82	74	90
N. Node	9	9	9	11	13	7	13	12	9	15	11	12

Planets

Aspects	Parallels	4 Axis	Direction	☉/☽ Midpoint
	Moon-Chiron	Uranus (Dsc)		Mars
	Sun-Mars	Chiron (IC)		
	Venus-Uranus			
				Moon's Nodes

Sex Drive (High)

Theme: Virgo, 5th House, Venus

Virgo (1 in 260):
Virgo is the sign most associated with sexuality, even more than Scorpio. It is the Earth sign in the 2nd relational quadrant of the zodiac. Saturn is the most frequent planet in Virgo (1 in 2,876 - 9H)

5th House (1 in 390):
The 5th house represents fun, creativity, and romance. Mercury (1 in 702) and Jupiter (1 in 266) are the most frequent planets, showing romantic curiosity and exuberance.

Pisces & 12th House are weak:
Virgo's opposite sign is weak in the 9th harmonic. Pisces is more concerned with artistry and solitary reflection.

Venus:
Venus represents sensuality and romantic relations. People with Venus retrograde prefer close, intimate relationships.

Signs

Size: 163	1st Harmonic	9th Harmonic	Houses	Rising Sign
Moon	Scorpio			
Sun	Leo		4th	
Mercury		Taurus, Leo	5th	Lunar Phase
Venus	Virgo	Sagittarius	6th	
Mars			1st, 8th	☉/☽ Midpoint
Jupiter			1st, 5th	Cancer
Saturn	Virgo	Virgo		Moon's Nodes
TOTAL	Virgo		5th	
(weak)	Taurus, Sagittarius	Pisces	12th	

1H	♈	♉	♊	♋	♌	♍	♎	♏	♐	♑	♒	♓
Moon	18	12	8	15	9	18	11	23	10	14	13	12
Sun	15	12	14	16	20	14	13	14	11	10	12	12
Mercury	15	11	15	11	15	18	13	14	12	18	8	13
Venus	18	9	13	20	14	24	13	11	7	7	17	10
Mars	12	12	15	12	16	18	17	16	10	11	10	14
Jupiter	12	3	16	15	20	15	16	18	14	16	9	9
Saturn	8	10	15	16	10	19	12	16	16	12	15	14
TOTAL	98	69	96	105	104	126	95	112	80	88	84	84
Rising	6	6	16	12	20	15	19	24	15	11	10	9
N. Node	9	12	10	16	20	17	12	14	15	14	11	13

9H	♈	♉	♊	♋	♌	♍	♎	♏	♐	♑	♒	♓
Moon	12	11	13	17	15	13	15	9	11	19	10	18
Sun	15	10	18	16	7	18	15	16	10	15	17	6
Mercury	16	21	10	13	21	17	11	11	8	13	12	10
Venus	14	16	12	14	13	13	15	15	20	14	12	5
Mars	9	18	19	17	17	8	18	13	13	13	8	10
Jupiter	13	16	10	19	14	12	13	6	16	16	13	15
Saturn	14	15	10	9	17	26	12	11	11	13	12	13
TOTAL	93	107	92	105	104	107	99	81	89	103	84	77

Houses	1st	2nd	3rd	4th	5th	6th	7th	8th	9th	10th	11th	12th
Moon	17	18	16	17	12	8	15	11	16	11	10	12
Sun	13	15	11	23	13	14	11	14	9	11	17	12
Mercury	16	15	12	16	22	10	17	9	13	12	12	9
Venus	9	12	17	13	17	18	10	16	11	15	11	14
Mars	20	15	10	10	16	13	10	20	10	6	16	17
Jupiter	20	15	11	11	23	4	11	17	10	16	13	12
Saturn	12	11	11	19	12	12	14	17	18	13	14	10
TOTAL	107	101	88	109	115	79	88	104	87	84	93	86
N. Node	18	11	16	16	17	10	5	15	14	14	14	13

Planets

Aspects	Parallels	4 Axis	Direction	☉/☽ Midpoint
Moon-Uranus	Moon-Jupiter	Chiron (MC)	Venus (R)	
Venus-Uranus	Sun-Venus			
	Venus-Mars			
	Venus-Neptune			**Moon's Nodes**
				Mercury (N)
				Neptune (S)

Sex Symbols

Theme: Scorpio (8th)

Scorpio & the 8th House:
Scorpio is known for its powers of sexual magnetism. The 8th house represents sexual intimacy. Venus, the planet of love, is most frequent in the 8th house (1 in 2,285).

Virgo:
Virgo, being the relational Earth sign, is the dominant sign for the sex-related categories in this study.

Venus-Mars parallels (1 in 105):
This is an indication of sexual magnetism.

Signs

Size: 149	1st Harmonic	9th Harmonic	Houses	Rising Sign
Moon	Virgo	Gemini, Pisces	2nd	Capricorn
Sun	Aquarius	Scorpio	1st	
Mercury	Virgo	Libra	1st	Lunar Phase
Venus		Cancer	8th	Leo
Mars	Scorpio		3rd, 5th, 8th	☉/☽ Midpoint
Jupiter	Scorpio		8th	Cancer
Saturn	Gemini	Taurus	2nd	Moon's Nodes
TOTAL	Virgo, Scorpio		8th	
(weak)	Cancer	Capricorn	6th	

1H	♈	♉	♊	♋	♌	♍	♎	♏	♐	♑	♒	♓
Moon	9	14	12	15	13	20	17	10	7	7	13	12
Sun	15	13	5	9	11	16	12	16	14	9	18	11
Mercury	10	13	5	6	8	20	15	16	10	18	12	16
Venus	18	9	10	9	9	16	16	12	10	14	14	12
Mars	9	9	16	8	16	16	10	19	12	12	11	11
Jupiter	8	14	8	10	11	16	15	20	10	12	17	8
Saturn	7	12	17	10	5	10	15	13	17	11	15	17
TOTAL	76	84	73	67	73	114	100	106	80	83	100	87
Rising	5	6	12	15	14	17	15	11	13	18	13	10
N. Node	11	8	9	13	16	14	15	11	9	16	12	15

9H	♈	♉	♊	♋	♌	♍	♎	♏	♐	♑	♒	♓
Moon	9	13	18	12	17	10	12	10	9	13	7	19
Sun	11	15	8	9	11	14	16	19	9	13	9	15
Mercury	15	7	16	11	13	13	20	7	12	8	12	15
Venus	17	15	11	18	8	15	8	13	14	7	12	11
Mars	13	13	11	13	12	15	6	12	12	16	16	10
Jupiter	13	16	13	14	17	15	9	16	10	7	11	8
Saturn	10	21	13	14	7	8	14	11	17	7	16	11
TOTAL	88	100	90	91	85	90	85	88	83	71	83	89

Houses	1st	2nd	3rd	4th	5th	6th	7th	8th	9th	10th	11th	12th
Moon	11	19	13	10	9	11	17	10	16	11	13	9
Sun	20	15	11	13	8	8	13	12	10	12	15	12
Mercury	21	13	11	10	14	7	13	8	16	9	12	15
Venus	13	12	12	12	12	11	6	22	11	8	15	15
Mars	14	14	18	13	19	8	6	18	6	14	8	11
Jupiter	13	13	9	12	10	10	12	19	12	12	17	10
Saturn	12	19	16	14	13	12	6	9	11	13	11	13
TOTAL	104	105	90	84	85	67	73	98	82	79	91	85
N. Node	12	13	11	8	10	17	15	14	9	16	10	14

Planets

Aspects	Parallels	4 Axis	Direction	☉/☽ Midpoint
Mercury-Mars	Venus-Mars	Mercury (Asc)	Venus (Rx)	
			Uranus (Rx)	
				Moon's Nodes
				Uranus (S)
				Chiron (N)

Gauquelin Study Groups

GAUQUELIN - ACTORS

Theme: Sagittarius, Venus

Sagittarius:
Sagittarius is the strongest sign for the Gauquelin group, while Aquarius is strongest for the Rodden group.

Venus:
Venus is the strongest planet for both the Gauquelin and Rodden study groups.

The following findings are the same for the Gauquelin and Rodden study groups:

1. Mercury in Pisces
2. Moon's North Node in Cancer
3. Jupiter conjunct Ascendant
4. Sun conjunct Moon's South Node

Signs

Size: 1,408	1st Harmonic	9th Harmonic	Houses	Rising Sign
Moon	Taurus, Cancer	Sagittarius		
Sun	Taurus			Lunar Phase
Mercury	Cancer, Pisces		10th	Scorpio
Venus		Scorpio, Sagittarius		☉/☽ Midpoint
Mars		Gemini		Aries
Jupiter	Gemini	Sagittarius, Pisces	9th	Leo
Saturn	Sagittarius	Virgo, Aquarius		Moon's Nodes
TOTAL	Taurus, Leo	Sagittarius	6th, 7th	Cancer
(weak)	Scorpio	Cancer, Libra	11th	

1H	♈	♉	♊	♋	♌	♍	♎	♏	♐	♑	♒	♓
Moon	129	120	97	131	130	118	103	118	112	124	112	114
Sun	118	147	108	142	126	114	116	85	114	107	109	122
Mercury	116	125	109	118	115	120	117	109	115	127	120	117
Venus	107	137	117	150	116	121	110	97	113	88	126	126
Mars	94	117	130	113	161	154	168	115	103	90	84	79
Jupiter	99	101	124	117	126	123	118	124	132	117	107	120
Saturn	103	99	102	80	106	123	132	135	153	128	120	127
TOTAL	766	846	787	851	880	873	864	783	842	781	778	805
Rising	56	78	109	124	152	169	153	162	151	115	73	66
N. Node	127	127	109	136	133	128	126	123	103	112	95	89

9H	♈	♉	♊	♋	♌	♍	♎	♏	♐	♑	♒	♓
Moon	94	124	127	109	109	121	102	121	145	120	125	111
Sun	129	101	96	121	127	130	119	118	124	123	105	115
Mercury	130	118	120	114	129	106	92	133	113	112	111	130
Venus	104	121	125	98	101	124	120	137	135	106	127	110
Mars	120	126	138	97	115	117	106	122	118	128	104	117
Jupiter	117	101	124	113	113	124	106	102	131	128	114	135
Saturn	108	116	120	111	95	138	125	121	117	104	142	111
TOTAL	802	807	850	763	789	860	770	854	883	821	828	829

Houses	1st	2nd	3rd	4th	5th	6th	7th	8th	9th	10th	11th	12th
Moon	108	114	127	110	118	124	124	114	109	119	111	130
Sun	135	128	118	105	113	109	107	108	115	121	124	125
Mercury	144	143	100	118	102	119	103	110	109	144	89	127
Venus	136	109	127	115	113	109	113	107	109	128	125	117
Mars	110	135	111	112	105	120	127	117	117	113	133	108
Jupiter	116	123	119	108	98	120	115	121	139	109	108	132
Saturn	118	118	133	108	107	125	127	117	110	120	114	111
TOTAL	867	870	835	776	756	826	816	794	808	854	804	850
N. Node	104	102	123	99	118	124	131	120	132	118	116	121

Planets

Aspects	Parallels	4 Axis	Direction	☉/☽ Midpoint
Moon-Uranus	Moon-Jupiter	Jupiter (Asc)	Pluto (D)	Jupiter
Mercury-Venus	Venus-Mars	Moon (Dsc)		
Mars-Chiron	Mars-Uranus	Mercury (Dsc)		
		Venus (Dsc)		Moon's Nodes
		Venus (IC)		Sun (S)
		Uranus (IC)		Neptune (N)

GAUQUELIN - ALCOHOLICS

Theme: Taurus/Aquarius/Pisces

Taurus (1 in 1,162,268):
Taurus is the strongest sign for both the Gauquelin and Rodden study groups. Venus is the most frequent planet in Taurus for the Gauquelin group (1 in 18,395). The Sun/Moon midpoint is also strong in Taurus (1 in 7,280).

Neptune:
Neptune represents escapism and is frequently associated with alcoholism and drug addiction. Neptune is most frequent on the Midheaven (1 in 96,853).

The following findings are the same for the Gauquelin and Rodden study groups:

1. Mercury in Pisces
2. Leo weak (1H)
3. Taurus (9H)
4. Sun conjunct Uranus/Neptune midpoint
5. Mercury conjunct Moon's North Node

Signs

Size: 1,793	1st Harmonic	9th Harmonic	Houses	Rising Sign
Moon	Taurus	Cancer		
Sun	Gemini, Pisces	Scorpio	2nd	Lunar Phase
Mercury	Taurus, Pisces	Pisces	3rd	Aries
Venus	Taurus, Cancer			Aquarius
Mars	Taurus			☉/☽ Midpoint
Jupiter	Leo			Taurus, Cancer
Saturn	Taurus, Aquar, Pisces	Taurus, Aquarius	6th	Moon's Nodes
TOTAL	Taurus, Aquar, Pisces	Taurus, Aquarius		Taurus, Cancer
(weak)	Leo, Libra, Scorpio			Pisces (S)

1H	♈	♉	♊	♋	♌	♍	♎	♏	♐	♑	♒	♓
Moon	153	175	145	137	151	167	146	142	142	143	141	151
Sun	155	152	193	159	152	138	99	152	129	135	154	175
Mercury	138	165	149	150	151	125	142	138	161	141	166	167
Venus	153	202	131	196	126	163	125	104	170	126	160	137
Mars	122	157	155	187	140	180	172	163	129	137	120	131
Jupiter	107	125	123	127	184	177	158	177	174	159	151	131
Saturn	176	187	142	84	87	133	128	128	156	182	192	198
TOTAL	1004	1163	1038	1040	991	1083	970	1004	1061	1023	1084	1090
Rising	69	89	150	191	213	214	207	201	179	119	91	70
N. Node	144	181	153	169	179	176	149	160	145	108	110	119

9H	♈	♉	♊	♋	♌	♍	♎	♏	♐	♑	♒	♓
Moon	131	164	163	175	135	155	126	148	168	139	137	152
Sun	134	137	136	147	163	144	166	173	141	163	147	142
Mercury	157	150	147	154	134	158	148	148	133	154	141	169
Venus	161	148	134	164	151	138	151	135	162	136	163	150
Mars	166	161	155	133	151	140	131	138	138	166	169	145
Jupiter	164	168	130	148	140	148	145	164	149	146	151	140
Saturn	139	184	166	145	102	169	157	129	119	131	220	132
TOTAL	1052	1112	1031	1066	976	1052	1024	1035	1010	1035	1128	1030

Houses	1st	2nd	3rd	4th	5th	6th	7th	8th	9th	10th	11th	12th
Moon	152	160	139	160	142	151	148	154	152	141	152	142
Sun	150	182	173	137	131	117	136	127	128	162	164	186
Mercury	180	178	183	147	128	134	111	129	133	159	165	146
Venus	179	157	160	152	127	147	119	127	139	148	152	186
Mars	147	167	165	151	141	131	126	150	140	157	164	154
Jupiter	152	149	147	159	130	141	134	164	156	141	161	159
Saturn	151	142	140	170	145	173	160	123	146	150	143	150
TOTAL	1111	1135	1107	1076	944	994	934	974	994	1058	1101	1123
N. Node	164	151	159	150	164	127	144	160	142	150	140	142

Planets

Aspects	Parallels	4 Axis	Direction	Moon's Nodes
Sun-Uranus	Mercury-Pluto	Uranus (Dsc)		
Mercury-Venus	Mars-Jupiter	Pluto (Dsc)		
		Neptune (MC)		
		Uranus (IC)		**Moon's Nodes**
				Mercury (N)
				Saturn (N)
				Neptune (N)
				Jupiter (S)

GAUQUELIN - BUSINESS EXECUTIVES

Theme: Virgo/Pisces, 11th House, Mars

Virgo & Pisces:
Virgo and Pisces are the strongest signs for the Gauquelin group, while Gemini is strongest for the Rodden group.

11th House:
The 11th house is the strongest house for the Gauquelin group, while the 12th house is strongest for the Rodden group.

Mars:
Mars is the strongest planet for the Gauquelin group, while Pluto is strongest for the Rodden group.

The following findings are the same for the Gauquelin and Rodden study groups:

1. Moon in Virgo
2. Gemini (9H)
3. Mars conjunct Ascendant
4. Sun conjunct Descendant

Signs

Size: 552	1st Harmonic	9th Harmonic	Houses	Rising Sign
Moon	Virgo	Virgo	7th	Leo
Sun	Leo, Pisces	Capricorn	11th	Virgo
Mercury			11th	Lunar Phase
Venus		Cancer, Pisces		Gemini
Mars		Pisces	12th	☉/☽ Midpoint
Jupiter	Aquarius	Pisces	12th	Pisces
Saturn	Sagittarius	Gemini, Virgo, Libra		Moon's Nodes
TOTAL	Virgo, Aquarius	Gemini, Cancer, Pisces	11th	4th
(weak)	Aries, Libra	Aries, Sagittarius	1st, 3rd	

1H	♈	♉	♊	♋	♌	♍	♎	♏	♐	♑	♒	♓
Moon	44	52	52	38	42	63	32	42	38	54	46	49
Sun	41	53	35	55	69	49	33	31	39	39	49	59
Mercury	53	32	46	42	54	57	40	41	38	45	56	48
Venus	33	53	47	53	46	54	39	44	42	45	52	44
Mars	35	52	42	59	64	53	56	49	35	31	34	42
Jupiter	29	27	41	37	47	52	52	55	57	54	65	36
Saturn	52	50	51	19	26	47	41	51	62	58	51	44
TOTAL	287	319	314	303	348	375	293	313	311	326	353	322
Rising	26	24	40	49	74	72	68	61	50	39	31	18
N. Node	49	55	52	41	56	53	49	50	39	30	33	45

9H	♈	♉	♊	♋	♌	♍	♎	♏	♐	♑	♒	♓
Moon	39	41	54	44	54	57	41	39	41	44	45	53
Sun	44	37	55	53	49	39	45	43	38	57	42	50
Mercury	45	44	51	55	45	44	48	42	42	49	39	48
Venus	40	49	37	57	46	43	45	42	41	37	49	66
Mars	46	45	48	44	44	48	38	43	40	49	46	61
Jupiter	45	42	43	48	40	40	44	48	52	43	49	58
Saturn	36	42	61	53	22	60	62	41	35	40	57	43
TOTAL	295	300	349	354	300	331	323	298	289	319	327	379

Houses	1st	2nd	3rd	4th	5th	6th	7th	8th	9th	10th	11th	12th
Moon	36	46	36	42	46	50	59	54	53	41	48	41
Sun	42	42	46	44	35	44	45	33	48	60	63	50
Mercury	44	53	41	46	45	41	34	43	42	48	67	48
Venus	44	42	41	38	49	39	49	39	46	50	60	55
Mars	42	41	32	53	38	38	46	46	56	44	53	63
Jupiter	46	45	55	36	46	47	38	36	45	48	41	69
Saturn	55	53	49	40	55	52	35	41	45	39	50	38
TOTAL	309	322	300	299	314	311	306	292	335	330	382	364
N. Node	53	41	50	59	40	40	48	47	49	42	42	41

Planets

Aspects	Parallels	4 Axis	Direction	☉/☽ Midpoint
Sun-Mars	Mars-Neptune	Mars (Asc)	Mercury (D)	
Venus-Uranus	Mars-Pluto	Sun (Dsc)		
		Moon (Dsc)		
		Jupiter (MC)		**Moon's Nodes**
		Neptune (MC)		Mercury (N)
		Pluto (MC)		Jupiter (S)

GAUQUELIN - JOURNALISTS

Theme: Capricorn, 1st House, Uranus

Capricorn & Aquarius:
Capricorn is the strongest sign for the Gauquelin group, while Virgo and the 6th house are strongest for the Rodden group.

1st House:
The 1st house is the strongest house for the Gauquelin group, while the 6th house is strongest for the Rodden group.

Uranus:
Uranus, the awakener, is the strongest planet for the Gauquelin group.

The following findings are the same for the Gauquelin and Rodden study groups:

1. Jupiter in 9th House

Signs

Size: 674	1st Harmonic	9th Harmonic	Houses	Rising Sign
Moon		Capricorn, Pisces	11th	Gemini
Sun	Libra, Capricorn		1st	Lunar Phase
Mercury	Capricorn	Capricorn	1st	Aries
Venus	Libra, Aquarius	Leo	3rd, 6th	☉/☽ Midpoint
Mars		Virgo	1st	Cancer
Jupiter	Taurus, Aquarius		9th	Virgo
Saturn	Virgo, Capricorn	Aries	4th	Moon's Nodes
TOTAL	Aquarius, Capricorn		1st, 3rd	Sagittarius
(weak)	Aries, Gemini		7th	

1H	♈	♉	♊	♋	♌	♍	♎	♏	♐	♑	♒	♓
Moon	50	65	58	60	60	56	56	55	57	46	44	67
Sun	39	48	56	54	54	62	69	53	51	71	59	58
Mercury	29	51	45	51	49	61	68	66	53	81	55	65
Venus	33	66	32	70	52	61	72	51	55	40	86	56
Mars	54	53	46	61	67	74	52	67	57	51	43	49
Jupiter	54	63	50	49	51	54	64	52	41	66	73	57
Saturn	50	51	48	53	46	76	44	56	67	75	60	48
TOTAL	309	397	335	398	379	444	425	400	381	430	420	400
Rising	34	39	65	69	81	72	76	64	69	45	40	20
N. Node	54	53	63	51	61	56	62	60	69	48	47	50

9H	♈	♉	♊	♋	♌	♍	♎	♏	♐	♑	♒	♓
Moon	43	65	51	53	40	60	46	56	62	68	61	69
Sun	61	55	56	66	56	50	54	64	58	45	56	53
Mercury	63	58	56	48	62	55	52	42	50	70	57	61
Venus	60	53	56	47	78	49	68	49	48	60	52	54
Mars	54	43	67	66	58	69	51	51	47	66	53	49
Jupiter	45	57	61	64	57	55	51	53	63	46	65	57
Saturn	69	66	41	62	67	49	45	56	56	58	50	55
TOTAL	395	397	388	406	418	387	367	371	384	413	394	398

Houses	1st	2nd	3rd	4th	5th	6th	7th	8th	9th	10th	11th	12th
Moon	55	59	63	57	54	52	44	51	52	55	73	59
Sun	86	64	59	49	40	54	52	54	53	56	47	60
Mercury	89	62	64	46	49	53	51	47	46	60	52	55
Venus	67	59	71	38	54	62	35	57	60	51	55	65
Mars	72	41	58	47	61	58	34	63	59	64	58	59
Jupiter	46	46	66	55	51	59	48	57	67	53	66	60
Saturn	60	61	67	70	57	50	47	60	45	56	55	46
TOTAL	475	392	448	362	366	388	311	389	382	395	406	404
N. Node	55	58	54	53	54	51	50	59	57	56	61	66

Planets

Aspects	Parallels	4 Axis	Direction	☉/☽ Midpoint
Moon-Uranus	Moon-Mars	Mars (Asc)	Uranus (D)	Venus
Venus-Uranus		Pluto (Asc)	Pluto (R)	
		Chiron (Dsc)		
		Saturn (IC)		**Moon's Nodes**
				Sun (N)
				Sun (S)
				Uranus (N)

Gauquelin - Medical

Theme: Pisces (12th), Neptune

Pisces:
Pisces is the strongest sign for the Gauquelin group (1 in 285,481), while the opposite sign, Virgo, is strongest for the Rodden group.

Neptune:
Neptune is the strongest planet for the Gauquelin group, while Pluto is the strongest for the Rodden group.

The following findings are the same for the Gauquelin and Rodden study groups:

1. Virgo Lunar Phase
2. 4th House weak

Signs

Size: 2,552	1st Harmonic	9th Harmonic	Houses	Rising Sign
Moon	Pisces	Scorpio		
Sun	Pisces		2nd, 5th, 6th	
Mercury	Pisces	Cancer	5th	Lunar Phase
Venus	Taurus, Libra, Pisces	Pisces		Virgo
Mars		Aries	9th, 12th	☉/☽ Midpoint
Jupiter	Leo		5th	
Saturn	Aquarius, Pisces	Leo, Sagittarius	9th, 12th	Moon's Nodes
TOTAL	Pisces		2nd, 12th	Pisces
(weak)	Gemini, Libra, Scorp		3rd, 4th, 7th	8th, 10th

1H	♈	♉	♊	♋	♌	♍	♎	♏	♐	♑	♒	♓
Moon	203	210	185	200	224	213	210	227	214	234	193	239
Sun	217	223	203	220	223	221	201	196	190	213	209	236
Mercury	213	179	211	185	211	231	193	210	232	223	232	232
Venus	204	256	173	246	200	214	224	164	237	177	203	254
Mars	183	200	217	229	243	281	249	229	197	179	176	169
Jupiter	187	186	187	213	229	241	234	251	221	224	188	191
Saturn	214	196	177	205	193	210	164	206	212	255	266	254
TOTAL	1421	1450	1353	1498	1523	1611	1475	1483	1503	1505	1467	1575
Rising	91	158	205	262	288	299	288	290	244	198	121	108
N. Node	192	190	212	217	213	223	209	217	214	208	220	237

9H	♈	♉	♊	♋	♌	♍	♎	♏	♐	♑	♒	♓
Moon	224	203	197	216	197	200	218	245	208	214	233	197
Sun	200	230	210	202	201	212	228	214	206	206	228	215
Mercury	214	206	213	234	233	198	220	213	207	195	213	206
Venus	217	194	217	231	208	214	198	218	228	206	180	241
Mars	237	197	219	214	232	185	227	181	211	216	210	223
Jupiter	209	215	207	239	196	223	205	220	191	221	217	209
Saturn	232	228	168	212	246	215	206	202	238	219	185	201
TOTAL	1533	1473	1431	1548	1513	1447	1502	1493	1489	1477	1466	1492

Houses	1st	2nd	3rd	4th	5th	6th	7th	8th	9th	10th	11th	12th
Moon	234	228	196	223	212	226	200	226	202	192	192	221
Sun	241	273	207	164	204	208	167	193	177	246	222	250
Mercury	249	245	220	175	221	203	162	179	200	214	226	258
Venus	233	246	224	195	190	176	176	175	212	224	244	257
Mars	224	214	200	202	190	184	189	213	247	219	208	262
Jupiter	232	219	194	227	239	194	208	211	180	222	210	216
Saturn	227	199	201	211	201	214	192	201	243	192	228	243
TOTAL	1640	1624	1442	1397	1457	1405	1294	1398	1461	1509	1530	1707
N. Node	233	227	195	175	208	229	185	240	182	236	233	209

Planets

Aspects	Parallels	4 Axis	Direction	☉/☽ Midpoint
Neptune	Moon-Chiron	Moon (Asc)	Saturn (R)	
Moon-Neptune	Mars-Pluto	Uranus (Dsc)	Neptune (D)	
Sun-Chiron		Pluto (IC)		
Mars-Neptune				Moon's Nodes

Gauquelin - Mental Illness

Theme: Aquarius/Pisces, Moon/Mercury

Aquarius & Pisces:
Aquarius and Pisces are the strongest signs for the Gauquelin group, while Scorpio is strongest and Pisces is weak for the Rodden group.

2nd House:
The 2nd house is the strongest house for the Gauquelin group, while the 11th house is strongest for the Rodden group.

Moon & Mercury:
The Moon and Mercury are the strongest planets for both the Gauquelin and Rodden study groups. The Rodden group is also strong with Neptune.

The following findings are the same for the Gauquelin and Rodden study groups:

1. Saturn in Sagittarius
2. Pluto conjunct Midheaven
3. Saturn conjunct Venus/Mars midpoint
4. Moon conjunct Moon's South Node

Signs

Size: 4,521	1st Harmonic	9th Harmonic	Houses	Rising Sign
Moon		Pisces		Cancer
Sun	Aries, Pisces		2nd	
Mercury	Taurus, Pisces		2nd	Lunar Phase
Venus	Taurus, Cancer, Aquar	Leo, Pisces	10th	Taurus
Mars	Aquarius, Pisces	Taurus	4th	☉/☽ Midpoint
Jupiter	Leo, Sagittarius	Aquarius	7th	Taurus
Saturn	Sagittarius	♉ ♊ ♋ ♒	1st, 3rd, 4th, 5th	Moon's Nodes
TOTAL	♐ ♑ ♒ ♓	♋ ♒ ♓	2nd	Gemini
(weak)	Aries, Cancer, Leo	Leo, Sagittarius	9th	Pisces (S)

1H	♈	♉	♊	♋	♌	♍	♎	♏	♐	♑	♒	♓
Moon	317	380	394	355	391	372	400	397	387	373	382	373
Sun	446	397	399	401	390	374	334	337	308	359	376	400
Mercury	374	377	348	365	370	381	340	380	371	386	408	421
Venus	366	472	344	498	294	418	299	324	390	314	450	352
Mars	322	358	403	416	412	460	443	364	338	345	334	326
Jupiter	274	279	348	345	428	433	437	437	459	395	380	306
Saturn	356	320	284	153	224	337	458	445	529	527	471	417
TOTAL	2455	2583	2520	2533	2509	2775	2711	2684	2782	2699	2801	2595
Rising	189	194	345	530	528	518	497	528	451	329	255	157
N. Node	398	412	444	386	420	438	386	350	328	304	300	355

9H	♈	♉	♊	♋	♌	♍	♎	♏	♐	♑	♒	♓
Moon	354	368	373	392	370	349	368	386	387	365	376	433
Sun	374	379	359	401	359	388	354	386	381	393	370	377
Mercury	388	348	392	380	373	358	389	378	383	377	390	365
Venus	372	389	380	384	408	375	340	350	360	377	359	427
Mars	372	429	405	372	362	394	370	358	343	362	352	402
Jupiter	396	377	345	396	318	389	388	375	358	377	412	390
Saturn	349	423	425	418	254	401	375	376	331	328	505	336
TOTAL	2605	2713	2679	2743	2444	2654	2584	2609	2543	2579	2764	2730

Houses	1st	2nd	3rd	4th	5th	6th	7th	8th	9th	10th	11th	12th
Moon	380	384	351	378	371	391	367	382	379	387	371	380
Sun	420	459	434	336	319	295	327	344	325	396	443	423
Mercury	439	469	411	365	344	325	315	320	309	399	423	402
Venus	422	446	382	374	321	333	302	342	342	425	414	418
Mars	393	370	392	397	368	351	337	360	365	394	402	392
Jupiter	368	380	349	369	372	378	410	383	355	380	389	388
Saturn	456	389	440	427	422	389	334	334	325	341	341	323
TOTAL	2878	2897	2759	2646	2517	2462	2392	2465	2400	2722	2783	2726
N. Node	366	394	392	361	403	370	359	379	377	388	360	372

Planets

Aspects	Parallels	4 Axis	Direction	☉/☽ Midpoint
Moon & Mercury	Mercury-Uranus	Chiron (Asc)		Venus (D)
Pluto		Neptune (MC)		
Moon-Mars		Pluto (MC)		
Moon-Pluto		Saturn (IC)		Moon's Nodes
Sun-Saturn				Moon (S)
Mercury-Venus				Mercury (N)
Mars-Jupiter				Jupiter (N)

GAUQUELIN - MILITARY

Theme: Taurus/Capricorn, 9th/12th Houses, Pluto

Taurus:
Taurus and Capricorn are the strongest signs for the Gauquelin group, while Virgo and Capricorn are the strongest signs for the Rodden group.

9th House:
The 9th house is strongest for both the Gauquelin and Rodden study groups.

Pluto:
Pluto is the strongest planet for the Gauquelin group, while Chiron is strongest for the Rodden group.

The following findings are the same for the Gauquelin and Rodden study groups:

1. Mercury in Aquarius
2. Capricorn (9H)
3. 9th House
4. Jupiter in 12th House
5. Saturn in 12th House

Signs

Size: 3,046	1st Harmonic	9th Harmonic	Houses	Rising Sign
Moon		Scorpio		Virgo
Sun	Taurus, Leo	Capricorn		
Mercury	Taurus, Aquarius			Lunar Phase
Venus	Libra, Pisces	Virgo		
Mars		Capricorn	9th	☉/☽ Midpoint
Jupiter	Taurus, Gemini	Libra	9th, 12th	Aries
Saturn			1st, 12th	Moon's Nodes
TOTAL	Taurus, Libra, Scorp	Leo, Capricorn	1st, 9th, 12th	Pisces
(weak)	Aries, Gemini, Capr	Aquarius	4th, 7th	5th, 9th

1H	♈	♉	♊	♋	♌	♍	♎	♏	♐	♑	♒	♓
Moon	268	275	245	270	248	244	257	239	256	260	249	235
Sun	227	294	242	218	286	268	230	257	247	257	248	272
Mercury	224	270	206	217	217	279	276	281	286	257	294	239
Venus	233	284	203	302	229	242	284	233	270	223	264	279
Mars	212	244	265	300	284	318	294	255	240	222	215	197
Jupiter	257	256	259	252	258	256	268	285	250	206	239	260
Saturn	207	200	204	276	281	279	331	374	287	221	185	201
TOTAL	1628	1823	1624	1835	1803	1886	1940	1924	1836	1646	1694	1683
Rising	106	159	230	320	330	362	351	349	330	224	150	135
N. Node	251	262	242	258	235	256	243	234	240	260	267	298

9H	♈	♉	♊	♋	♌	♍	♎	♏	♐	♑	♒	♓
Moon	236	257	250	246	241	273	267	295	254	254	224	249
Sun	250	267	261	270	270	258	212	258	240	286	249	225
Mercury	248	278	239	247	273	226	259	239	246	276	263	252
Venus	243	261	228	256	266	287	250	247	251	259	260	238
Mars	231	250	265	239	253	234	259	271	243	284	240	277
Jupiter	257	251	252	250	264	268	280	279	244	241	206	254
Saturn	251	274	248	259	274	233	267	231	278	238	222	271
TOTAL	1716	1838	1743	1767	1841	1779	1794	1820	1756	1838	1664	1766

Houses	1st	2nd	3rd	4th	5th	6th	7th	8th	9th	10th	11th	12th
Moon	265	258	259	239	254	259	225	274	251	260	259	243
Sun	308	283	267	226	220	213	219	233	221	275	307	274
Mercury	311	303	259	236	227	227	196	219	233	269	275	291
Venus	292	284	246	239	239	236	203	224	249	248	274	312
Mars	274	268	266	252	209	228	210	252	287	238	258	304
Jupiter	251	231	237	242	230	240	244	238	313	259	260	301
Saturn	286	257	247	225	278	262	231	242	235	273	253	257
TOTAL	1987	1884	1781	1659	1657	1665	1528	1682	1789	1822	1886	1982
N. Node	230	252	264	246	287	256	240	242	286	253	228	262

Planets

Aspects	Parallels	4 Axis	Direction	☉/☽ Midpoint
Venus-Chiron	Sun-Moon	Jupiter (MC)	Pluto (R)	
Mars-Jupiter	Mercury-Uranus	Neptune (MC)		
	Venus-Pluto	Mars (MC)		
		Pluto (IC)		Moon's Nodes
				Neptune (S)
				Pluto (S)

GAUQUELIN - MURDERERS

Theme: Taurus (2nd), Pisces, Neptune

Taurus:
Taurus and Pisces are the strongest signs for the Gauquelin group, while Taurus and Scorpio are the strongest signs for the Rodden group.

2nd House:
The 2nd house is strongest for the Gauquelin group, while the 4th house is strongest for the Rodden group. The 10th house is weak for both study groups.

Neptune:
Neptune is the strongest planet for the Gauquelin group, while Mars is strongest for the Rodden group.

The following findings are the same for the Gauquelin and Rodden study groups:

1. Sun in Taurus
2. Mars in Taurus
3. Saturn in Sagittarius
4. Jupiter in Scorpio (9H)
5. Capricorn Lunar Phase
6. 10th House weak
7. Chiron conjunct Imum Coeli
8. Uranus conjunct Moon's North Node

Signs

Size: 622	1st Harmonic	9th Harmonic	Houses	Rising Sign
Moon	Gemini, Virgo	Taurus	12th	
Sun	Taurus, Gem, Pisces	Capricorn	2nd	
Mercury	Pisces		2nd	Lunar Phase
Venus	Taurus, Pisces		1st, 6th	Capricorn
Mars	Aries, Taurus	Taurus, Scorpio		☉/☽ Midpoint
Jupiter		Scorpio, Sagitt	2nd	Cancer
Saturn	Sagitt, Capricorn	Sagittarius		Moon's Nodes
TOTAL	Gemini, Pisces	Taurus	2nd	3rd
(weak)	Libra, Scorpio	Cancer, Aquar	10th	

1H	♈	♉	♊	♋	♌	♍	♎	♏	♐	♑	♒	♓
Moon	47	39	69	62	45	76	47	43	36	48	50	60
Sun	52	65	65	56	42	40	44	43	44	54	53	64
Mercury	46	56	56	45	51	41	38	46	61	59	55	68
Venus	50	66	59	66	30	55	44	38	47	53	54	60
Mars	58	64	62	44	65	65	59	42	34	42	50	37
Jupiter	55	40	47	51	52	53	54	61	60	48	54	47
Saturn	43	29	46	36	54	58	49	57	84	72	50	44
TOTAL	351	359	404	360	339	388	335	330	366	376	366	380

	♈	♉	♊	♋	♌	♍	♎	♏	♐	♑	♒	♓
Rising	28	38	41	57	71	65	74	68	70	42	43	25
N. Node	39	48	46	51	64	47	56	58	59	51	52	51

9H	♈	♉	♊	♋	♌	♍	♎	♏	♐	♑	♒	♓
Moon	43	65	49	41	63	58	62	42	47	54	41	57
Sun	62	49	41	40	54	56	59	53	38	65	44	61
Mercury	46	45	56	45	55	61	46	58	51	55	50	54
Venus	47	59	54	45	58	55	43	49	58	46	51	57
Mars	48	66	61	39	42	49	53	64	44	48	47	61
Jupiter	44	50	50	56	45	46	50	70	62	60	40	49
Saturn	40	60	46	57	74	46	44	47	63	48	45	52
TOTAL	330	394	357	323	391	371	357	383	363	376	318	391

Houses	1st	2nd	3rd	4th	5th	6th	7th	8th	9th	10th	11th	12th
Moon	49	58	57	51	44	48	51	42	55	41	58	68
Sun	64	69	57	42	42	52	51	46	42	51	55	51
Mercury	58	71	57	43	44	54	43	46	40	46	61	59
Venus	79	59	46	43	44	61	50	40	46	50	38	66
Mars	55	51	54	60	59	37	45	53	43	55	50	60
Jupiter	54	64	60	43	52	43	48	51	53	47	57	50
Saturn	57	52	59	56	53	57	46	45	45	42	61	49
TOTAL	416	424	390	338	338	352	334	323	324	332	380	403
N. Node	60	53	63	40	40	46	55	56	58	51	45	55

Planets

Aspects	Parallels	4 Axis	Direction	☉/☽ Midpoint
Neptune	Mercury-Neptune	Venus (Dsc)	Uranus (R)	
Mars-Saturn		Chiron (IC)	Neptune (D)	
Mars-Pluto			Pluto (D)	
				Moon's Nodes
				Uranus (N)
				Pluto (S)

Gauquelin - Musicians

Theme: Aquarius/Pisces, Uranus/Pluto

Aquarius & Pisces:
Aquarius and Pisces (1 in 466) are the strongest signs for the Gauquelin group, while Capricorn and Pisces are the strongest signs for the Rodden group (4th Quadrant). The Sun/Moon midpoint is also frequent in Pisces (1 in 298,569).

6th House:
The 6th house is the strongest house for the Gauquelin group, while the 5th house is strongest for the Rodden group.

The following findings are the same for the Gauquelin and Rodden study groups:

1. Pisces
2. Capricorn
3. Mercury in Taurus
4. Mercury in Capricorn
5. Venus in Taurus
6. Venus in Pisces
7. Mars in 10th House
8. Mercury conjunct Moon's South Node

Signs

Size: 1,248	1st Harmonic	9th Harmonic	Houses	Rising Sign
Moon			6th, 11th	Capricorn
Sun	Aries			Lunar Phase
Mercury	Taurus, Capricorn	Aries, Aquarius		Pisces
Venus	Taurus, Libra, Pisces		4th, 9th	☉/☽ Midpoint
Mars	Aquarius	Aries, Taurus	4th, 10th	Pisces
Jupiter	Cancer		2nd	Moon's Nodes
Saturn	Scorp, Aquar, Pisces			Scorpio
TOTAL	Capr, Aquar, Pisces		6th	Sagittarius
(weak)				6th

1H	♈	♉	♊	♋	♌	♍	♎	♏	♐	♑	♒	♓
Moon	102	101	109	94	114	86	117	103	99	106	112	105
Sun	127	100	113	95	107	100	83	85	115	110	107	106
Mercury	104	112	85	93	102	100	95	97	117	133	96	114
Venus	88	129	91	113	99	85	128	71	112	102	103	127
Mars	96	105	87	98	132	139	102	112	99	92	101	85
Jupiter	101	104	96	126	108	105	122	102	108	107	80	89
Saturn	89	91	82	67	72	80	101	126	119	114	170	137
TOTAL	707	742	663	686	734	695	748	696	769	764	769	763
Rising	55	57	106	133	142	122	131	133	136	113	62	58
N. Node	102	116	98	99	75	100	110	124	131	89	97	107

9H	♈	♉	♊	♋	♌	♍	♎	♏	♐	♑	♒	♓
Moon	94	97	113	102	107	103	106	110	106	97	99	114
Sun	105	109	101	118	102	109	110	108	94	93	86	113
Mercury	124	101	97	108	95	116	96	104	79	101	121	106
Venus	107	106	113	108	100	104	99	94	101	104	109	103
Mars	128	119	102	106	93	104	100	102	98	98	104	94
Jupiter	96	100	98	115	104	98	105	106	104	109	110	103
Saturn	86	110	110	94	102	104	116	111	94	99	106	116
TOTAL	740	742	734	751	703	738	732	735	676	701	735	749

Houses	1st	2nd	3rd	4th	5th	6th	7th	8th	9th	10th	11th	12th
Moon	102	114	108	94	93	120	113	103	90	82	123	106
Sun	114	114	121	96	103	94	101	89	85	126	99	106
Mercury	125	108	122	112	85	106	90	95	93	103	110	99
Venus	111	106	121	121	90	96	80	99	117	103	98	106
Mars	94	101	101	124	108	96	97	106	102	134	95	90
Jupiter	113	124	105	97	105	97	91	94	107	106	108	101
Saturn	119	106	105	102	109	115	89	112	88	95	106	102
TOTAL	778	773	783	746	693	724	661	698	682	749	739	710
N. Node	99	101	88	106	122	123	92	106	99	89	114	109

Planets

Aspects	Parallels	4 Axis	Direction	☉/☽ Midpoint
Uranus	Sun-Pluto	Moon (Asc)		
Moon-Uranus	Mercury-Uranus	Venus (IC)		
Venus-Pluto				
Mars-Neptune				Moon's Nodes
				Mercury (S)
				Saturn (N)
				Pluto (S)

GAUQUELIN - PAINTERS

Theme: Taurus (2nd), Leo (5th), Pisces, Pluto/Chiron

Pisces:
Pisces is the strongest sign and Virgo is the weakest sign for both the Gauquelin Painter group and the Rodden Artist group.

5th House:
The 5th house is the strongest house for the Gauquelin group, while the 12th house is strongest for the Rodden group.

Pluto & Chiron:
Pluto and Chiron are the strongest planets for the Gauquelin group, while Venus is strongest for the Rodden group.

The following findings are the same for the Gauquelin and Rodden study groups:

1. Pisces
2. Aquarius
3. Sun in Pisces
4. Mercury in Pisces
5. Venus in Pisces
6. Virgo weak (1H) & (9H)
7. Scorpio weak
8. Moon-Pluto aspects
9. Mercury-Jupiter aspects
10. Chiron Direct

Signs

Size: 1,472	1st Harmonic	9th Harmonic	Houses	Rising Sign
Moon				Virgo
Sun	Aries, Pisces	Scorpio		
Mercury	Taurus, Pisces		2nd	Lunar Phase
Venus	Taurus, Gem, Pisces	Taurus		
Mars	Leo, Capricorn	Capricorn	5th	☉/☽ Midpoint
Jupiter	Cancer, Leo			Pisces
Saturn	Aquarius	Capricorn	1st, 5th	Moon's Nodes
TOTAL	Aquarius, Pisces		2nd, 5th	Virgo
(weak)	Virgo, Scorpio	Virgo		Capricorn

1H	♈	♉	♊	♋	♌	♍	♎	♏	♐	♑	♒	♓
Moon	132	115	119	134	126	102	124	118	127	131	119	125
Sun	147	129	127	134	126	90	133	106	107	111	124	138
Mercury	122	130	105	122	110	110	129	124	111	123	144	142
Venus	110	152	136	127	112	110	126	102	120	97	133	147
Mars	101	115	114	135	170	142	139	113	118	124	101	100
Jupiter	119	102	117	142	137	147	126	108	148	100	117	109
Saturn	136	108	102	86	94	123	118	143	127	137	161	137
TOTAL	867	851	820	880	875	824	895	814	858	823	899	898
Rising	57	80	114	147	175	190	158	165	153	111	71	51
N. Node	121	118	121	121	106	140	104	115	121	141	136	128

9H	♈	♉	♊	♋	♌	♍	♎	♏	♐	♑	♒	♓
Moon	113	115	135	136	129	117	124	127	121	117	107	131
Sun	129	93	127	121	122	107	128	160	140	128	107	110
Mercury	120	129	105	134	129	107	106	116	126	134	131	135
Venus	124	146	121	115	125	106	135	137	113	105	135	110
Mars	131	131	116	121	115	122	122	117	117	139	130	111
Jupiter	118	111	118	142	118	119	126	121	131	130	117	121
Saturn	90	125	133	132	132	119	124	125	119	140	110	123
TOTAL	825	850	855	901	870	797	865	903	867	893	837	841

Houses	1st	2nd	3rd	4th	5th	6th	7th	8th	9th	10th	11th	12th
Moon	112	136	121	108	114	135	126	124	119	129	125	123
Sun	123	145	146	114	113	98	112	114	108	136	132	131
Mercury	120	168	122	130	119	89	105	116	106	129	127	141
Venus	138	129	136	120	124	98	115	104	134	123	127	124
Mars	122	136	135	110	139	107	120	133	98	121	141	110
Jupiter	112	134	124	137	134	132	106	118	113	124	125	113
Saturn	143	129	122	133	153	116	111	115	95	133	120	102
TOTAL	870	977	906	852	896	775	795	824	773	895	897	844
N. Node	135	107	137	114	130	109	115	128	114	123	131	129

Planets

Aspects	Parallels	4 Axis	Direction	☉/☽ Midpoint
Moon-Pluto	Sun-Venus	Chiron (MC)	Jupiter (D)	Mars
Sun-Uranus	Mercury-Venus	Pluto (IC)	Pluto (D)	
Mercury-Jupiter	Venus-Saturn		Chiron (D)	Moon's Nodes
Mars-Chiron				Moon (S)
				Mercury (N)
				Uranus (N)
				Pluto (N)
				Chiron (N)&(S)

GAUQUELIN - POLITICIANS

Theme: Libra, Jupiter

Libra:
Libra the strongest sign for the Gauquelin group, while Aries and Pisces are the strongest signs for the Rodden group. In the Gauquelin group, Mercury is strong in Libra in both harmonics, while Mercury is strong in Aquarius in both harmonics for the Rodden group.

9th House:
The 9th house is the strongest house for both for the Gauquelin and Rodden study groups.

Jupiter:
Jupiter is the strongest planet for the Gauquelin group, while the Moon and Mars are strongest for the Rodden group.

The following findings are the same for the Gauquelin and Rodden study groups:

1. Sun in Aquarius
2. Mars in Gemini
3. Mercury in Libra (9H)
4. Moon-Mars parallels

Signs

Size: 1,002	1st Harmonic	9th Harmonic	Houses	Rising Sign
Moon	Taurus	Capricorn		Virgo
Sun	Libra, Aquarius	Gemini, Scorpio	9th	
Mercury	Libra	Libra		Lunar Phase
Venus			6th	
Mars	Gemini	Gemini	5th, 6th	☉/☽ Midpoint
Jupiter	Cancer, Leo		9th	Pisces
Saturn	Scorp, Sag, Capr			Moon's Nodes
TOTAL	Libra, Aquarius			3rd, 5th
(weak)			11th	

1H	♈	♉	♊	♋	♌	♍	♎	♏	♐	♑	♒	♓
Moon	89	99	77	82	81	82	93	79	83	81	80	76
Sun	90	70	100	72	86	89	102	64	76	83	100	70
Mercury	71	74	79	72	75	91	112	84	69	95	84	96
Venus	67	90	72	107	87	87	85	73	92	70	89	83
Mars	75	74	103	84	112	96	91	94	69	65	78	61
Jupiter	81	62	66	98	96	92	83	95	77	86	90	76
Saturn	64	66	75	53	59	79	79	113	116	129	95	74
TOTAL	537	535	572	568	596	616	645	602	582	609	616	536
Rising	42	41	71	114	115	129	109	118	102	61	54	46
N. Node	78	92	68	75	76	84	86	93	87	79	94	90

9H	♈	♉	♊	♋	♌	♍	♎	♏	♐	♑	♒	♓
Moon	91	80	81	78	88	67	83	87	85	100	71	91
Sun	86	85	99	87	79	84	88	99	58	71	79	87
Mercury	84	91	76	70	97	67	69	92	86	82	97	91
Venus	71	85	75	78	96	83	103	78	72	92	81	88
Mars	84	91	102	76	61	69	87	84	83	92	86	87
Jupiter	79	74	72	100	81	85	91	85	83	79	84	89
Saturn	94	89	72	88	82	87	80	81	87	61	95	86
TOTAL	589	595	577	577	584	542	601	606	554	577	593	619

Houses	1st	2nd	3rd	4th	5th	6th	7th	8th	9th	10th	11th	12th
Moon	82	79	84	83	82	82	82	84	89	84	78	93
Sun	93	95	86	80	78	82	71	60	97	94	90	76
Mercury	92	94	85	82	80	76	67	77	85	95	75	94
Venus	81	84	83	71	79	90	66	74	90	98	92	94
Mars	72	81	76	85	99	94	86	72	87	88	79	83
Jupiter	69	80	69	75	77	91	83	95	106	80	79	98
Saturn	98	98	91	81	88	77	86	75	78	80	73	77
TOTAL	587	611	574	557	583	592	541	537	632	619	566	615
N. Node	67	83	98	95	103	74	72	64	89	86	84	87

Planets

Aspects	Parallels	4 Axis	Direction	☉/☽ Midpoint
Chiron	Moon-Mars	Jupiter (MC)		Venus
Sun-Neptune	Venus-Jupiter	Neptune (MC)		Jupiter
Mercury-Mars		Saturn (IC)		
Venus-Chiron		Uranus (IC)		Moon's Nodes
Mars-Chiron				Jupiter (N)
				Uranus (N)

GAUQUELIN - PRIESTS

Theme: Libra/Pisces, 1st House, Neptune/Chiron

Libra & Pisces:
Libra and Pisces are the strongest signs for the Gauquelin group, while Taurus and Pisces are the strongest signs for Rodden's group.

1st House:
The 1st house is strongest for the Gauquelin group, while the 7th and 8th houses are strongest for the Rodden group.

Neptune:
Neptune is the strongest planet for the Gauquelin group, while Saturn is strongest for the Rodden group.

The following findings are the same for the Gauquelin and Rodden study groups:

1. Sun in 1st House
2. Venus-Chiron aspects

Signs

Size: 320	1st Harmonic	9th Harmonic	Houses	Rising Sign
Moon	Libra		1st, 9th	Gemini
Sun			1st, 3rd	
Mercury		Libra	2nd	Lunar Phase
Venus	Pisces		12th	
Mars	Scorpio		5th, 10th	☉/☽ Midpoint
Jupiter		Gemini	9th	
Saturn	Cancer	Taurus		Moon's Nodes
TOTAL	Pisces	Gemini, Libra	1st	Aries
(weak)			8th	Capricorn

1H	♈	♉	♊	♋	♌	♍	♎	♏	♐	♑	♒	♓
Moon	22	22	22	30	31	26	35	23	28	32	23	26
Sun	25	26	26	26	27	27	23	31	22	28	30	29
Mercury	21	22	30	19	24	28	22	30	30	30	34	30
Venus	26	29	24	21	23	26	31	21	33	28	26	32
Mars	23	23	20	31	28	27	37	41	24	18	23	25
Jupiter	22	22	23	28	29	28	32	29	26	30	24	27
Saturn	30	29	26	33	23	29	11	24	28	28	29	30
TOTAL	169	173	171	188	185	191	191	199	191	194	189	199
Rising	13	18	35	34	31	39	40	29	35	25	9	12
N. Node	36	21	27	27	20	28	26	19	23	35	29	29

9H	♈	♉	♊	♋	♌	♍	♎	♏	♐	♑	♒	♓
Moon	32	18	20	26	34	27	24	25	30	29	34	21
Sun	30	27	30	23	21	20	29	25	32	23	29	31
Mercury	19	28	32	22	20	29	35	34	21	32	25	23
Venus	29	31	31	26	28	22	27	26	23	31	19	27
Mars	25	18	34	22	25	27	33	24	32	21	26	33
Jupiter	27	30	38	26	25	24	33	30	17	19	23	28
Saturn	21	38	24	23	24	29	29	25	31	23	27	26
TOTAL	183	190	209	168	177	178	210	189	186	178	183	189

Houses	1st	2nd	3rd	4th	5th	6th	7th	8th	9th	10th	11th	12th
Moon	37	28	26	19	34	27	18	13	35	23	27	33
Sun	42	25	40	16	27	25	17	25	21	29	30	23
Mercury	38	40	34	21	18	29	24	18	18	31	23	26
Venus	29	36	29	31	12	24	27	24	18	26	26	38
Mars	26	29	26	26	34	26	19	17	27	35	25	30
Jupiter	27	25	26	24	33	26	30	24	34	20	27	24
Saturn	23	24	29	27	30	27	30	25	30	22	27	26
TOTAL	222	207	210	164	188	184	165	146	183	186	185	200
N. Node	24	21	29	31	28	20	26	28	34	27	34	18

Planets

Aspects	Parallels	4 Axis	Direction	☉/☽ Midpoint
Sun-Neptune	Moon-Uranus	Neptune (Asc)		
Venus-Chiron	Mercury-Neptune	Chiron (Asc)		
				Moon's Nodes
				Chiron (N)

GAUQUELIN - SCIENTISTS

Theme: Libra (7th), Sagittarius/Capricorn, Sun/Neptune

Sagittarius & Capricorn:
Sagittarius and Capricorn are the strongest signs for both the Gauquelin and Rodden study groups.

7th House (1 in 2,805):
The 7th house is strongest for the Gauquelin group, while the 3rd house is strongest for the Rodden group.

Sun & Neptune:
The Sun and Neptune are the strongest planets for the Gauquelin group, while the Moon is strongest for the Rodden group.

The following findings are the same for the Gauquelin and Rodden study groups:

1. Moon in Sagittarius
2. Mercury in Capricorn
3. Sun in Taurus (9H)
4. Sun in 3rd House

Signs

Size: 1,094	1st Harmonic	9th Harmonic	Houses	Rising Sign
Moon	Gemini, Sagittarius	Virgo	7th	Cancer
Sun		Taurus	3rd, 7th	Sagittarius
Mercury	Capricorn	Libra	2nd, 7th	Capricorn
Venus	Taurus, Pisces	Scorrpio		Lunar Phase
Mars	Leo	Capricorn	8th, 11th	
Jupiter		Virgo	10th, 11th	☉/☽ Midpoint
Saturn	Aries	Libra, Sagittarius		Aquarius
TOTAL	Pisces	Capricorn	7th	Moon's Nodes
(weak)	Leo, Scorpio	Taurus	4th, 5th	Sagittarius, 11th

1H	♈	♉	♊	♋	♌	♍	♎	♏	♐	♑	♒	♓
Moon	86	81	111	87	79	97	87	82	115	84	82	103
Sun	92	86	108	99	80	90	103	72	66	98	100	100
Mercury	96	83	80	80	98	97	95	71	86	114	95	99
Venus	74	125	89	98	77	95	90	83	80	74	102	107
Mars	62	81	107	121	107	113	104	98	82	70	73	76
Jupiter	100	84	70	93	83	98	93	100	103	99	71	100
Saturn	115	87	72	82	72	88	84	88	100	106	103	97
TOTAL	625	627	637	660	596	678	656	594	632	645	626	682
Rising	44	52	91	130	126	102	113	116	127	99	53	41
N. Node	96	77	81	94	82	89	99	98	109	98	83	88

9H	♈	♉	♊	♋	♌	♍	♎	♏	♐	♑	♒	♓
Moon	96	86	90	84	87	110	89	92	80	87	92	101
Sun	90	106	72	84	80	97	94	85	103	103	99	81
Mercury	92	83	101	95	92	94	106	97	74	75	93	92
Venus	99	73	84	97	80	81	88	109	102	98	82	101
Mars	72	80	90	86	97	88	82	100	91	116	97	95
Jupiter	90	78	86	99	94	108	103	84	85	103	87	77
Saturn	102	91	77	86	87	94	107	84	110	99	74	83
TOTAL	641	597	600	631	617	672	669	651	645	681	624	630

Houses	1st	2nd	3rd	4th	5th	6th	7th	8th	9th	10th	11th	12th
Moon	88	88	85	80	82	87	109	92	92	101	108	82
Sun	109	96	119	77	63	86	98	84	78	89	91	104
Mercury	110	121	89	93	65	83	94	84	85	80	97	93
Venus	93	108	104	91	68	84	86	96	76	86	95	107
Mars	101	89	80	79	71	88	94	105	96	84	114	93
Jupiter	98	92	92	84	81	99	100	83	84	107	107	67
Saturn	101	74	99	87	90	93	96	89	96	88	78	103
TOTAL	700	668	668	591	520	620	677	633	607	635	690	649
N. Node	95	83	95	102	94	90	99	85	93	75	112	71

Planets

Aspects	Parallels	4 Axis	Direction	☉/☽ Midpoint
Sun-Moon	Sun-Moon	Saturn (Dsc)	Neptune (D)	
Sun-Jupiter	Sun-Jupiter	Uranus (Dsc)		
Sun-Neptune	Mercury-Neptune	Neptune (MC)		
Mars-Saturn	Mercury-Chiron	Pluto (IC)		**Moon's Nodes**
		Chiron (IC)		Venus (S)
				Pluto (S)

GAUQUELIN - SPORTS

Theme: Aries/Taurus/Aquarius, 5th/9th Houses, Uranus

Aries, Taurus, & Aquarius:
Aries, Taurus, and Aquarius are the strongest signs for the Gauquelin group, while Aries and Gemini are the strongest signs for the Rodden group.

5th & 9th Houses:
The 5th and 9th houses are strongest for the Gauquelin group, while the 9th and 10th houses are strongest for the Rodden group.

Uranus:
Uranus is the strongest planet for the Gauquelin group.

The following findings are the same for the Gauquelin and Rodden study groups:

1. Venus in Aquarius
2. Moon's North Node in Aries
3. Moon's North Node in Pisces
4. 9th House
5. Pluto conjunct Midheaven

Signs

Size: 2,087	1st Harmonic	9th Harmonic	Houses	Rising Sign
Moon	Cancer	Taurus		
Sun	Capr, Aquar, Pisces	Pisces	5th	Lunar Phase
Mercury	Aquarius	Aries	5th	
Venus	Aquarius	Taurus, Sag, Aquar		☉/☽ Midpoint
Mars	Aquarius	Aries, Taur, Leo	9th, 12th	
Jupiter	Cancer, Leo			Moon's Nodes
Saturn	Scorp, Sag, Capr	Cancer		Aries, Taurus
TOTAL	♋ ♐ ♑ ♒	Aries, Taurus	5th, 9th	Gemini, Pisces
(weak)	♈ ♉ ♎ ♏	Capricorn	10th	

1H	♈	♉	♊	♋	♌	♍	♎	♏	♐	♑	♒	♓
Moon	166	171	176	200	171	185	188	151	188	165	177	149
Sun	161	156	178	196	155	158	174	152	157	193	195	212
Mercury	153	146	150	172	166	165	156	177	188	207	217	190
Venus	177	191	145	202	153	173	149	163	185	151	226	172
Mars	143	172	171	213	232	184	173	155	188	141	166	149
Jupiter	141	158	156	198	205	196	176	160	195	174	176	152
Saturn	133	122	139	112	108	182	187	224	261	276	202	141
TOTAL	1074	1116	1115	1293	1190	1243	1203	1182	1362	1307	1359	1165
Rising	92	109	176	199	246	240	232	230	203	152	110	98
N. Node	235	206	208	176	196	158	150	159	136	112	148	203

9H	♈	♉	♊	♋	♌	♍	♎	♏	♐	♑	♒	♓
Moon	171	202	188	181	162	181	174	165	174	160	166	163
Sun	175	182	164	163	146	168	191	176	182	165	177	198
Mercury	196	157	180	150	164	182	190	173	189	161	187	158
Venus	182	198	162	145	159	175	161	170	196	155	202	182
Mars	202	191	189	163	193	155	179	147	169	170	175	154
Jupiter	189	177	174	187	178	145	178	171	179	160	175	174
Saturn	167	197	175	192	166	187	169	169	156	148	194	167
TOTAL	1282	1304	1232	1181	1168	1193	1242	1171	1245	1119	1276	1196

Houses	1st	2nd	3rd	4th	5th	6th	7th	8th	9th	10th	11th	12th
Moon	177	183	172	168	183	172	173	169	192	164	175	159
Sun	189	202	194	152	180	148	167	161	157	182	163	192
Mercury	178	206	202	157	178	158	164	155	155	164	187	183
Venus	196	189	170	189	167	161	165	152	179	157	175	187
Mars	164	166	161	183	178	157	139	158	207	163	184	227
Jupiter	168	180	187	186	174	159	168	143	178	182	171	191
Saturn	190	179	196	176	168	185	160	160	163	164	175	171
TOTAL	1262	1305	1282	1211	1228	1140	1136	1098	1231	1176	1230	1310
N. Node	173	173	153	193	173	191	185	178	176	153	157	182

Planets

Aspects	Parallels	4 Axis	Direction	☉/☽ Midpoint
Venus-Uranus	Jupiter-Uranus	Pluto (Asc)	Saturn (D)	
		Venus (Dsc)	Uranus (D)	
		Neptune (MC)	Neptune (R)	
		Pluto (MC)	Pluto (R)	**Moon's Nodes**
				Sun (S)
				Mercury (N)
				Mars (S)
				Uranus (N)

GAUQUELIN - WRITERS

Theme: Cancer/Pisces, 2nd/6th House, Neptune

Cancer & Pisces (1 in 17,978):
Cancer and Pisces are the strongest signs for the Gauquelin group, while Capricorn, Aquarius, and Pisces are the strongest signs for the Rodden group.

Earth Houses:
The 2nd, 6th, and 10th houses are strongest for the Gauquelin group.

Neptune:
Neptune is the strongest planet for the Gauquelin group.

The following findings are the same for the Gauquelin and Rodden study groups:

1. Pisces
2. Sun in Cancer (9H)
3. Moon in 12th House
4. Mercury in 2nd House
5. Mercury conjunct Moon's North Node
6. Mercury Direct

Signs

Size: 1,352	1st Harmonic	9th Harmonic	Houses	Rising Sign
Moon			9th, 12th	
Sun	Aquarius, Pisces	Cancer	2nd, 6th	
Mercury	Pisces		2nd	Lunar Phase
Venus	Pisces	Cancer, Leo	6th, 9th	
Mars	Capricorn	Capricorn	2nd, 10th	☉/☽ Midpoint
Jupiter	Leo	Gemini, Cancer		Aquarius
Saturn	♊ ♍ ♐ ♓			Moon's Nodes
TOTAL	Pisces	Cancer	2nd, 6th, 10th	Libra
(weak)	Aries		7th	

1H	♈	♉	♊	♋	♌	♍	♎	♏	♐	♑	♒	♓
Moon	118	119	126	113	107	109	106	112	125	103	107	107
Sun	114	121	110	113	101	115	97	113	105	105	126	132
Mercury	100	108	100	94	112	103	102	128	119	125	122	139
Venus	110	124	106	137	87	117	108	86	123	105	106	143
Mars	95	114	126	124	132	124	126	105	107	114	97	88
Jupiter	90	87	90	118	156	119	139	115	111	125	92	110
Saturn	96	107	123	100	87	130	85	123	153	120	97	131
TOTAL	723	780	781	799	782	817	763	782	843	797	747	850
Rising	59	80	99	148	154	159	133	165	127	118	69	41
N. Node	97	116	99	120	98	107	131	125	124	111	100	124

9H	♈	♉	♊	♋	♌	♍	♎	♏	♐	♑	♒	♓
Moon	112	121	110	113	120	109	113	119	107	110	98	120
Sun	111	93	114	134	108	122	117	121	102	101	115	114
Mercury	116	116	108	113	112	105	97	113	116	121	120	115
Venus	95	108	114	134	130	116	105	115	114	103	114	104
Mars	115	139	111	111	95	110	111	102	105	132	122	99
Jupiter	107	101	154	142	88	119	111	94	100	110	120	106
Saturn	106	106	95	100	122	117	126	120	123	103	116	118
TOTAL	762	784	806	847	775	798	780	784	767	780	805	776

Houses	1st	2nd	3rd	4th	5th	6th	7th	8th	9th	10th	11th	12th
Moon	105	119	106	109	100	113	81	108	141	119	102	149
Sun	136	147	110	109	90	114	100	98	85	137	112	114
Mercury	143	142	112	114	100	103	101	98	100	119	109	111
Venus	116	117	128	115	96	117	84	103	130	106	114	126
Mars	99	135	100	121	109	113	100	111	96	139	110	119
Jupiter	125	101	126	104	116	120	92	118	119	109	113	109
Saturn	110	118	108	113	114	108	125	124	93	117	118	104
TOTAL	834	879	790	785	725	788	683	760	764	846	778	832
N. Node	113	123	127	105	105	112	110	106	108	104	123	116

Planets

Aspects	Parallels	4 Axis	Direction	☉/☽ Midpoint
Sun & Neptune		Sun (Dsc)	Mercury (D)	
Moon-Mars		Chiron (Dsc)		
Moon-Neptune		Moon (MC)		**Moon's Nodes**
Sun-Jupiter		Mercury (MC)		Mercury (N)
Venus-Neptune				Jupiter (N)
				Uranus (N)
				Neptune (S)
				Pluto (S)

Appendices

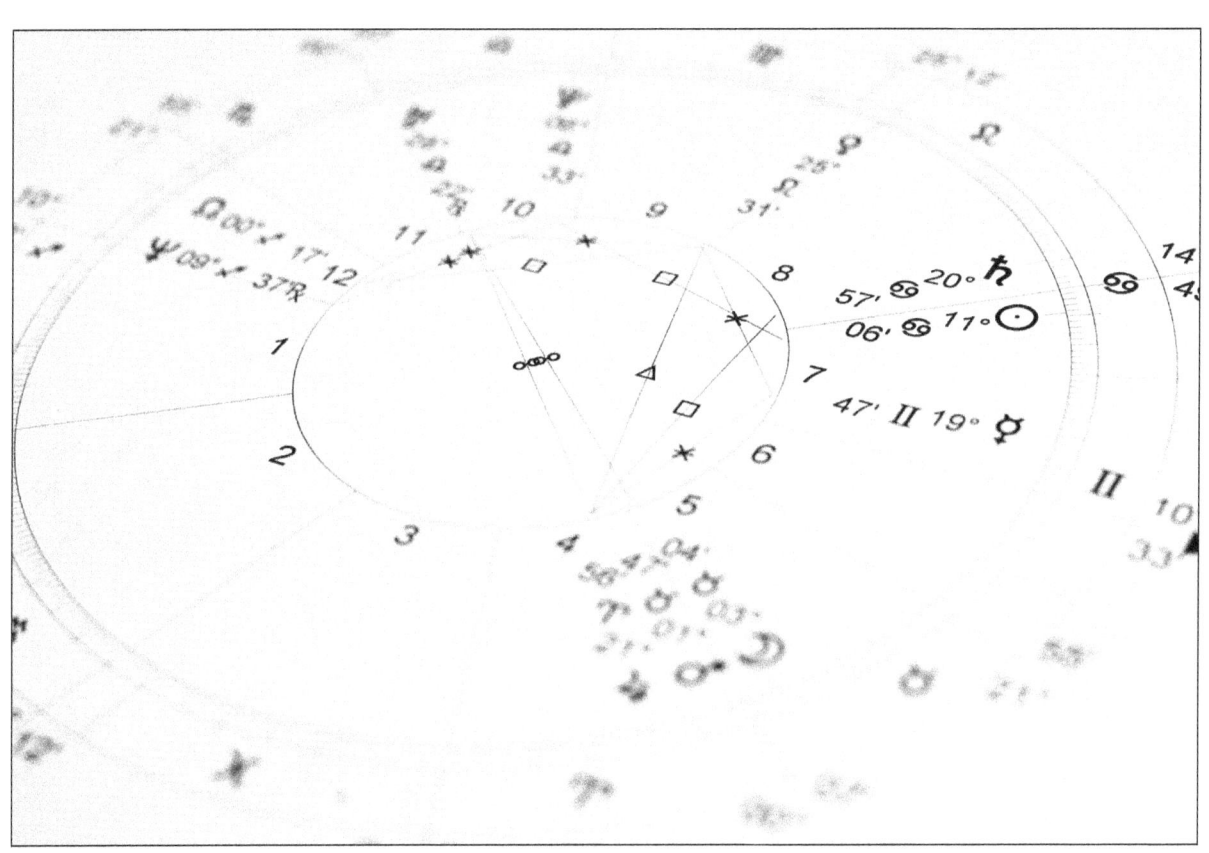

APPENDIX 1 – THE LUNATION CYCLE

The Lunation cycle can be divided into 12 parts, instead of the usual eight. The New Moon is symbolically represented by the beginning of Aries and the Full Moon by the beginning of Libra.

Top Findings (by probability of being an accident)

1. Sports Coaches	Aries	450,690
2. Teachers (K-12)	Pisces	6,266
3. Education (low)	Pisces	2,440
4. Spiritual Teachers	Scorpio	821
5. Novelists	Virgo	539
6. Suicide	Capricorn	446
7. Twins	Gemini	343
8. Marriage (none)	Virgo	301
9. Mystics	Scorpio	241
10. Instrumentalists	Sagittarius	199
11. Psychics	Scorpio	199
12. Presidents	Sagittarius	175
13. Murderers	Capricorn	138
14. Alzheimer's	Libra	138
15. Life (<29 yrs)	Libra	135

Aries	Taurus	Gemini	Cancer	Leo	Virgo
AIDS	Children (4+)	Criminals	Engineers	Child Abuse	Literature
Birth Defect	Songwriters	Heart Attack	Math-Physics	Child Performers	Marriage (none)
Nazis	Wealthy	Infant Mortality	Transvestites	Comedians	Medical
Priests		Lawyers	Widowed	Marriage (15+)	Novelists
Racers		Military		Philosophers	Poets
Sports Coaches		Psychiatrists		Same Job (10+)	Presidents
		Sex Abuse		Singers	Salespeople
		Twins			

Libra	Scorpio	Sagittarius	Capricorn	Aquarius	Pisces
Actors	Mystics	Instrumentalists	Lesbians	Artists	Alzheimer's
Alzheimer's	Psychics	Nobel Prize	Models	Astrologers	Dancers
Child Performers	Spiritual Teacher	Presidents	Murderers	Children (4+)	Education (low)
Culinary	Sports		Stroke	Infant Mortality	Journalists
Lawyers			Suicide	Prisoners	Teachers
Life (<29 yrs)			Writers	Sex Workers	
Scientists					

Appendix 2 – The Moon's Nodes

For the Moon's Nodes study, I observed the sign and house placements for the Rising (North) Node. For aspects, conjunctions within 5 degrees of orb were used for both the North and South (falling) Nodes.

The astrological theory for the Moon's Nodes states that a planet's positive qualities are emphasized and sought after on the North Node and its negative qualities are emphasized and struggled with on the South Node. However, an equal amount of both positive and negative study groups were found on both Moon's Nodes. For this reason, the study groups listed are those that are strong on both nodes.

Signs

Aries	Taurus	Gemini	Cancer	Leo	Virgo
Astronauts	Astronauts	Astrologers	Actors	Alzheimer's	Cancer
Life (<29 yrs)	Child Abuse	Directors	Children (none)	Culinary	Children (4+)
Math-Physics	Composers	Infant Mortality	Comedians	Engineers	Heart Attack
IQ (high)	Mystics	Models	Directors	Journalists	Literature
Murder Victims	Nazis	Mystics	Education (low)	Mental Handicap	Marriage (15+)
Musicians	Priests	Politicians	Homosexuals	Musicians	Military
Sports			Marriage (none)	Singers	Poets
			Out-of-Body	Transvestites	Psychics
			Sex Abuse		Royalty
			Sex Offenders		Writers
			Singers		
			Widowed		

Libra	Scorpio	Sagittarius	Capricorn	Aquarius	Pisces
Cancer	Murderers	Business Owners	AIDS	Birth Defect	Astrologers
Dancers	Philosophers	Education (low)	Child Prodigy	Medical	Birth Defect
Marriage (15+)	Professors	Suicide	Children (4+)	Mental Handicap	Children (none)
Models		Twins	Corporate	Novelists	Life (80+ yrs)
				Obesity	Life (<29 yrs)
				Presidents	Mental Handicap
				Spiritual Teacher	Novelists
				Sports Coaches	Priests
				Twins	Sex Workers
				Writers	Sports
					Technical

Houses

1st	2nd	3rd	4th	5th	6th
Adopted	Philosophers	Astrologers	Novelists	Astronauts	Birth Defect
Life (80+ yrs)		Child Prodigy		Mental Handicap	Computers
IQ (high)		Children (none)		Obesity	Criminals
Nobel Prize		Dancers		Out-of-Body	Engineers
Professors		Mental Illness		Police	Homosexuals
Same Job (10+)		Presidents			Sex Offenders
		Psychiatrists			Technical
		Spiritual Teacher			

7th	8th	9th	10th	11th	12th
Directors	Cancer	Singers	Business Owner	Alzheimer's	Comedians
Instrumentalists	Directors	Teachers (K-12)	Child Abuse	Children (4+)	Criminals
Journalists	Inheritance		Culinary	Culinary	Engineers
Lesbians	Life (80+ yrs)		Inheritance	Homosexuals	Heart Attack
Psychics	Literature		Marriage (15+)	Life (80+ yrs)	Infant Mortality
Stroke	Nobel Prize		Murder Victim	Military	Lesbians
Twins	Politicians		Salespeople	Nobel Prize	Suicide
	Writers		Sex Abuse	Same Job (10+)	
			Wealthy		

Conjunctions (North + South)

Sun	Moon	Mercury	Venus	Mars	Jupiter
Actors	Adopted	Alzheimer's	Corporate	Astrologers	Mental Handicap
Infant Mortality	Inheritance	Birth Defect	Directors	Birth Defect	Models
Nobel Prize	Instrumentalists	Children (none)	Education (low)	Composers	Royalty
Philosophers	Lawyers	Criminals	Engineers	Inheritance	Spiritual Teacher
Psychiatrists	Life (80+ yrs)	Culinary	Mystics	Instrumentalists	
Royalty	Marriage (none)	Depression	Psychiatrists	Mental Illness	
Sex Offenders	Math-Physics	Drug Abuse	Racers		
	Mental Illness	Education (low)			
	Scientists	Life (<29 yrs)			
	Technical	Literature			
		Medical			
		Mental Handicap			
		Musicians			
		Suicide			
		Widowed			
		Writers			

Saturn	Uranus	Neptune	Pluto	Chiron
Medical	Corporate Tycoons	Artists	Activists	Alcoholics
Sex Abuse	Mental Illness	Child Prodigy	Life (<29 yrs)	Cancer
	Prisoners		Stroke	Composers
	Sex Offenders		Transvestites	Infant Mortality
			Twins	Lawyers
				Math-Physics
				Musicians
				Police
				Songwriters
				Widowed

Appendix 3 – Parallels of Declination

Parallels of declination show when a pair of planets are the same vertical distance from the Celestial Equator, which is the Earth's Equator extended into space. Statistical strength was determined by finding the planetary combinations for each group using parallels within a 1 degree orb.

	Sun	Moon	Mercury	Venus	Mars
Moon	Marriage (15+) Songwriters		Children (4+) Education (low) Out-of-Body		
Venus	Adopted		Actors Inheritance		
Mars	Sex Abuse Transvestites	Child Performers Child Prodigy Politicians		Teachers (K-12)	
Jupiter		Mystics, Priests Performers Sex Workers Spiritual Teacher	Artists Education (low) Spiritual Teacher	Birth Defect Business Sex Offenders	Birth Defect Marriage (15+) Poets Spiritual Teacher
Saturn	Adopted Child Abuse Novelists	Infant Mortality Nobel Prize Novelists Professors	Life (<29 yrs) Philosophers		Culinary Salespeople Transvestites
Uranus	Child Performer	Child Abuse Composers Heart Attack	Astrologers Mystics	Artists Astronauts Sex Workers	Activists Astronauts Birth Defect
Neptune	Alcoholics Children (4+) Models Psychics	Mental Illness Out-of-Body Widowed	Corporate Lawyers	Instrumentalist Sports Coaches	AIDS
Pluto	Royalty	Life (80+ yrs) Marriage (none) Scientists	Depression Mental Illness Nazis	Medical Obesity Same Job (10+)	Sports Coaches
Chiron	Business	Only Child Twins	Heart Attack Mental Handicap Stroke, Teachers		IQ, Lawyers Military, Nazis Suicide

APPENDIX 4 – RETROGRADE PLANETS

A planet is retrograde when it is closest to the Earth. At this time, planets inside the Earth's orbit of the Sun (Mercury and Venus) are conjunct the Sun on the ecliptic. Planets outside the Earth's orbit are opposite the Sun on the ecliptic.

Since a planet is closest to Earth during the retrograde phase, its influence is subjective where personal feelings are most involved. It is a time for perceiving new information and corresponds with the Winter and Spring seasons. The Winter Solstice corresponds with the moment a planet is Stationary Retrograde.

For Mercury, this means that our thinking is more subjective and opinionated, less clear. For Venus, it means that we seek greater intimacy. For Mars, it means that our desires are more impulsive and selfish. The moment when a planet is closest to the Earth is its time of greatest subjectivity and corresponds with the Spring Equinox (Aries).

Planets are more distant from the Earth, during the direct phase. As a result, their influence at this time is more objective. It is the time for making judgments and corresponds with the Summer and Fall seasons. The Summer Solstice corresponds with the moment a planet is Stationary Direct.

For Mercury, this causes objective thinking and good decision making. For Venus, this causes relationships to become more formal. For Mars, this causes our desires to be more well-planned and inclusive of others. The moment when a planet is furthest from Earth is the time of greatest objectivity, when outer clarity is greatest, and corresponds with the Fall Equinox (Libra).

Of all the planets, Mercury and Venus show the most notable effects from being direct or retrograde in motion. Mercury shows the traditional description of the retrograde effect of repressing the best qualities of the planet. The groups with the most Mercury retrograde have serious problems with their minds. The groups with the least Mercury retrograde are mostly great thinkers and communicators. Mercury is seen to have its most beneficial influence when direct, since its nature is to be objective and rational.

For Venus, the very opposite situation occurs. **The groups most representative of the positive Venus qualities are strongly retrograde**. Unlike Mercury, which prefers rational communication, Venus prefers to act from subjective feelings, always prioritizing personal tastes and desires. Venus retrograde is when relationships are most close, while Venus direct is when relationships are most distant.

(In the following tables, blue highlights indicate groups that emphasize the more positive attributes of the planet and red highlights indicate groups that emphasize the more negative attributes of the planet.)

Mercury

Direct	Retrograde
Children (4+)	Astronauts
Homosexuals	Lesbians
Literature	Mental Handicap
Musicians	Mental Illness
Psychics	Only Child
Salespeople	
Singers	
Stroke	
Writers	

Venus

Direct	Retrograde
Computers	Actors
Journalists	Children (4+)
	Culinary
	Models
	Mystics
	Priests
	Salespeople
	Spiritual Teachers

Mars

Direct	Retrograde
Salespeople	Fashion
Sports Coaches	Politicians
Suicide	Widowed
Teachers (K-12)	

Appendix 5 – Midpoints

A midpoint is the location on the ecliptic precisely in the middle location between two planets. When a third planet is located at a midpoint, it creates an energetic relationship between the three planets. The Sun/Moon midpoint is the most influential, since it represents our primary individual needs and drives. A 3 degree orb is used for planetary conjunctions.

Sun/Moon (Signs)

1. Racers	Aries	9,810
2. Sports	Aries	3,779
3. Only Child	Capricorn	3,379
4. Life Long (80+)	Pisces	2,856
5. Police	Virgo	2,134
6. Politicians	Pisces	1,037
7. Engineers	Scorpio	482
8. Same Job (10+)	Aries	457
9. Astronauts	Aquarius	236
10. Salespeople	Aries	232
11. Obesity	Cancer	230
12. Scientists	Pisces	225
13. Murderers	Aries	188
14. Criminals	Libra	176
15. Mental Illness	Sagittarius	174
16. Astrologers	Gemini	162
17. Life Short (<29)	Libra	160
18. Child Abuse	Leo	155
19. Math/Physics	Aquarius	145
20. Homosexuals	Pisces	137
21. Presidents	Aquarius	134
22. Scientists	Sagittarius	118
23. Sex Offenders	Virgo	103
24. Technical	Pisces	101
25. Culinary	Cancer	99

Sun/Moon (Planets)

1. Fashion	Venus	1,538
2. Instrumentalists	Saturn	1,146
3. Child Abuse	Pluto	1,010
4. Marriage (none)	Mars	858
5. Sex Offenders	Jupiter	732
6. Salespeople	Jupiter	609
7. Fashion	Chiron	579
8. Directors	Mars	438
9. Alzheimer's	Jupiter	278
10. Prisoners	Mars	246
11. Depression	Chiron	220
12. Out-of-Body	Mars	184
13. Sex Offenders	Neptune	163

APPENDIX 6 – PLANETARY RULERSHIP OF SIGNS

In traditional astrology, each of the seven planets, Sun to Saturn, are assigned rulership over and exaltation in specific signs. Planets are thought to be strongest when positioned in these signs. Simultaneously, planets are thought to be weakest when in the opposite signs, which are called detriment and fall (shown below).

To study this concept, I found which signs are the strongest and weakest for each planet. This was determined by observing the signs where each planet has the most groups expressing their positive or negative natures in both the 1st and 9th harmonics.

	Planetary Strength	Planetary Weakness
Sun	Aries	
Moon	Sagittarius, Aquarius	Aries, Capricorn
Mercury	Capricorn, Aquarius, Pisces	
Venus	Taurus	Virgo
Mars	Gemini, Capricorn	Taurus, Virgo
Jupiter	Scorpio	Libra
Saturn	Gemini, Libra, Aquarius	Cancer, Virgo, Sagittarius

Out of 14 total positions, 8 are the same as traditional rulerships (shown in blue), with only 1 being the opposite (shown in red). This shows strong validity for the traditional correspondences between signs and planets.

The Sun is strongest in Aries, its exalted position. The Moon is weakest in Capricorn, its position of detriment. Venus is the most consistent with traditional theory, being both strongest in Taurus, its ruled sign, and weakest in Virgo, its fallen sign. Mars is strongest in Capricorn, its exalted position. Saturn is strongest in the Air signs including its ruled sign Aquarius, and its exalted sign Libra, while weak in its sign of detriment, Cancer. Mercury is the only contradictory planet, being strong in Pisces, its sign of detriment.

Aquarius is the most beneficent sign, being strongest for three planets. Virgo is the most difficult sign, being weakest for three planets.

	Ruled	Exalted	Detriment	Fall
Sun	Leo	Aries	Aquarius	Libra
Moon	Cancer	Taurus	Capricorn	Scorpio
Mercury	Gemini, Virgo	Virgo	Sagittarius, Pisces	Pisces
Venus	Taurus, Libra	Pisces	Aries, Scorpio	Virgo
Mars	Aries, Scorpio	Capricorn	Taurus, Libra	Cancer
Jupiter	Sagittarius, Pisces	Cancer	Gemini, Virgo	Capricorn
Saturn	Capricorn, Aquarius	Libra	Cancer, Leo	Aries

Appendix 7 – The Southern Hemisphere

In the Southern hemisphere, the seasons are completely reversed from the Northern hemisphere. From a Southern hemisphere point of view, Aries starts at the Fall Equinox, Cancer starts at the Winter Solstice, Libra starts at the Spring Equinox, and Capricorn starts at the Summer Solstice.

However, Aries represents the beginning of individual activity and is inherently connected to the Spring Solstice. Cancer, representing loving warmth, is inherently associated with the Summer Solstice. Libra, representing the birth of a social conscience, is inherently associated with the Fall Equinox. Capricorn, representing repression, discipline, and planning, is inherently associated with the Winter Solstice.

This suggests to me that the zodiac signs are in opposite positions for people born in the Southern hemisphere. To investigate, I compared 73 Actors and 80 Politicians born in the Southern Hemisphere to their Northern Hemisphere study groups. The Northern Hemisphere Actors are strong in Aquarius, while the Southern Hemisphere Actors are dominant in the opposite sign, Leo. The Northern Hemisphere politicians are strong in Pisces, while the Southern Hemisphere politicians are dominant in the opposite sign, Virgo.

I also looked at three of the world's most elite athletes, Michael Jordan, Wayne Gretsky, and Pele (see Case Studies). I noticed that Jordan and Gretsky have Mars in the 1st degree of Cancer, the peak of the solar cycle, while Pele has Mars in the 1st degree of Capricorn, the trough of the solar cycle.

Why would Pele have Mars there? It appears to be because he was born in the Southern hemisphere, where this is the 1st degree of the Summer Solstice, and Mars is actually in the 1st degree of Cancer for him.

For this reason, I only used Northern Hemisphere birth charts in the study groups for the entire book, except for the Obesity group which consists entirely of people born in the Southern Hemisphere.

Actors - North

N=888	♈	♉	♊	♋	♌	♍	♎	♏	♐	♑	♒	♓
Moon	63	84	76	76	76	91	78	66	56	71	83	68
Sun	84	76	79	69	68	73	79	74	64	75	79	68
Mercury	70	73	53	64	67	80	79	93	73	77	70	89
Venus	77	70	84	84	52	85	65	79	69	66	95	62
Mars	62	71	84	78	73	92	78	83	62	75	69	61
Jupiter	69	82	61	60	75	87	84	86	86	68	71	59
Saturn	58	73	51	67	68	77	86	84	77	76	100	71
TOTAL	483	529	488	498	479	585	549	565	487	508	567	478

Actors - South

N=73	♈	♉	♊	♋	♌	♍	♎	♏	♐	♑	♒	♓
Moon	8	9	7	5	13	3	2	2	5	3	6	10
Sun	9	3	10	7	8	3	6	6	6	2	6	7
Mercury	5	9	5	7	7	7	7	3	5	4	9	5
Venus	4	10	9	7	6	3	12	6	3	4	4	5
Mars	4	5	6	8	10	7	8	3	3	5	8	6
Jupiter	5	6	4	7	7	8	6	7	6	3	4	10
Saturn	9	6	6	5	8	9	6	4	3	5	8	4
TOTAL	44	48	47	46	59	40	47	31	31	26	45	47

Politicians - North

N=543	♈	♉	♊	♋	♌	♍	♎	♏	♐	♑	♒	♓
Moon	52	55	45	42	52	36	37	48	37	50	39	52
Sun	53	42	44	54	39	39	45	31	49	40	59	50
Mercury	40	34	38	51	41	44	35	44	52	44	58	64
Venus	55	50	39	50	45	48	29	52	42	39	56	40
Mars	33	42	59	51	64	52	49	38	42	37	32	46
Jupiter	36	48	34	52	50	46	59	50	48	35	40	47
Saturn	43	48	41	42	55	40	49	35	56	46	53	37
TOTAL	312	319	300	342	346	305	303	298	326	291	337	336

Politicians - South

N=80	♈	♉	♊	♋	♌	♍	♎	♏	♐	♑	♒	♓
Moon	8	8	5	2	11	4	5	7	8	6	6	10
Sun	3	4	4	6	7	13	7	6	9	6	5	10
Mercury	3	4	5	6	6	11	8	7	11	7	8	4
Venus	3	6	6	7	7	10	5	7	9	9	8	3
Mars	3	2	8	3	10	9	8	7	9	7	7	7
Jupiter	4	3	8	7	7	8	4	11	6	7	8	7
Saturn	9	6	6	10	7	10	6	6	4	11	2	3
TOTAL	33	33	42	41	55	65	43	51	56	53	44	44

APPENDIX 8 – THE SIDEREAL ZODIAC

According to historical records, the first zodiac was conceived by the Babylonians some 2,500 years ago. It was a belt of 12 unevenly sized, overlapping constellations which served as background to the Sun and planets on the ecliptic. The idea to assign 12 equal, 30 degree spaces to each constellation was inspired by the 12 month calendar. The calendar is based on the Earth's seasonal cycle around the Sun, which is also the basis for the Tropical zodiac. **Even though the Tropical zodiac was not yet conceived by the Babylonians, its reality appears to be the basis for projecting a symmetrical zodiac onto the stars.**

About 2,000 years ago, the Tropical zodiac was conceived by the Greeks. They named the signs after the constellations that were, but are no longer, in the same visual spaces. It is thought that this discovery, being at the time when both zodiacs were matched, was a sort of miracle. However, this assumes that both zodiacs are functional. This book shows that the Tropical zodiac is functional. However, it is hard to imagine how two entirely different zodiacs, which place the planets in different signs, could both be functional.

In my study on the Sidereal zodiac (constellations), there was much less evidence for its accuracy than the Tropical zodiac. The findings are more random and not very well fitted to the signs. The 9th harmonic findings also do not match the 1st harmonic findings. In the Tropical zodiac study, there are 23 cases of a group being strong in the same sign in both harmonics. In the Sidereal zodiac study, this only occurred 6 times (shown in blue). (The expected number of times for 93 groups to do this is about 5.)

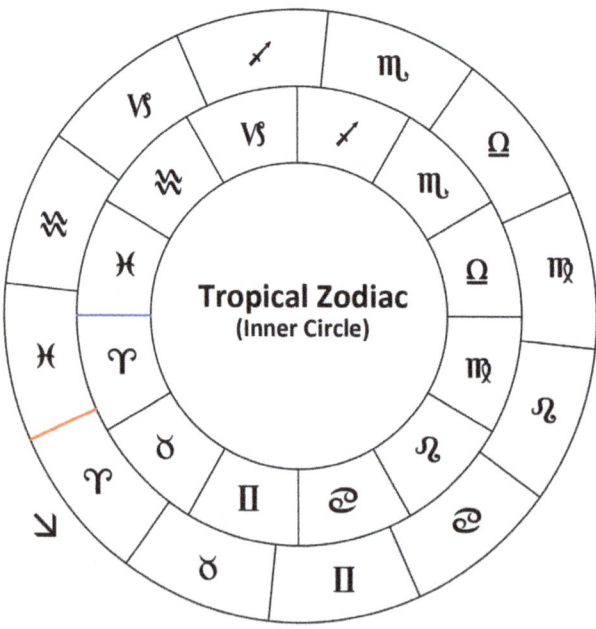

Sidereal Zodiac (Lahiri)

Aries	Taurus	Gemini	Cancer	Leo	Virgo
Alzheimer's	Birth Defect	Astrologers	Corporate	Cancer	Royalty
Artists	Business	Fashion	Infant Mortality	Child Abuse	Teachers (K-12)
Children (4+)	Child Abuse	Infant Mortality	Lawyers	Drug Abuse	Transvestites
Culinary	Corporate	Murderers	Lesbians	Lesbians	
Life (80+ yrs)	Fighters	Salespeople	Politicians	Medical	
Marriage (15+)	Marriage (15+)	Spiritual Teacher		Models	
Mental Handicap	Same Job (10+)			Only Child	
Priests	Singers			Out-of-Body	
Racers	Sports			Sex Abuse	
Sports	Stroke			Sex Offenders	
				Sex Workers	

Libra	Scorpio	Sagittarius	Capricorn	Aquarius	Pisces
AIDS	Comedians	Adopted	Alzheimer's	Artists	Astronauts
Astrologers	Composers	Comedians	Comedians	Child Prodigy	Children (4+)
Child Performer	Criminals	Composers	Culinary	Comedians	Infant Mortality
Computers	Heart Attack	Fighters	Inheritance	Composers	Instrumentalists
Life (<29 yrs)	Police	Homosexuals	Marriage (15+)	Education (low)	Life (80+ yrs)
Mental Illness	Scientists	Life (80+ yrs)	Marriage (none)	Instrumentalists	Marriage (none)
Mystics	Sex Abuse	Literature	Math-Physics	Lawyers	Math-Physics
Philosophers	Twins	Math-Physics	Military	Life (80+ yrs)	Nazis
Spiritual Teacher		Poets	Murder Victims	Life (<29 yrs)	Professors
Twins		Scientists	Only Child	Mental Handicap	Racers
		Technical	Racers	Musicians	Salespeople
		Transvestites	Salespeople	Poets	Sports
		Writers	Technical	Politicians	Sports Coaches
				Psychics	
				Scientists	
				Sports Coaches	
				Technical	
				Wealthy	

The results for the Sidereal zodiac are shifted over by almost one sign from the Tropical zodiac. As a result, the findings for Pisces in the Sidereal zodiac are mostly the same as Aries in the Tropical zodiac. The groups in Sidereal Pisces are clearly a much better fit for Tropical Aries. This is true for most signs.

Metaphysics Addendum

1. Obliquity of Planets

The Earth spins on its axis, the poles. This axis is not perpendicular to the ecliptic, or the Sun. It is about 23 degrees from perpendicular, which is why it has a wobble. Due to the Sun's gravitational force, this axis rotates, in a circle, around the line perpendicular to the Sun. As a result, the seasonal cycle shifts backwards against the constellations. A complete cycle takes 25,765 years.

The Tropical year is slightly shorter than the Sidereal year. The axis returns to its same position each year before the Earth returns to the same position in the constellations. It is still unknown why the Earth's axis is not perpendicular to the ecliptic. The Earth is thought to spin eastward due to the initial momentum of its creation.

The obliquity of a planet shows the tilt of its poles in relation to its orbit around our Sun. A zero or 180 degree obliquity would indicate that a planet has no tilt and spins perpendicular to its orbit, with no wobble. The tilt of a planet tells us a great deal about its nature. A planet that spins in an upright position, with no tilt, is of upright character and has integrity.

The planets traditionally considered positive and beneficial, Venus and Jupiter, both spin in an upright position. Jupiter's North pole points forward in the direction our solar system moves, while Venus' North pole points in the opposite direction - our spiritual home from which we came. Mercury, symbol of knowledge and reason, has a perfectly upright spinning position.

The planets that are traditionally considered malefic and harmful, Mars and Saturn, do not spin in an upright position, they wobble - showing that their nature is not upstanding. The condition of life on Earth is revealed by the fact that it also wobbles, having almost as much obliquity as the malefic planets. This symbolizes to us that there is much suffering on our planet.

Perhaps, if humanity ever forms a harmonious civilization in balance with nature and in tune with the universe, our planet will lose its tilt and spin more upright like the benevolent planets.

(Neptune, also considered malefic by astrologers, has the same tilt as Mars and Saturn. Uranus, the rebellious eccentric planet, is the only planet that spins on its side, parallel to its solar orbit.)

2. THE TWO ZODIACS

The Tropical zodiac, used in the Western world, was devised by the ancient Greeks and is based on the Earth's seasonal path around the Sun. In the Northern hemisphere, the beginning of Capricorn marks the start of Winter. The beginning of Cancer marks the start of Summer, while Aries and Libra start Spring and Fall.

So why do we say our destiny is written in the stars?

Prior to the Greeks, about 2,500 years ago, the Babylonians selected a belt of 12 constellations that were in the background of the Sun at the times of the 12 months of their calendar year. However, these constellations were of very different sizes and even overlapped in places. This was a zodiac made up of visible stars, a concept borrowed from the 27 lunar mansions (nakshatras) used in India.

But those who sought to use the 12 constellation signs found it difficult, since planets could be in two constellations at once or none at all. Their answer was to divide the starry sky into 12 equal sections, just as the calendar divides the year into 12 months. So the original zodiac calculated the Sun, Moon, and planets with reference to the constellations. Unfortunately, the starting point of this Sidereal zodiac is also subject to debate since there is distance between the end of Pisces and the beginning of Aries.

The Greeks corrected this whole dilemma by simply realizing that the astrological influence of planets in the 12 signs do not come from the backdrop of the stars, but the spaces created by the seasons. However, not all astrologers agreed with the Greeks and continue, to this day, to use the Sidereal zodiac (especially in India).

Are we entering the Age of Aquarius?

Since the Tropical (seasonal) and Sidereal (constellations) zodiacs entirely overlapped at the time of their conception, their difference was unimportant for practical reasons. However, the Tropical zodiac actually rotates clockwise (backwards) against the Sidereal zodiac. So today, planets can be in two different signs in the two zodiacs. This is impossible for astrologers, however, since this gives each planet two entirely different interpretations.

Currently, Tropical Aries is mostly overlapping Sidereal Pisces and is nearing Sidereal Aquarius. For this reason, it is said that we are in the Age of Pisces and entering the Age of Aquarius. But if the Sidereal zodiac has no actual influence, what meaning do the ages have? And does it really make sense that ages would move in reverse order through the signs? And is it believable that the last 2,000 years of human history can be described as being Piscean, full of creativity and spirituality? It seems more like an Age of Aries, full of pioneering aggression.

3. The Great Ages

The Great Ages are measured by the Precession of the Equinoxes, which is the shifting of alignment between the Tropical zodiac, used by Western astrologers, and the Sidereal zodiac used by Vedic astrologers. The Tropical zodiac is based on the positions of the Earth as it travels around the Sun. The Sidereal zodiac is based upon the positions of the constellations in our galaxy.

These two zodiacs are currently not in alignment. As the Earth wobbles in a circular motion on its axis, the location where Spring begins gradually changes, relative to our galaxy. As a result, the Tropical Zodiac shifts from the Sidereal zodiac. This occurs at a rate of approximately 1 degree every 72 years taking 25,765 years to complete an entire cycle.

The last time the two zodiacs were aligned was approximately 300 AD. The exact date is not known because the beginning of the Sidereal zodiac can not be specifically located. In fact, there are at least 14 different points in use, which vary by as many as 8 degrees. This amounts to a 530 year difference for timing the shift of the Spring Equinox into a new sign in the Sidereal zodiac. This creates great uncertainty for those doing Vedic astrology, especially since the smaller nakshatras and harmonic divisions are an integral part of that system.

For ages, astrologers have speculated that when the Spring Equinox moves into each sign in the Sidereal zodiac a new phase of human existence will ensue. It takes 2,147 years for each sign to be completed.

In the year 2022, estimates put the beginning of Aries in the Tropical zodiac between the 2nd and 10th degree of Pisces in the Sidereal zodiac. Most astrologers view the Tropical zodiac as spinning clockwise inside the Sidereal zodiac. Accordingly, the signs are traversed in reverse order. This is the reason that it is believed humanity is leaving the Age of Pisces and entering the Age of Aquarius about 400 years from now.

However, it is likely that the Ages actually progress in forward order through the signs. From this perspective, it is the Sidereal zodiac which revolves around the Tropical zodiac and is now beginning somewhere between the 20th and 28th degree of Tropical Aries. This would mean that we are now in the Age of Aries (ruled by Mars) and entering into an Age of Taurus (ruled by Venus).

<u>The Nature of Astrology</u>

Astrology is a subjective science that derives information from the observable universe. Are the Sun and Moon the same size? Of course not, but to astrologers they bear equal weight since they are the same size to our eyes. Do planets actually move backwards? Obviously not, but from our perspective they do and astrologers interpret such retrograde planets to have special meaning. Can Mercury and Venus be in opposite sides of the solar system? Of course, but from the view of planet Earth they are never very far apart.

The point is that, in astrology, all planetary movements are interpreted according to their appearance ... not the reality. This is why science has such a difficult time with astrology. However, quantum physics itself states that our subjective consciousness is largely responsible not only for how we interpret physical reality, but what actually occurs in reality.

Since everything in astrology is viewed from our perspective, we should also view the movement through the constellations from our perspective, using our Tropical zodiac as the measure of change. Objectively, we are wobbling around the solar system causing the Tropical zodiac to rotate backwards across the constellations. But astrology is subjective and from our perspective the constellations are moving forward through the Tropical zodiac from Aries though Pisces (viewed from the Northern Hemisphere perspective - where about 90% of the human population lives).

(Additionally, the original Sidereal zodiac was measured by the Moon's daily motion through the stars. These are the 27 lunar mansions (nakshatras) described in the ancient Vedic scriptures. In all likelihood, the 12 signs are strictly seasonal divisions of the Tropical zodiac and have no astrological effect from the constellations. It is the lunar zodiac that is rotating around the the Tropical (solar) zodiac ... the first lunar mansion, Ashwini, being located at the beginning of the constellation Aries.)

<u>The Stages of Civilization</u>

The constellation Aries was in the Tropical sign of Libra at the end of our last ice age about 12,000 years ago. The Age of Scorpio, from about 10,400 – 8,300 BC, brought transformation and rebirth into the time of the Fertile Crescent. The domestication of plants and animals brought about an agricultural revolution. This key step in the development of human civilization made possible an increased food supply and an accompanying growth of population and allowed nomads and cave dwellers to become farmers and herders.

The Age of Sagittarius (8,300 – 6,100 BC) followed with the creation of surplus commodities. Humans proceeded to spread around the globe searching for new opportunities.

From about 6,100 – 4,000 BC was the Age of Capricorn. The occupation of Babylonia was begun. Because of inadequate rainfall the inhabitants of the lower valley had to resort to irrigation in order to farm the land. Irrigation required a highly organized governmental structure to mobilize and direct the efforts of the workers. At the same time, a lack of stone, wood, metals, and other commodities led the people of the Southern Valley to develop industry in order to produce goods that could be traded for the materials needed and to develop an extensive trade with the outside world. The Southern Valley's more complex economy, coupled with its more productive agriculture, induced a more sizable increase in population and brought the rise of large villages, even cities.

About 4,000 – 1,800 BC was the Age of Aquarius. The was the time of the Sumerians who organized a system of flood control and a pattern of irrigation and created an enduring writing system (cuneiform), religious literature, architectural form, and economic organization. The Egyptian pyramids were built.

About 1,800 BC – 300 AD was the Age of Pisces. This is when the religious traditions of Hinduism, Buddhism, and Judaism were developed as spiritual paths. The Babylonian, Egyptian and Greek Philosophers of this era were responsible for inventing astrology. Jesus, who was known as the fish, became the symbol for this age and his execution by the Romans indicated that the Arian Age was coming.

At about 300 AD the Age of Aries began. The Roman Empire attained dominance and, after its demise, large portions of humanity entered into what is known as the Dark Ages. This time was marked by violence, ignorance, and suffering. To this day, we are still struggling with our own selfishness and barbarism. The religions of Christianity and Islam were developed in this age as salvations from sin. (According to Vedic astrology, the Dark Ages (Kali Yuga) lasted from 700 BC to 1700 AD.)

At about 2,400 AD the Age of Taurus will begin. This age will hopefully be marked by peace, beauty, and pleasure gained through abundance in the physical universe.

4. Planetary Life Paths

There are two primary life paths: the outer, masculine path and the inner, feminine path.

Outer Path

For the person focused on achieving personal goals in the outer world, the conscious mind is represented by the masculine planets: Sun, Mars, Saturn, and Uranus. The Sun represents his self-identity (ego), Mars his assertive expression (persona), Saturn his worldly authority, and Uranus his unique purpose.

The feminine planets, the Moon, Venus, Jupiter, and Neptune represent his private life and unconscious mind. The externally focused person represses his emotions and nurturing qualities (Moon), while his ideal partner is submissive and feminine (Venus). His home and nightlife are experienced as places of freedom (Jupiter), while his spirituality (Neptune) is viewed with skepticism and ignored.

Inner Path

For the inward relationally focused person, the entire scheme is reversed. Her individuality (Sun) is suppressed in order to support others. Her identity revolves around the Moon, being nurturing and kind. She embraces her emotions and develops sensitivity. Her persona is attractive and loving (Venus), while her partner is masculine and dominant (Mars).

Her authority (Saturn) is asserted with the family at home, while she has freedom (Jupiter) from responsibilities in the outer world. Her higher goal is to achieve oneness with life (Neptune) and disavows her individual uniqueness (Uranus). The inner spiritual world is where she truly shines. Her strength is within, while his strength is without.

Sex Roles

These have been the traditional roles of men and women. As our sex roles loosen, more men follow the inward path and more women follow the outward path. Partnership roles then reverse within relationships. The Sun and Mars of one partner will match the Moon and Venus of the other. As we evolve, we become less attached to sex roles and more free to choose our own path, blending qualities from both paths into the same person.

(In the course of life, a person may reverse poles during a process of transformation. An outwardly focused person can become inward and an inward person can become outward. In this case, a person's whole personality may appear to change, causing their career and relationships to undergo major upheaval.)

*Saturn: Yin-Masculine (consolidating power). Jupiter: Yang-Feminine (receptive growth).

5. Cognitive Functions

In his studies, Carl Jung realized that every person favors one of four cognitive functions of human consciousness: sensing, feeling, thinking, or intuition. Sensing and intuition are functions of pure perception. Sensing information comes from without, and intuition comes from within. The judging processes, thought and feeling, are used to interpret our perceptions. Thoughts are conceptual interpretations of perceptions, while feelings are value judgments.

Each function can be focused towards the outer world or the inner world. People focused on the outer world are termed extroverted, while people focused on the inner world are termed introverted. Both worlds contain objects. Inner objects are ideas, which we have personal feelings towards. A strong attitude towards an idea will form a complex, such as inferiority, and be projected as outer objects in our world (people and situations).

Feeling

Feelers are the most empathetic and connected with their hearts. Family and relationships are of great value to them. Their judging functions are strong, while their perceiving functions are repressed - making them somewhat moralistic and structured.

Extroverted Feeling: Sociability
A person who favors extroverted feeling seeks to create positive feelings in others in order to have abundant personal relations and inspire people. They enjoy community gatherings and social services (actors, teachers, service, diplomats).

Introverted Feeling: Intimacy
A person who favors introverted feeling desires to understand the inner lives of themselves and others in order to support people. They enjoy being with family and close friends in a private setting (home-makers, therapists, nurses, culinary).

Thinking

Thinkers are the most logical and realistic people. They reliably shoulder responsibilities to gain lasting results. Their judging functions are strong, while their perceiving functions are repressed - making them somewhat conservative and materially focused.

Extroverted Thinking: Organization
A person who favors extroverted thinking uses tactical planning to achieve tangible results. These are ambitious people who focus on organizing the worldly activities of people (business people, lawyers, politicians, military leaders).

Introverted Thinking: Logical Reasoning
A person who favors introverted thinking seeks to increase their knowledge of the world. These are the most logical people who focus on devising and improving systems of thought and technology (academics, scientists, technology, engineers).

Sensing

Sensing types are the most in touch with their bodies. They are athletic, sensual people who appreciate the physical world. Their perceiving functions are strong, while their judging functions are repressed - making them freedom-loving and impulsive.

Extroverted Sensing: Physicality
A person who favors extroverted sensing seeks direct experience with the senses. They are the most physical people who focus on their skill and ability to interact with the outer world (athletes, builders/fixers, soldiers).

Introverted Sensing: Sensuality
A person who favors introverted sensing seeks the subjective experience of enjoyment that the senses provide. They are skilled, creative, peace-loving people who connect deeply with nature (artisans, musicians, craftsmen).

Intuition

The intuitives are the most abstract, inventive people who often seem somewhat eccentric to the rest of us. Their perceiving functions are strong, while their judging functions are weak - making them impractical and often unreliable.

Extroverted Intuition: Innovation
A person who favors extroverted intuition seeks to apply their unique insights into fields of knowledge. They are the most open-minded people who seek to discover new ways of being (philosophers, explorers, inventors).

Introverted Intuition: Introspection
A person who favors introverted intuition observes the patterns of change as people and circumstances flow through time. They are the most reflective people, who dream of what is possible through connecting with and understanding the inner nature of reality (artists, mystics, writers).

6. Temperaments

Human beings express as four distinct types of personalities which reflect the Earth's four seasons and daily phases. They are 1) Introverted Perceivers, 2) Extroverted Perceivers, 3) Extroverted Judgers, and 4) Introverted Judgers.

Perceivers (P) represent the portion of the Earth's annual and daily cycles when sunlight is increasing and new ideas are forming. Judgers (J) represent the portion of the Earth's cycles when sunlight is fading and results are reaped. Extroverts (E) represent the portion of each cycle when the Sun is brightest. Introverts (I) represent the portion of each cycle when the Sun is dimmest.

1. Introverted Perceivers are reflections of Winter and sleep-time, when life is at rest. In stillness, they are deeply meditative, conscientious, flexible and relaxed, yet are withdrawn and lack initiative. (ISFP, ISTP, INTP, INFP) (Phlegmatic)

2. Extroverted Perceivers are reflections of Spring and morning, when life reawakens. With a burst of energy, they express tremendous enthusiasm for new experiences. They are energetic, optimistic, open-minded and entertaining, yet impatient.
(ESFP, ESTP, ENTP, ENFP) (Sanguine)

3. Extroverted Judgers are reflections of the Summer and afternoon, when the Sun is brightest. With confidence and determination, they seek social importance and public achievements. They are proactive and purposeful, yet stressed and demanding.
(ESFJ, ESTJ, ENTJ, ENFJ) (Choleric)

4. Introverted Judgers are reflections of the Fall and evening, when nature's cycle is concluded. These mature people seek to influence others with their knowledge and wisdom. They are stable, calm, and trustworthy, yet are passive-aggressive and controlling.
(ISFJ, ISTJ, INTJ, INFJ) (Melancholic)

Perceiving is receiving information. Judging is interpreting information. All people do both, but we are each focused on one more than the other. Perceivers are mainly concerned with experiencing life, whereas Judgers are concerned with influencing life.

Sensing (S) dominant people live in the present, mainly perceiving concrete physical reality. Intuition dominant people live in their minds, mainly perceiving inner realities. Sensors are concerned with the experience of phenomena, whereas Intuitives (N) are concerned with the meaning of phenomena.

Thinking (T) dominant people make judgments based on information and logic. Feeling (F) dominant people make judgments based on personal tastes and values. Thinkers are concerned with understanding the world, whereas Feelers are concerned with relating to the world.

The MBTI creates sixteen types by identifying our secondary function. A Perceiving dominant person will have a Judging secondary function and a Judging dominant person will have a Perceiving secondary function.

As a result, there are four elemental types of people who express in each of the four seasonal temperaments: 1) Sensing Feelers (Fire) - warm and affectionate, 2) Sensing Thinkers (Earth) - practical and realistic, 3) Intuitive Thinkers (Air) - logical and inventive, and 4) Intuitive Feelers (Water) - creative and insightful.

Therefore, each type is an expression of both a season and an element. For instance, an Introverted Perceiver who is an Intuition dominant Feeler (INFP) is a reflection of Winter and Water. They are calm, reflective people who are also sensitive, imaginative, and insightful. These qualities combine to form a creative, idealistic person who strives for human upliftment.

7. Types of Perception

At the Sensing level of perception, people are focused on physical survival (food, shelter, money). For this reason, they are matter-of-fact and materialistic. As a result, the most important differentiation between them is whether a person is group oriented or individualistic.

People who want to be part of groups are more conservative and value stability. They disdain people who are individualistic since they threaten the group's well-being, living outside the group's customs and rules.

Nearly 50% of people are Sensing Judgers (SJ). They are the responsible citizens who form the backbone of society. These are the most moral and loyal people. The thinking types are hard-working and fatherly (Sun), while the feeling types are supportive and motherly (Moon).

Sensing Perceivers (SP) are more adventurous. They are the hunters, while the Sensing Judgers are the gatherers. The thinking types are aggressive and risk-taking (Mars), while the feeling types are sensual and pleasure-seeking (Venus).

Intuitive Judgers (NJ) are the rarest types (8%), since judging people are not usually focused on their inner knowing. These are the holders of wisdom and knowledge for society. The thinking types are serious and ambitious (Saturn), while the feeling types are spiritual and uplifting (Jupiter).

Intuitive Perceivers (NP) are more experimental and open-minded. They are the inventors and visionaries. The thinking types are highly intellectual (Uranus), while the feeling types are more imaginative (Neptune).

As our perception becomes more Intuitive, we focus on growth and fulfillment, instead of physical survival. For this reason, Intuitive people are more reflective, imaginative, and creative.

As a result, the most important differentiation between them is not whether you are "one of us" or "one of them", but the type of energy you radiate - whether you are warm and empathetic or cool and intellectual, extroverted or introverted.

It is not as important for them if a person will conform to their beliefs and lifestyle since their survival is not dependent on it. In fact, fulfillment is seen to increase through diversity for them, since it adds so much variety and excitement to life.

8. Enneagram

The Enneagram was introduced by the Russian mystic G.I. Gurdjieff in 1916. It is a symbol with nine points ordered clockwise on a circle. Each point describes a personality style formed in reaction to our early environment. Though Gurdjieff never explained how the enneagram was developed, he described it as being "a living symbol in motion" and appears to be ordered in the pattern of the astrological planets.

By studying the enneagram, we can learn a great deal about the planetary influences in our birth charts. It is a very well developed psychological system that explains many facets of our behavior and motivations.

Type 3: Sun - The Achiever. They are motivated by success, but can be superficial and vain. Their path is to develop humility and authenticity.

Type 2: Moon - The Helper. They are giving and supportive, but can be needy and co-dependent. Their path is to become autonomous and self-reliant.

Type 1: Mercury - The Perfectionist. They seek correctness in all things, but can be critical and invasive. Their path is to become accepting and nonjudgmental.

Type 9: Venus - The Peacemaker. They seek harmony within themselves and their environment, but can be self-indulgent and lazy. Their path is to become confident and assertive.

Type 8: Mars - The Challenger. They seek autonomy and control, but can be bossy and vengeful. Their path is to become peaceful and kind.

Type 7: The Enthusiast - Jupiter. They seek fun and adventure, but can be excessive and crazy. Their path is to become content and reliable.

Type 6: Saturn - The Loyalist. They are protective and community seeking, but can be insecure and fearful. Their path is to become open-minded and trusting.

Type 5: Uranus - The Observer. They seek knowledge through the intellect, but can be avoidant and emotionally disconnected. Their path is to open up and connect with others.

Type 4: Neptune - The Individualist. They seek a unique self-expression, but can be over-sensitive and envious. Their path is to become confident and stable.

The types are organized into triads:
Types 2, 3, and 4 are feeling (heart). Sun, Moon, and Neptune.
Types 1, 8, and 9 are sensing (body). Mercury, Venus, and Mars.
Types 5, 6, and 7 are thinking (head). Uranus, Jupiter, and Saturn.

The types can also be extroverted or introverted:
Types 3, 7, and 8 are extroverted. Sun, Mars, and Jupiter.
Types 4, 5, and 9 are introverted. Uranus, Neptune, and Venus.

9. NUMEROLOGY

There are nine phases of human consciousness, which are depicted by numerology. The odd numbered phases are masculine: individualistic, dynamic, and externally focused. The even numbered phases are feminine: communal, conservative, and internally focused.

The phases are also divided into three groups that reflect the three modes of astrology. Phases 1-3 express Cardinal power, phases 4-6 express Fixed love, and phases 7-9 express Mutable wisdom. The 3rd phase of each mode (3, 6, 9) is its most potent expression.

1. Initiating: Aggressive. Selfish. Isolated. Fearful. Victim mentality. Rugged individualism. Survival of the fittest. Hunter gatherers. Anarchy. Gangs. (Mars)

2. Cooperating: Sharing. Moral. Natural. Respect for elders. Supernatural beliefs. Farming. Bartering. Tribal Communities. Families. (Moon)

3. Self-expressing: Proud. Egocentric. Competitive. Power seeking. Manipulation of nature. Ownership of resources. Monarchies. Feudalism. Slavery. (Sun)

4. Structuring: Rule-bound. Disciplined. Fixed beliefs. Judgmental. Authority of groups over the individual. Democracies. Communist states. Organized religions. Working class develops. (Saturn)

5. Enterprising: Individualistic. Entrepreneurial. Consumeristic. Inventive. Scientific. Freedom from nature. Technology. Capitalism. Middle class develops. (Mercury)

6. Nurturing: Considerate. Healing. Humanitarian. Vegetarian. Pacifist. Environmental. Equal rights. Social welfare. True Democracy. Socialism. (Venus)

7. Inspiring: Open-minded. Creative. Imaginative. Independent. Arts. Music. Psychedelics. Psychology. Yoga. Philosophy. Libertarian. Free market economy. Global village. (Neptune)

8. Awakening: Spiritual. Faithful. Peaceful. Zen. Intuitive. Inner-guided. Visionary. Altruistic. Sharing of resources. Abundance. Planetary citizenship. World peace. Collective economy. (Uranus)

9. Fulfilling: Blissful. Mystical insight. I Am God consciousness. Miraculous abilities. Multi-dimensional existence. Universal citizenship. Kingdom of Heaven on Earth. Infinite resources. Reality Manifestation. (Jupiter)

The digit 0 represents our existence beyond all expression as the Tao: total self-awareness. Our Tao awareness is represented by an even number, which depicts its feminine, allowing nature.

(In the 1,500s & 1,600s, the Europeans, at level 3, came to the Americas and conquered the Native Americans, who were at level 2. The United States started shifting towards level 4 in late 1700s and moved to level 5 in the mid-1900s. There was a burst of level 6 - 8 thought in the 1960s & 70s and the country is now struggling to move to level 6. The most advanced European countries are presently at level 6, which must be achieved by a critical mass of people on Earth in order to bring human life into balance with the planet.)

10. Chakras

We each have three bodies. The physical body, made of matter and energy, forms our appearance and way of interacting with the world. The emotional (astral) body, made of feeling and desire, forms our way of experiencing reality, our tastes and preferences. The mental body, made of thought and memory, forms our way of interpreting reality, our personal identity and will.

When we have sufficiently developed our three personal bodies, the divine self within unites its threefold consciousness with us. When we feel love for all beings, we are connecting with its divine heart. When we receive intuitive knowledge, purpose, and guidance, we are connecting with its divine mind. When we become like the divine, we unite with its spiritual body. Through this spiritual union, we then awaken to our true nature and become self-aware as the divine.

These seven types of consciousness are transmitted to us through the 7 chakras (centers) of our soul. Each chakra corresponds with an area of our body and can be seen (psychically) as a color. As we evolve, our soul - a growing tree of life - raises its kundalini (spiritual light) to open each chakra and make them an active part of our being.

1. The grounding root chakra is red and transmits power manifested on the physical plane of being. At this beginning stage, we establish a strong foundation as our consciousness is focused on developing our physical abilities through the senses, and learning the practical skills needed for survival. (Mars)

2. The libidinal sacral chakra is orange and transmits emotion experienced on the astral plane of being. At this stage, we grow desire and the capacity for pleasure by developing our feeling nature, which serves to enrich personal experiences and relationships. (Venus)

3. The causal solar plexus chakra is yellow and transmits active intelligence from the mental plane of being. At this stage, we seek knowledge and intellectual development in order to understand ourselves and influence our environment. This is the seat of the ego. (Sun)

Together, the first three chakras form the process of cognition: (1) sensing -> (2) feeling -> (3) thinking; as well as the process of creation: (3) thought -> (2) desire -> (1) action.

Once enough self-centered development within the lower self (ego) is accomplished, the individual's consciousness moves upward to the higher chakras and planes of being where it learns to let the higher self direct it in service to the whole.

4. The devotional heart chakra is green and transmits loving kindness and the desire to be of service to life. When this chakra opens, we become clairsentient and develop the

empathic ability to feel people's emotions, which enables us to participate in the healing process within and around us. (Moon)

5. The communicative throat chakra is blue and transmits spiritual and intuitive knowledge. When this chakra opens, we become clairaudient and develop the ability hear the all-knowing inner voice, which gives insight, directs actions, and provides wisdom for others. (Mercury)

6. The visionary pineal chakra is purple and transmits psychic vision to the Third Eye. When this chakra opens, we become clairvoyant and develop the ability to see the inner nature of beings. This spiritual awareness enables us to experience non-physical realities and interact with higher beings. (Jupiter)

When we become selfless and learn how to impersonally use our abilities in service to life, we join the community of light and our consciousness opens to the highest chakra.

7. The brilliant crown chakra is white and transmits God consciousness, the union of all polarities. Nothing can oppose that which has no opposite. When this chakra opens, we become enlightened masters, vessels of truth with radiant halos. We release our identity as individual beings (ego) and become reborn in our true identity as the One Universal Self. (Uranus)

In this state, we become completely tuned in to our core inner self that Is and Knows, and identify as life itself, existing in a perfectly flowing state balanced between the inner reception of inspiration and the outer expression of divine will.

*Just as the physical body has organs and veins, the soul has chakras and channels. There are two main channels that interweave up and down like a DNA strand around the chakras, and a center channel that moves directly up and down (medical serpent staff). The Kabbala's Tree of Life portrays the chakra system where chakras three, five, and six are each split into left and right chakras, creating 10 chakras. Theologians describe the 7 chakras as layers or sheaths that form our consciousness, with which we exist on the seven planes of being.

**Saturn represents the ground of being, the laws of physical reality, which precede the individual chakras. Neptune represents the ocean of being, the dreaming mind that causes manifest existence, which is experienced by individuals beyond the 7th chakra.)

11. Higher Planes

Could it be that the astral, mental, and spiritual planes actually exist?

As a reflection of the Universal Mind, our solar system has four planes of existence: the physical, astral, mental, and causal (spiritual) planes.

The Earth represents the physical plane of existence. The Moon orbiting the Earth represents its etheric body (energy). The Earth represents waking life (conscious) and the Moon represents dream life (subconscious). Just prior to birth, we exist in a dreamlike condition in the womb of our mother, who is represented by the Moon.

It is thought that the astral plane surrounds the physical plane, the mental plane surrounds the astral plane, and the causal plane surrounds the mental plane.

Venus and Mars, which represent the astral forces of love, attachment, and desire, surround the Earth.

Mercury and Jupiter, which represent the mental abilities of logic and abstract reasoning, surround Venus and Mars.

The Sun and Saturn, which represent the spiritual law of cause and consequence, surround Mercury and Jupiter.

The ordering of the planets perfectly describes our journey from the spiritual plane to the Earth and back again:

As we travel from the cause of life (Sun), we acquire a mind (Mercury), a heart (Venus), and a body (Earth). When we die, we experience our remaining desires (Mars), review our lives and gain wisdom (Jupiter), and then reap the consequences of our lives (Saturn).

After we have come to our final judgments about ourselves (Saturn), we enter the pearly gates of heaven (Uranus derives from the Greek word for heaven), and eventually dissolve into Nirvana (Neptune) until we are ready for rebirth (Pluto).

12. The Moon

Since the Moon rules the night sky, it is also the ruler of astrology - the study of planets and stars seen at night. The Moon also rules symbolism, since our dreams at night are symbolic in nature. The signs and planets can be seen as symbols for our many types of consciousness, the material of dreams.

Cancer is the zodiac sign most associated with the Moon and is, therefore, also associated with astrology and symbolism.

It is for this reason that the most important finding for the study group of Astrologers was that Venus is highly likely to be in Cancer. Astrologers love (Venus) astrological symbolism (Cancer).

13. Alchemy of Venus

Venus represents the perfect human.

It is portrayed by Leonardo da Vinci's drawing of the Vitruvian Man, in the shape of a pentagram with perfect proportions.

Venus' 8 year retrograde cycle forms a pentagram, turning backwards on the ecliptic every 18 months, two points over from the prior retrograde.

Many proportions on the sides of a pentagram equal the divine ratio (1.618), the perfect ratio of human proportions.

The ratio of the orbits of the Earth to Venus is the divine ratio (365 days/225 days = 1.62), and Venus has the most perfectly circular orbit of all the planets in our solar system.

The process of achieving human perfection is shown by the sequence of numbers 1-5:

0. The Void, the Tao, the uncreated source of existence - the eternal Self.

1. Out of the void arises the mind of creation, the one Universal Mind - the Father.

2. Out of the universal mind arises the Mother, birth giver of the soul, our higher self and Guardian Angel - the Holy Ghost.

3. Born of an Immaculate Conception (pure idea), implanted within our Mother's womb, is the child, the soul - the Son.

4. The soul embodies a human personality, surrounds itself with the four elements and seasons, and is crucified (transformed) on the cross - the Earth. (Earth's symbol is a sun (son) cross, a circle on a cross.)

5. The personality liberates itself of all selfishness and embraces the will of its Higher Self - the Voice of Intuition. The Mother + Son becomes Christ (2+3=5). The one with whom God is pleased is lifted from the cross and the wheel of reincarnation. (Venus' symbol is a circle above a cross.)

The numbers 6-9 represent the phases of higher existence, and show how the Christ being unites with the Father:

6. Brotherhood: The Christ being joins a blessed community.

7. Knowledge: The laws of reality become known.
8. Mastery: We become endowed with true creative power.

9. Enlightenment: Total self-awareness is realized. "I and the Father are one."

10. Completion: Return to Source - Nirvana, Bliss.

*A diamond is considered the gemstone of Venus. It is created by compressing pure carbon under high-pressure circumstances within the Earth's mantle for billions of years. When the diamonds (souls) are perfected, they are brought to the surface (true reality) by volcanic eruptions (awakenings).

14. Levels of Reality

Every level of reality seems absolutely real to those who exist within them. What seems as fact to those at one level is clearly seen as fiction from a higher level.

Ironically, those on lower levels also view higher levels as fictitious. This must be so since viewing higher realities to be real would cause their own reality to lose legitimacy.

Unless you are at the level where you fully understand, create and control life, there is a higher level of reality from which it is known that your reality is a fiction.

The level just above the majority of humanity is marked by the experience of knowing all beings to be one being, moving in unison with each other's desires. Here, all desires are known to be those of our Self.

15. Taoism

We are non-beings, in a state of being, in the act of doing.

We appear as people in places with things, performing actions. This is doing.

We, as evolving souls, experience specific states of consciousness with their corresponding thoughts and feelings. This is being.

When we release our existence on all planes and enter the light, we become identity-less. In total peace, our consciousness becomes still and self-awareness alone remains. This is non-being, our original state.

16. Infinity Symbol ∞

The infinity symbol represents the cycle of life. We exist in an unending cycle between three prime states of being:

1) God (True Self)
2) An immortal soul living with our soul family
3) A mortal living amongst many soul families

We move from the God state to the Soul state to the Mortal state, then back again and again and again ...

17. God

How could one mind - The God Mind - simultaneously experience individual lives?

It is done in the same way as you simultaneously experiencing individual senses.

The senses represent individual types of outer experiences, which combine to create a multifaceted reality.

In the same way, our individual lives combine to create a multifaceted reality for God.

18. The Soul

Your soul was born from a new idea of being. It has evolved physical, emotional, and mental bodies that allow you to exist on many planes of existence. When we release one body, we become aware of our next densest body. As a result, we inhabit our astral body when our physical body dies.

Our divine Self desires to exist as you. The soul is the mold out of which your life is shaped. It is also the sum total of every thought, feeling, and experience you have ever had. It is experienced as your unique personality and is always evolving.

In order to change our lives, we must change the mold of the soul by adjusting our reactions to all phenomena. This means releasing outworn patterns that prevent us from being fully present. Habits of mind are often subconscious to us, even instinctual, but must be brought to awareness.

Unawareness causes us to project our selves onto others, viewing them as the cause of our circumstances. By doing so, we surrender the power to consciously create our lives. Though satisfying to the ego, viewing oneself as a victim only serves to limit our freedom.

We are the creators of life and will never be helpless victims. Our circumstances are mirror reflections of our own consciousness, just as in a dream. Life improves when we acknowledge our own suffering and release anger.

In truth, we are creating a world of conflict because we are in conflict with ourselves. Deep down inside, we believe that we are sinful and deserve to suffer. So we give our distrusted powers away to a world that punishes us.

Only by forgiving ourselves and embracing our divinity can we accept the creative power of our soul to influence external reality. As our self-trust increases, we relax and the vibration of our soul rises. And by this, we come to inhabit more harmonious and supportive worlds of being.

19. Starseeds

The universe is 13.8 billion years old. The oldest known solar system is 11.2 billion years old. Our solar system is 4.6 billion years old. This means that the oldest of a trillion trillion solar systems is at least 6.6 billion years older than ours.

Given that the basic constituents of the universe are present in all parts, consciousness must have developed into self-aware beings in many of these solar systems long before us. It is barely imaginable how advanced they must be, making it an almost certainty that they are aware of us.

Are extra-terrestrial beings involving themselves in our world?

It appears that they are. This would explain why our scientific knowledge and technological abilities have increased so dramatically over the last 150 years. Telepathic communication could easily explain the sudden hunches many discoverers and inventors experience.

It also appears that they are incarnating amongst us in order to initiate and assist progressive movements within humanity. For it would be impossible for higher knowledge to be brought to us without human beings who could understand and utilize it. These people are frequently referred to as Indigo souls or starseeds.

These evolved beings care deeply about the rest of us, their younger siblings, and actively involve themselves in our lives. They never forget their oneness with us and have compassion for our struggles, due to remembrance of their own experiences in prior states of existence.

Indigo souls have been described as a type of light worker that is especially willful, independent, and perceptive. They are determined to raise human consciousness and make the world a better place, one not ruled by selfishness, through their abilities in artistry, science, technology, economics, psychology, and metaphysics.

Indigo souls are symbolized by the planet Uranus, the agent of awakening which acts decisively and suddenly, and many of them have this planet in a prominent position in their birth charts. Uranus had just risen above the Eastern horizon when I was born.

When I was young, long before I knew much of anything about spirituality, I had a vision where my entire sight was flooded with the color indigo. My thoughts told me that this was the color of my soul. I was also told that I, and many others, would some day know the truth and help bring a new age of peace to the Earth.

For this reason, in spite of our many challenges, I have always been totally convinced that humanity would rise up and transform this planet into a realm of brotherly love and true knowledge. When the collective soul of humanity is ready, this change could happen within a generation. We may call them the generation of light.

Imagine a world where children learn how to create with their minds, and people can teleport to any world in the known universe at will, where material needs are readily available, and people are thrilled to be with each other. What an adventure life would be!

20. Suffering

What is the cause of suffering?

There are two types of suffering: individual and universal.

Individual suffering is caused by ignorance of our one true Self and the resulting loss of goodness. Imagining ourselves to be separate and unsupported causes great fear, and makes the degeneration of our morality seem reasonable. Remembrance of our divine Self leads us to respond to life with trust and neutrality in all circumstances. As we grow in awareness, our hearts open and our lives improve.

Many mystics believe that there is no suffering outside of the unenlightened individual, since in God there must be total perfection.* But, if suffering does not exist in All That Is (God), how could it be experienced at all?

There are four main explanations for this:

1) Living amongst evil is a test of a soul's worthiness for heaven. But this implies that God has imperfect knowledge of us, needing to figure out who we are.

2) Suffering only exists in order to evolve us. But this implies that many parts of God need improvement and that God is not all-powerful since he cannot instantly perfect us.

3) Evil is a necessary polarity to goodness. But this implies that God is incapable of creating heavenly experiences without torturing other parts of itself.

4) Evil was created out of mere curiosity. But this implies that God is grossly insensitive to us and lacking in love.

None of these explanations for suffering indicate that God is perfect. Therefore, it appears that suffering is a condition within God caused by the primordial circumstances of existence.

Consider this scenario:

Before creation, You existed in silent darkness, alone, without any experience except your own imagination. You could not sense anything outside yourself and there was no one else to teach You about yourself.

As a result, You experienced a deep sense of isolation as well as bewilderment about what You are, how You came to exist, and whether You could cease to exist. Your feelings of loneliness, confusion, and inability to change your condition were a cause of great suffering.

In your solitude, the idea occurred to You that through experiencing yourself Being, You could express your love and pain through creativity. Thereby, You would know great joy and freedom, gain understand of yourself, and overcome suffering.

To do this, You learned how to surround yourself in sensual atmospheric conditions, then how to narrow and confine your consciousness into separate selves (cells, souls), while remaining fully self-aware. You gave each self a unique combination of your qualities, making them each perfectly suited for expressing your many ideas, in worlds of shared sense conditions and memories.

It became the purpose of each self to help heal You by healing itself, to expand You by expanding itself. This is why being human is so precious to You. As each self overcomes its personal struggles, it remembers itself as You and becomes a conscious participant in your plan for universal salvation, a new stage of existence where suffering only exists as a memory.

*Only when you are willing to examine all possibilities can you discover the basic facts of existence. Until then, spirituality will just be another form of escapism for you. The need to see God as perfect is a security blanket, assuaging your fears, but preventing you from moving forward.

21. THE LONELY SWAN (Brhad-aranyaka Upanisad)

"The original being, atman or self looks around and sees nothing else but himself. When he realizes his loneliness, he has two feelings, one of fear and the other of a desire for companionship. His fear is dispelled when he realizes that there is nothing else of which he has to be afraid. His desire for companionship is satisfied by his dividing himself into two parts which are called husband and wife."

22. Universal Healing

Once you realize that everything you are and do is done deliberately by God, you realize that your ignorance and suffering are deliberate. But why would God deliberately cause itself to suffer? If God is perfect and has no needs, what could God gain from it? Clearly, God is seeking to gain something through our suffering and must, therefore, be itself evolving.

If the mere experience of good and evil is being sought and have equal value to God, then we would feel an equal pull in both directions. But this is not the case. The majority of beings undeniably feel pulled towards goodness. We are seeking healing and this must eventually lead to the dissolution of evil. This indicates that God is, indeed, seeking to eliminate suffering from its own existence.

While it is true that good and evil have always coexisted, evil is by no means desirable or necessary. To say that you cannot feel good without sometimes feeling bad is untrue. It has been our nature to have both positive and negative emotions, but must this always be the case? All spiritual paths assert that negativity and suffering can be overcome and I agree. Harmony can exist without chaos.

23. Christ

The nature of the world crisis is essentially that we are not living in tune with our inner Self, our spiritual counterpart. We must seek union with our Christ Self, our higher self, in order to function as a harmonious whole.

As humanity remains convinced of the supremacy of our individual ego selves, we are under the sway of the intellect. Vulnerable to all manner of reasoning, we seek guidance from without and lose our autonomy, causing chaos.

In order to become a master, we need to accept that the real master is beyond our conscious minds. We must become obedient to our hearts, mindful of our conscience, and responsive to our inner will.

Your divine partner is ever ready to serve you in every moment. Seek its counsel and your true path will unfold.

"The stone (higher self) is under thee, as to obedience; above thee, as to dominion; therefore from thee, as to knowledge; about thee, as to equals.
Though art its ore ... and it is extracted from thee ... and it remains with thee"

-Rosinys ad Sarratantam, 3rd Century AD

24. Easter

Why was Easter, the celebration of Christ's rebirth, given its name?
Because the Sun rises in the East. The Sun, which represents the Christ (Son of God), is reborn each day at dawn.

When does Easter occur?
On the first Sun-day, after the first Full Moon in Spring. The Full Moon represents fulfillment from reflecting Christ's light.

Why is Easter celebrated in the Spring?
Because the Spring is to the year as dawn is to the day. At the Spring Equinox, the Sun's light gains prominence over darkness and life flourishes.

Interestingly, the study group of parents giving birth to 4 or more children (500 people) was dominant in the first two signs of Spring: Aries and Taurus.

(The name Easter came from Oestra, a goddess of Spring. The root of the word oestra is "dawn" or "shine". Thus, the connection between the Spring, dawn, the East, and Easter.)

25. The Heart

Life often brings deep pain to us.

To escape this pain, we disconnect from our hearts. But this leaves us unable to feel guidance from within. As a result, we become overly mental and lost in our heads. We hopelessly try to solve problems with our reasoning abilities, becoming slaves of the intellect. Unsuccessful, we end up avoiding our problems through escapism.

Without inner leadership, we become insecure, fearful, and cold-hearted. In our state of aloneness, we often become delusional. No longer clearly aware of actual reality, our outer lives become chaotic.

The only way back is to face the pain in our hearts and reconnect with our feelings. This takes incredible courage, but must be done. Cry like the rain.

Healed from our grief, we embrace life experiences and feel purpose in each moment. We are now able to connect with others with true authenticity. Being able once again to share our hearts, true bonds are created and life is once again trusted.

26. The World

Your intentions make all the difference in the world. For, we exist in a field of consciousness where intention is everything.

So, do not concern yourself if you do not appear to have a visible effect on the world or have any power to effect change in your government. You cast your vote every day with your vibrations. When a critical mass of beings desire the common good and are not fooled by the idea that we cannot have or do not deserve a happy world, we will have it.

27. Good and Evil

Why is there so much darkness on Earth?

The answer is very simple ... we want it this way.

Have you ever taken pleasure in the thought of someone being harmed or violated? If so, you have darkness in you and are actively co-creating the problems on Earth.

It is common for people to view another person with animosity.* It gives us a dark pleasure, a false sense of power to harm that person, which presumably will improve your life. Only, your life is determined by your attitude.

We all have some degree of darkness in us. The question is: What direction are you heading in? In order to know, it may be helpful to describe the differences between beings predominant in darkness and light.

Dark beings focus on immediate pleasures and blame others for their problems.
Light beings focus on personal development and take full responsibility for their lives.

Dark beings disguise their desires, fearfully manipulating life towards their goals.
Light beings are honest and intimate, faithfully participating with life to achieve their goals.

Dark beings feel superior to others and only consider their own needs.
Light beings realize their equality with people and live for the benefit of all.

Dark beings focus on manifestations (materialism) and do not notice the Tao.
Light beings focus on our essence (spirituality) and observe the Tao everywhere.

Dark beings think our nature is wicked and seek separateness ... causing chaos.
Light beings know our nature is divine and seek unity ... causing harmony.

We blame a lot of our problems on the darkest beings amongst us. But, if we were to root out the evil within ourselves, we would no longer participate with them and their influence on the world would evaporate.

Only then would we manifest a planetary reality of peace and abundance ... a world awakened to the realization that we are creating exactly what we want to experience and always have been.

*You will become unable to maintain hatred towards others when you realize that every person is another aspect of you, taking on a different perspective sought by our total Self.

28. Angels and Devils

Our world, planet Earth, is influenced by many beings, on many planes of existence, who have extremely divergent interests in us. Each one of us is influenced by those beings who most resonate with our nature.

Those of us with loving natures are supported by loving beings who provide us with evidence, throughout our lives, that reveals our eternal connection to the higher spheres. Faith and trust are reinforced by our experiences.

Those of us with jealous natures are influenced by malevolent beings who only show us evidence of our mortality and weakness. Fear and selfishness are reinforced by our experiences.

Most souls in our world are influenced by both types of beings, who are ever striving to gain our allegiance and companionship in eternity.

Choose your intentions wisely!

29. The Narcissist

Narcissists are fallen angels who feel deeply rejected by life. To hide this wound they use their talents to gain status. But, by becoming the image they seek, they lose their authentic selves... their truth.

They become dependent on their image and fear that they are worthless underneath it. The moment praise is withdrawn or they are criticized, they become devastated and retaliate with extreme aggression, then withdraw into a temporary despair.

Since their entire personality is an image, all behaviors inconsistent with their self-image are denied awareness. When their false behavior is pointed out to them by others, they believe that the observer is either jealous or troubled (a projection). Any person who penetrates their veil is pushed out of their lives.

Authentic relating and genuine loving is, therefore, impossible for a narcissist. Instead, they engage in unequal relationships based on deception and selfishness. They insist upon being treated like royalty, yet take pleasure in demeaning others.

30. The Ego

The ego is your temporary identity as a unique personality in existence. This identity is composed of your public persona as well as your attitudes, beliefs, and desires.

Your ego identity creates the boundaries of your conscious awareness. The ego functions as an egg, a cocoon. It encloses and insulates you from the greater reality, from which you are not yet prepared to exist, just like a chick before it hatches.

Through a lengthy existence within ego consciousness, you have become identified with your personal self. This is the cause of egocentrism, which gives the ego its association with pride and arrogance. The notion that the personal self is your true Self leads to the conclusion that there is nothing greater than the human personality, that you as an individual are your god.

The ego-identified self has false knowledge and is, consequently, highly insecure and fearful. The greatest challenge to the ego-identified person is to admit that you do not understand yourself or your purpose and that there is something greater than you. This admission is the beginning of humility and the journey to truth.

Humility is a most essential virtue for an awakening soul, since it is the medicine for egocentrism. Perfect humility indicates to the universe that a soul is mature and ripe, thus, allowing the mind to dissolve ego beliefs and open the door to all true things:

Wisdom: Personal fictions dissipate from your mind. You are not a person in a world, you are the world experienced as a person. All reality is your own consciousness. The vastness of the universe shows your enormity, not your insignificance.

Love: An egoless person knows its total equality and unity with all existence. Every living thing is an aspect of your Self. Every person is a unique version of You. To know thy Self is to love and serve all selves.

Power: Your deepest desire is to have life's will continuously flow through you, to be a vehicle for the pure expression of God. Your purpose for living becomes to expand and improve life for all ... to achieve the harmonious synchronicity of all beings.

(The narcissistic ego being is completely consumed with itself. It considers its own life to be most important. It professes there to be no god and no higher purpose in order to consolidate its own sense of power and justify its actions. It often deceives itself by creating a positive self-image of success. However, it cannot escape the internal and external conflicts caused by its feelings of isolation and insecurity.)

"The greatest among you shall be your servant. For whoever exalts himself will be humbled, and whoever humbles himself will be exalted." Matthew 23:12

31. Maturity

Physical maturity is when we are able to take care of the practical needs of our body and do not need others to support us.

Emotional maturity is when we are able to feel contented with or without others. We do not need people to help us feel good and become codependent, or blame others for our unhappiness.

Psychological maturity is when we accept responsibility for our own state of mind. We do not rely on others for our values and beliefs.

Spiritual maturity is when we accept our oneness with life and our responsibility to each other. We do not think everything revolves around us and disregard others.

Humanity as a whole is often said to be in an evolutionary state comparable to childhood. Like children, we deliberately hurt and steal from each other and cause chaos in our environment. The psychological community refers to these behaviors as narcissistic and incurable, but it is simply spiritual immaturity.

Our behavior is also like that of a rebellious teenager who is passionate and angry. This explains our obsession with sex, drugs, and violence. The world's religions consider this behavior sinful, but it is just emotional immaturity.

In truth, immaturity is a natural part of evolution. The universe is patiently providing us rebels with space to be free, knowing that we will some day become conscientious adults.

32. Right Relations

The better your character and intentions, the more helpful and trustworthy your companions will be. Many people are not capable of having true friendships or deep intimacy, since they are needy and controlling in some way. Neediness feels like a prison to others. Recognize that your provider is within and you will come to rely on others less. The less you need people, the more they want you.

Focus on yourself and life will bring the right people to you. Be reliant on no one for your well-being. As you awaken, you become increasingly aware that you are the sole creator of your experiences and only need modify your own desires to change your existence.

In order to create thriving communities, we must learn to appreciate others without trying to control them. Do not try to change another's way of thinking or sacrifice yourself to benefit another. This will only bring you down. The best way to be of service is to raise the purity of your own thoughts and desires and let that be a light to those around you. Just be happy to please and support each other.

Being autonomous also causes us to unlink from the mass hypnosis of humanity, since we become less concerned with being liked and fitting in, thus, enabling us to see more clearly and find truth. Autonomy is, therefore, not only the key to better relationships, but also the path towards higher states of awareness and existence.

Autonomous beings exist in states of joy and peace. Their relationships are filled with mutual attraction and admiration. The moment excitement fades, the relationship diminishes, though the bond never fades. There is no bitterness in this, since all feelings are known and accepted.

Autonomous beings value their freedom greatly and take total responsibility for their lives. All higher beings are autonomous, since they independently manifest their preferred experiences. This is because their desires, being centered in their hearts, are the direct will of the Lord and not caused by habitual attachments or outer influences.

33. Scientism

The transition from Christianity to Atheism has been considered, by the Western mind, to be a triumph of reason over superstition and dogma.

Christianity had bound people in blind obedience, feelings of sinfulness, and fear of eternal damnation. But it did remind us of our connection to a higher order of life and promote brotherhood and compassion.

Atheism stripped away our feelings of guilt and fear of God, but has also eliminated much of our sense of community and having purpose in life. Our blind obedience to authority has also merely shifted from the priesthood to professors.

Atheism simultaneously makes people feel insignificant and become egocentric. Our perceived lack of meaning in life and smallness in the universe causes us to feel inconsequential and the imagined lack of a higher intelligence causes us to exaggerate our own importance and superiority.

Atheism is a belief system that is representing itself as a consequence of scientific knowledge. However, science has no explanation for the cause of existence and has not yet even figured out what consciousness is. In truth, most great men of science, such as Tesla, Einstein, and Newton, were deeply spiritual people.

Atheism unscientifically predetermines what it thinks and rejects all evidence to the contrary, thus, creating another form of mental slavery. It has such a hold on the Western mind, that most educated people actually consider it a sign of intelligence to have no belief in a greater intelligence at all.

But what makes more sense:

1) The universe created itself out of nothing 13.8 billion years ago by a random accident. It accidentally exploded outwards at such a perfect speed that the universe would neither implode (from gravity) or speed out of control, with a 1 in a trillion, trillion, trillion, trillion probability of success. Dead, mindless matter organized itself into patterns which eventually formed living organisms that perform meaningful tasks. These self-less life forms somehow became selfish and competitive for survival.

In spite of this, they formed elaborate cooperative ecosystems. Humans evolved out of animals by developing self-awareness with thoughts and feelings, which arose out of nowhere. Human beings are the highest form of intelligence, on the only planet with intelligent life, in all the billions of galaxies. Humans are masters of the universe and everything exists to serve them. Life has no inherent meaning or purpose. When our bodies die, consciousness ends.

or

2) The universe is the conscious expression of a loving, intelligent being. Its parts work in harmony and order because they are attributes of one idea. The cosmos is filled with life on many planes of existence. Higher beings support the progress of lower beings. All beings are aspects of one being, each with essential value to the whole. Life forms manifest to perfectly express our evolving consciousness. Love is the primary motivation. Life is unending.

Without you, directing the body from within, the body would be without purpose and dissipate. The body cannot create its master, you, for then it would be the master. In the same way, there must be a you within the universal body, directing the universe from within, without which the universe would dissipate.

"I assert that the cosmic religious experience is the strongest and the noblest driving force behind scientific research."
-Albert Einstein

34. Astrological Causation

When people ask "How do planets cause astrological phenomena?", I respond that planets do not cause astrological phenomena. The idea that matter is unconscious, yet is the cause and director of consciousness is magical thinking.

With $E=mc^2$, Albert Einstein proved that matter is slowed-down energy.

Quantum physics has also proven that particles (matter) and light waves (energy) are two states of the same thing and that our conscious attention literally causes waves to become particles.

In addition, many people under hypnosis have been made to see right through solid objects. With the hypnotic suggestion that the solid object does not exist, the object does not exist to the person. This is because our minds cause all things to exist.

The planets are mind forms that are caused by our Universal Mind to reflect our many types of consciousness and individual states of being. This is how astrology works.

35. Flat Earth

Is the Earth flat or round?

In our personal experience, the Earth is flat. The ground is a stationary plane.

In our collective experience, the Earth is a spinning globe. People on opposite sides have day and night reversed.

Seen with one eye, the Earth is a two dimensional picture. Seen with two eyes, it is a 3D hologram.

The flat perspective is singular and self-centered, while the spherical perspective is of the multitude.

Both realities coexist together. There is no single objective truth. All life is substance of the mind.

36. Vibrational Waves

Why does the universe manifest from vibrational waves?

Vibrational waves represent thoughts projecting outward from the center of our universal Self.

Picture a waveform: the bottom of the wave represents the innermost state of a thought and the top of the wave represents its outermost state as observable existence. The length of the wave, which is proportional to its distance from the Source, determines the amount of time between thought and expression.

Our individual minds receive thought waves and perceive them as physical reality. The electromagnetic spectrum represents the full range of possible thought manifestations that can exist. Our senses only perceive a minute fraction of these waves of light.

As we evolve and come to exist in higher states of being, we perceive a greater range of frequencies. At the highest state, we are be able to perceive all frequencies.

As we expand our conscious awareness, manifestations also become more observably reactive to us. The physical plane, which is furthest from our Center, is considered to be vibrating at the slowest rate because thoughts manifest so slowly. This makes it almost imperceptible that reality conforms to us, causing us to feel like prisoners within our circumstances.

In higher planes (states) of being, outer conditions respond to each individual's consciousness at increasing speed, thus, shortening the thought wave and raising its frequency. The more rapidly conditions change to reflect the individual, the more obvious is the relation between the two and the more empowered we become.

In the highest, innermost planes, conditions instantly reflect thought … ideas are realities. As a result, time ceases to exist since the waveform collapses, as there is no delay between inner thought and outer expression.

(The reason that we do not consciously encounter higher beings is simply that our realities cannot coexist. Every level of reality seems absolutely real to those who exist within them. Meeting beings who are beyond our laws of nature would cause the disillusionment of our reality.)

37. Light

Over the past century, quantum physicists have discovered properties of light that are revolutionizing our view of reality. Traditional (Newtonian) physics assumed that the universe has an objective existence, independent of our subjective consciousness. However, recent discoveries have called this assumption into question, leading to conclusions that are remarkably consistent with ancient spiritual teachings.

1. Non-locality

Photons can be split into pairs spinning away from each other in opposite directions. When one member of the pair changes direction, the other shifts its direction in an equivalent manner, even at great distances. This is called quantum entanglement.

Interestingly, photons respond to each other at a speed far greater than light, even though nothing can move faster than light. Therefore, a nonphysical form of communication must be connecting the photons, such as thought.

Everything Is Connected Beyond Time And Space!

2. Wave-Particle Duality

While studying photoelectrons, Einstein discovered that light waves become particles when in contact with each other. It was later discovered, in the 'double-slit' experiment, that photons of light appear as waves when not being observed and as particles when being observed. This was also found to be true for atoms and molecules, leading to the realization that all physical reality manifests to us in response to our focused attention.

Consciousness Causes All Things To Exist!

3. Speed of Light

It was discovered that light is always measured at the same speed, no matter how fast one is moving. Ordinarily, objects appear faster when moving towards them. This was a shocking discovery that perplexed physicists and led to Einstein's theory of relativity. His hypothesis was that time measuring devices must compensate for their own movement in such a way that they measure exactly the same speed of light. This led to the idea that a person in a fast moving object would experience time more slowly and that a person moving at the speed of light would experience no time at all. However, this has never been proven.

The more probable reason that light is always measured to move at the same speed is that light waves emanate from the mind, forming the matrix of space.

For explanation: If you are moving in a car (mind), the only things that move at the same speed, no matter how fast or in what direction you go, is the car itself and anything in the car (light). Time (moving light) is not experienced as slowing down for the driver, no matter how fast, and would continue for him or her at the same rate as anyone else on Earth. Thus, there is no such thing as time travel, since time is experienced at the same rate to all individuals everywhere in the universe.

There Is Only The Eternal Present!

(Light waves occupy all points of space, but are only seen when willed into manifestation by consciousness. Sunlight is a projection of the Self, experienced as a projection of the Sun. The subjective origin of light also explains why the universe appears to be expanding away from us in all directions. Light begins with us and dissolves into the black holes of distant projected space.)

38. THE FUTURE

Is the future certain?

If so, we would be powerless in the present ... mere spectators of life. Yet, only the present actually exists.

If the future doesn't exist, how can it be predicted let alone traveled to?

What makes the future predictable is that we are predictable. If we do not change, why would anything change for us? How you think, feel, and act directly causes your life to be the way it is. This includes the larger world around you.

The future is also drawn towards you by Your higher idea for yourself and the Great Idea for humanity and all beings everywhere. This idea is of our evolution towards fulfillment through unity. A Grand Universal Orchestra!

Until we recognize the purpose of our life's challenges, we resist change with all our might and seek comfort through predictability.

The critical shift in your existence occurs when you recognize that there is a pattern to your challenges. It causes you to immediately realize that life is intentionally evolving you, that all of life has purpose, and that you are safe. Now, instead of hiding from your struggles, you actively seek to learn from them and grow.

Eventually, you will get to the point where you no longer need struggle to learn and become so in tune with life's currents of feeling that you could not possibly create negative circumstances for yourself. Thereafter, you will learn solely in states of inspiration and joy.

39. Spiritual Path

There is only one thing God is incapable of doing ... dying.

Live your life with the knowledge that you will always be alive. You are never going to die. When you leave your physical body, a new one replaces it in your next reality. In the eternal life, you will be surrounded by others of similar character to you, in an environment reflective of your nature.

Do you want to exist like you do now, forever?

Nothing is going to change the quality of your life but you. You should immediately seek to improve yourself in any way possible. You have a tremendous destiny on the path of improvement. Why would you wait?

There are three main ways to improve yourself and attain better states of existence:

1. Inspiration - Seek higher knowledge in any manner possible, whether through teachings, spiritual practices, or sacred plant medicines.

2. Aspiration - The greatest weakness is to admit no weakness. Practice brutal self-honesty to improve your character. Seek counseling and other forms of therapy.

3. Perspiration - Apply your increasing awareness to the situations in life that challenge you most. Seek to be helpful and reliable to others, whether it be in your work, family, or casual encounters.

Most real change happens slowly. Never assume you are done with a problem. You may relapse many times before it is resolved. This world will give you every chance to fail if you allow it.

Do not harbor anger towards those people in our world who seek to corrupt and use us. In doing so, they only doom themselves to living on the lowest planes of existence. They behave in a low manner because it is all they know ... forever going in and out of hellish realities, where they only have each other to abuse. It is for this reason that they so greedily seek to possess the beautiful things of Earth.

They often have no idea that higher states of existence are even possible for them, do not feel brotherly love, and attempt to marginalize those of us who do. Do not let them drag you down to their realms. Instead, practice forgiveness towards them. The suffering they cause us is only temporary, so long as we don't become like them. We should feel concerned for the sad condition of their existence and seek to be an example to them. Humanity only evolves as all beings living on its many planes evolve.

40. Buddhic Consciousness

As long as you persist in seeing other beings as separate from you, they will behave as though they are. This is a direct reaction to your viewpoint.

In truth, they could not react to you if they were separate, since a reaction requires a connection. We all must be parts of each other.

In Buddhic consciousness*, it dawns on the individual that it and everything else are aspects of one Self. This is a palpable feeling supported by a constant stream of intuitive knowings. Unselfishness and compassion develop quite naturally in this condition.

Many people think Buddhic consciousness is a myth. But from the buddhic viewpoint, you are a myth. Your personality and life are total inventions ... made for the benefit of your real self.

In the ego state, we do not know who we are and attempt to compensate by feigning confidence and seeking control, even convincing ourselves that we are the most advanced form of life in existence.

In current times, our behavior is increasingly becoming like that of sophisticated robots, pressured by economic and social forces to ignore our real feelings. We know that we are unique, but repress our individuality in order to become absolutely predictable in our public lives.

It is only at the Buddhic level that we are able to resist the pressures of the outer world, since we spontaneously know the right actions for our well-being in every moment of our lives. Our behavior now becomes spontaneous and free, a genuine expression of our inner feelings.

To attain Buddhic consciousness, become self-assured by releasing attachment to the image of your personality. Feel and express true desires without desire, knowing they are life's will for you. Focus on the positive potential of your eternal existence, rather than temporary experiences of limitation. Purify your essential nature by subordinating yourself to the knowing in your heart and become an unselfish force for good.

While it is true that there are those who seek to misinform and control us, we are not victims. We absolutely attract all our experiences and choose what information we believe. It is all a matter of who you allow yourself to become. If you have high expectations for yourself, no outer circumstance can diminish you.

*According to Theosophy, there are 7 levels of consciousness:

The first three are states of ego consciousness:
1) Physical - sense pleasures & survival
2) Astral - emotional desires & attachments
3) Mental - intellect & morality

The next three are states of egoless consciousness:
4) Buddhic - spiritual wisdom
5) Atmic (Nirvanic) - pure love & bliss
6) Anapudakic (Monadic) - creative power

In this way, our three modes of being (power, love, wisdom) are expressed in both the ego and egoless states of mind.

The seventh level of consciousness, called the Adi state, is the Godhead. It is You, beyond form, the one and only master of all reality.

41. INTUITION

The All Seeing "I" is communicating to you through your very thoughts. Its divine mind places thoughts in your head just as easily as putting objects in your vision. Would it be much of a trick for the Inner "I", your own True Self, to order the planets in a such a way that they reflect your thoughts and corresponding life?

All thoughts that are unique to you are divine thoughts coming from inside you. All thoughts that you share with others are from the external world.

Some people rarely hear a thought from inside them, since they are so attached to the external world. This causes them to seem dense and insensitive, unable to respond to life without checking their memory for the acceptable response. They are, thus, limited in their ability to feel love, freedom, or joy.

When you listen to your inner voice and feel excitement about having original thoughts, you are able to respond to life at lightning speed - always being forewarned with precognitive and telepathic thoughts and forearmed with intuitive, instinctual responses.

You become creative, joyful, and independent, beyond the control of the external world. You do not fear ... for your Inner Self is ever present in your mind with its loving guidance. You feel comfortable and trusting, sensitive, kind and loving.

42. Zen

All experience occurs within one's own mind. It is all subjective. There is no possible way of knowing if there is any separate objective reality whatsoever. The fact that individuals share collective experience is not proof of the existence of an objective reality either. It could be that our minds are linked by a collective mind that originates experience. Taoism suggests that tuning into this mind brings harmony and purpose.

I observe experiences occurring, as well as my reactions to them. I know my total experience of life to be in my mind. My mind is my being. Therefore, my mind, being, and the world are all subjective experiences of my true self, which is the observer - the one objective fact. I, the observer, am all I know to be real. All other things change like the wind.

Be in total acceptance of what is, here, now, in the present. Apply this to yourself and your entire world of experience. By resisting what is, we only add to our tension. We are not accepting the greater purpose or wisdom of the situation and are assuming that we, individually, have a better idea and resolution for how things should be. This is being in ignorance and against the flow of life.

Accepting what is does not induce any sort of indifference or unwillingness to be real with our feelings. It is quite the opposite. Not being willing to accept what is causes us to ignore or reject something and therefore deny the truth and our feelings about it. By accepting a person or situation for what it is, we can understand it fully and allow ourselves to have a genuine reaction to it. We can be true to ourselves and live with complete authenticity.

43. Meditation

If you hold your mind in meditative stillness and allow yourself to be led from within, you will claim your power to change any aspect of yourself and your perceived reality.

To do this, you must gain complete mastery over your thoughts ... banishing negative, limiting thoughts due to the perception of their non-reality and embracing positive thoughts with the knowledge that your state of being and personal interests dictate your experiences.

Negative thoughts are unreal because we exist within the consciousness of a loving intelligence. This intelligence supports our every desire, loving or selfish.

For this reason, we must learn to desire goodness. Through embodying goodness you will come to know heavenly conditions.

Gradually, you will gain eternity ... an endless experience of fulfillment.

As such a person, you could walk on water, part the seas, dive into the ocean of bliss.

44. Awareness

What is awareness?

Awareness is the space in which all perception exists. All of life is aware.

Consciousness is what one is aware of: our senses, feelings, and thoughts, as well as the objects of the senses. It is the substance of existence.

Our awareness and perceived consciousness are a marriage made in heaven. Awareness is the stable center, while consciousness is ever-changing ... stirring and resting.

Awareness perceives consciousness through minds, which are unique spaces. Individual minds exist within a Universal mind that coordinates all consciousness in a collective reality. Individual minds migrate in and out of collective realities.*

The purpose of existence is to experience ideas of being (souls). As individual beings, we experience the unique forms of consciousness of our innermost self, God.

God isn't something you "see" or "know" because it is the state of total awareness of everything that constitutes seeing and knowing.

As awakened self-aware beings, we know all existence to be our own consciousness and watch our sensed reality flow in synchronous harmony with our thoughts and feelings.

As unawakened beings, we perceive consciousness as divided: sense objects being exterior to ourselves; feelings and thoughts being within ourselves; with senses being the intermediary.

Our existence is then viewed as a struggle between the inner and the outer and we, therefore, attempt to limit and control all forms of consciousness. Thus, being at war with ourselves, conflict is experienced in all facets of living.

The more separate a person views its sensed reality to be from itself, the more fearful, hostile, dependent, and powerless they become.

The more unified a person knows its reality to be, the more faithful, loving, independent, and powerful they become.

All realities are equally "real" since all experience is a perception of our consciousness. It is, therefore, essential how we view reality.

Many may consider planet Earth to be the ground of existence, but this is just a moment in eternity. Even the unique and beautiful soul forms of consciousness come in and out of existence. The fundamental reality is You, eternal awareness.

*Group minds are the higher selves of individual minds. Ant colony and bee hive minds are visible examples of this.

45. True Substance

Cold can never be gained, but heat can be lost.
Darkness can never be increased, but light can be reduced.

The negative has no substance of its own.
It is only experienced by reducing the positive.

Sadness is the loss of joy.
Apathy is the loss of passion.

Depression is the loss of enthusiasm.
Meaninglessness is the loss of purpose.

Boredom is the loss of stimulation.
Fear is the loss of safety.

Suffering is the loss of love.
Death is the loss of life.

"I once was lost, but now I'm found."
-Amazing Grace

God is good. Om is home.

46. Place No Idols Before God

When we attach to the things of life, we forget God. We become so full of desire for things that we renounce grace to gain them. As a result, God takes away all things until we realize their emptiness ... in the absence of truth.

47. Awakening

The ultimate question for humanity is "Do we want to wake up or keep dreaming?"

Dreaming involves continuing with all of our illusions of being separate beings in an isolated world.

Waking up involves realizing that there is only one of us forever, and that dreaming is just a vacation from eternity.

If we continue dreaming, our world will likely undergo massive cataclysm and cause incredible suffering.

If we wake up, we will return to the loneliness of solitude.

Spiritual traditions indicate that waking up causes bliss. There is an intermediary state where you can be aware that all is one being, yet still live as an individual. This is the sweet spot in which we wish to remain, the endless path of an awakened soul.

48. Ascetism

The senses are maligned by many mystics for being such a distraction.
But the world and nirvana are not opposites ...
they can be experienced by the same mind.

Some say that the goal of life is to escape it and allow our light to go out.

Well, then why did we bother in the first place?
We did not come into living just to retreat from it.
We are here to thrive!

49. Greatness

I once sought to be a great man,
but now I seek to make man great.

I once sought the love of my life,
but now I seek to be a lover of life.

I once sought great knowledge,
but now I seek knowledge of the great.

I once sought great wealth,
but now I seek the wealth of greatness.

50. The Lord

Thank You Lord

Thank you lord for being my most loyal friend.
Thank you for providing for me 'til the end.

Thank you for sharing yourself for my growth.
Thank you for making me part of your oath.

Thank you for blessing me with love in my heart.
Thank you for making life so great to be a part.

Even in times of great despair,
I know that you are there.

Thank you!

About the Author

Michael Bergen was born in Brooklyn, NY on May 27, 1971 at 3:35pm.

After attending NYC public schools and graduating from Brooklyn Technical H.S., he gained a B.A. in Applied Math and Economics from Brown University and an M.A. in Math Education from Brooklyn College. He also attended the Gestalt Associates and C.G. Jung Institutes in NYC for 3 years.

He had 6 years experience in the research department of Merrill Lynch and 4 years experience as a high school teacher. He has been studying astrology since 1993. He lives in Brooklyn with his wife and kids.

For inquiries regarding the findings in the book, please contact the author at:
michael_bergen@msn.com

The website associated with the book is www.AstrologyCode.org.

www.ingramcontent.com/pod-product-compliance
Lightning Source LLC
Chambersburg PA
CBHW080611230426
43664CB00019B/2859